# Landscape Balance and Landscape Assessment

Springer

*Berlin*
*Heidelberg*
*New York*
*Barcelona*
*Hong Kong*
*London*
*Milan*
*Paris*
*Singapore*
*Tokyo*

Rudolf Krönert • Uta Steinhardt •
Martin Volk (Eds.)

# Landscape Balance and Landscape Assessment

With 57 Figures and 44 Tables

Springer

EDITORS:
Professor Dr. Rudolf Krönert
Dr. Uta Steinhardt
Dr. Martin Volk
UFZ Centre for Environmental Research
Department of Applied Landscape Ecology
Permoserstr. 15
04318 Leipzig
Germany

Library of Congress Cataloging-in-Publication Data
Landscape balance and landscape assessment / ed.: Rudolf Krönert....- Berlin; Heidelberg; New York; Barcelona; Hong Kong; London; Milan; Paris; Singapore; Tokyo: Springer 2001



http://www.springer.de

ISBN 978-3-642-08678-6 ISBN 978-3-662-04532-9 (eBook)
DOI 10.1007/978-3-662-04532-9



Originally published by Springer-Verlag Berlin Heidelberg in 2001.

Softcover reprint of the hardcover 1st edition 2001



Cover Design: Erich Kirchner, Heidelberg
Typesetting: Camera-ready by the editors

SPIN: 10764761 30/3130/xz – 5 4 3 2 1 0 – Printed on acid free paper

# Contents

# 1 Introduction: Landscape balance and landscape assessment

Rudolf Krönert, Martin Volk, Uta Steinhardt

## 1.1 Aims and topics of the book

The articles in this book were written mainly[1] by colleagues working in the Department of Applied Landscape Ecology at the UFZ Centre for Environmental Research Leipzig-Halle. It is designed to improve the application of landscape-ecological methods in research and spatial planning as a contribution to sustainable landscape development, and pays particular attention to our related methodological procedures. The most important basis for achieving this aim is a better understanding of the interactions between landscape structures and processes within the landscape. In the face of the overwhelming number of books on landscape-ecological topics, this book represents the experiences and views of our department regarding the development, realization and application of the following landscape-ecological topics, with special consideration of German approaches to landscape ecology:

- History of landscape assessment *(Chap. 2)*
- Databases, data organization and data processing *(Chap. 3)*
- Research into state and dynamics in landscapes using remote sensing methods *(Chap. 4, 5)*
- Scales and spatial dimensions in landscape research *(Chap. 6)*
- Landscape balance *(Chap. 7)*
- Methods of landscape assessment *(Chap.8)*
- Integration of landscape assessment methods into spatial planning - processes as a basis for sustainable development *(Chap.9)*
- Decision support systems for land use changes: a comparison of methods *(Chap. 10)*

These topics in this logical order represent our typical methodological procedure in analysing and assessing the interactions within landscape systems and the human influence to which they are subject. It should be repeated that our general aim is to derive concepts for sustainable development. This requires

[1] The co-author of Chap. 5 works at the Department of Ecological Modelling of the UFZ, whereas Chap. 10 was written in close cooperation with our colleagues from the UFZ's Department of Economy, Sociology and Law.

cooperation and discussion among all colleagues working within the different branches of landscape ecology in our department to achieve the required integrated syntheses. Obviously, these colleagues hold partly different views due to their different backgrounds in geography, biology, agriculture, cartography or economics, as well as their different approaches to landscape ecology as a discipline - something which makes for very creative collaboration (as becomes apparent from the following chapters). In our opinion, this has to be considered as another positive effect within the transdisciplinary science of landscape ecology that stimulates discussion about a very complex object of research - the landscape with all its natural and societal impacts and interactions.

For landscape analysis and landscape diagnosis on different spatio-hierarchical levels, combinations of more 'classical' methods with geographic information systems (GIS) and related remote sensing methods are used. Besides cartographic methods, the application of GIS-based models plays an important role in landscape-ecological analysis. Integrated landscape-ecological analysis is essential for identifying land use conflicts resulting from an overlapping of different societal demands on landscapes. This forms the basis for the realization of landscape assessment methods, which should be used to conclude land use variants with a positive influence on the protection of our natural resources such as water and soil. The problem of using integrated landscape assessment methods in spatial planning has to be discussed among both researchers and planners. With this in mind, our work is concentrated on the protection of both the regulation functions and the multiple use of landscapes. This aim can only be achieved when account is taken of regionally different landscape conditions and the water-coupled fluxes of material within landscapes. The studies presented in this book were conducted in areas in eastern Germany, which is characterized by the strong influence of agriculture, industry and urbanization. In spite of the particular conditions of these regions, it should be possible to transfer our approaches to other landscapes with similar problems. As several problems are related to each of the individual topics, we include our proposals for solving these problems. We hope to contribute to the resolution of several open questions in landscape ecology. We would like to start with a general theoretical background as a basis for understanding our approaches. Although written with special reference to our approaches and opinions, this chapter tackles a number of open questions in order to provide the reader with enough scope for suggestions and discussions.

## 1.2 Theoretical background: Sustaining and improving of landscape regulation functions

The objects of landscape-ecological research are landscape elements, landscapes and landscape regions. The topic of landscape ecology is the analysis and explanation of structures and processes in landscapes and the landscape functions. In view of their interactions, this can all be defined as the landscape ecosystem. On the one hand, landscape functions can be directed 'internally' - to the functioning of the ecosystem itself. On the other hand, 'externally' directed landscape functions characterize the existence of potentials and the availability of natural resources of landscapes for human yields. As landscape functions ought to be realized in a sustainable manner, methods of landscape assessment aimed at preserving landscape functions play an important role in landscape ecology. The application of landscape assessment methods assumes knowledge of the role of structures and related processes in landscapes. It also requires the determination of worth and assessment rules and standards for stresses and capacities of landscapes. The extraordinary complexity and diversity of landscapes makes their integrated investigation and assessment difficult. Thus, over the past few decades, investigations have frequently addressed the landscape structure. Of the studies on the landscape's processes, the majority of (mostly small-scale) investigations focused on its vertical direction. By contrast, the analysis of horizontal processes, as well as the balancing of fluxes of water and material within landscapes and watersheds, were and still are one of the most serious problems of landscape-ecological research and the topic of many current studies. Knowledge and thus studies of the dynamics and interactions of landscape balance factors is essential as a basis for landscape assessment. Landscape research and landscape assessment bear mainly on the individual components of the landscape, such as soil, water, bios and relief. On the other hand, they aim at an integrated approach to the whole natural complex, considering the interactions between its various components. Finally, the whole landscape including the use of our natural resources and land use are at the focus of attention.

German landscape ecology has contributed greatly to these general ideas of landscape-related sciences during its development over the past 60 years. Unfortunately, in the majority of cases these theoretical insights are unknown to the international scientific community, the main reason being that the studies were published predominantly in German. Another reason is doubtless the theoretical complexity of these approaches involving several definitions and new linguistic creations complicating the English translation. Nevertheless, several approaches of German landscape ecology can be found today in the present integrated methods used in international landscape ecology. Thus, our work is mainly based on some of these traditional studies and is designed to build upon the knowledge, experiences and tendencies of international landscape ecology. Carl Troll's (1939) publication *Luftbildplan und ökologische Bodenforschung* can be regarded as the beginning of landscape ecology in Germany and other German-speaking countries. In it Troll (1939) pointed out the importance of interpreting aerial photographs for the integrated observation of landscapes as a whole. Similar to a "top-down" approach, a methodological procedure of mesoscale landscape ecology can be observed in the handbook for the classification of natural landscapes in Germany (Meynen & Schmithüsen 1953 - 1962, Schmithüsen 1953). In the German Democratic Republic (East Germany), the study by Neef, Schmidt & Lauckner (1961) dealing with landscape-ecological investigations into different physiotopes of north-west Saxony represents the beginning of process-oriented research into the microscale, whereas the doctoral thesis by G. Haase (1962) on landscape-ecological investigations in the uplands of the north-western Lausitz marks the start of landscape-related research into the micro- and mesoscale and provided a methodological framework for a generation of East and West German landscape ecologists. Unfortunately, Haase's (1968) work was only published in several individual manuscripts and writings. It should be mentioned that the former division into two German states (and the political division of the world) also highly influenced the scientific development of German landscape ecology. Despite partly intensive and comprehensive exchanges of ideas among East German landscape ecologists and scientists from the Soviet Union and countries of Eastern Europe (especially during the Sixties, Seventies and partly the Eighties), a truly common approach was not reached because even environmental science was affected by changing political conditions and the partial prevention of the global exchange of ideas. This is evident from the outstanding book by Naveh & Lieberman (1994) about the theory and application of landscape ecology, which contains hardly any citations or theoretical contributions from east German landscape ecologists.

The German approaches to landscape ecology are summarized and resumed by Leser (1976, 1991), due to the above-mentioned German tradition of developing theoretical system approaches at a very high level. The problem of transferring these complex ideas to the international scientific community is apparent from a remark by Naveh & Lieberman (1994) about this book: "As in other scientific disciplines, there exist a great many different approaches to terminology and classification of landscape ecology. This is especially the case in theoretical geography, in which an almost overwhelming amount of semantic discussion can bury the major methodological and practical issues, as in the recent (German) book on landscape ecology by Leser (1976, 1991)." Leser (1999) himself recently defined the position and state of the art of landscape ecology.

The essential results of East German landscape ecology are summarized in a book by Haase et al. (1991), which deals with problems of the investigation of natural landscapes and their relationships and interactions with land use (and whose title can be translated as "Investigation of natural landscapes and land use. Geochorological methods for the analysis, mapping and assessment of nature landscapes"). Bearing in mind all these different approaches and opinions in the interdisciplinary branch of research known as landscape ecology, the problem of compiling standarized methods is obvious. Nevertheless, Marks et al. (1992) suggested guidance for the integrated landscape assessment for scales up to 1:25,000. This book can be seen as a framework and an essential basis for landscape assessment and its application in research and, unfortunately no doubt to a lesser extent, in spatial planning in Germany. The significance of landscape ecology for spatial planning is discussed for instance by Finke (1994).

German landscape ecology - largely influenced by geography - has mainly dealt with the abiotic components of the landscape, whereas Bastian & Schreiber (1999) also consider biotic aspects in their book about the analysis and ecological assessment of landscapes. Nevertheless, landscapes comprehend both the abiotic and the biotic component, as well as land use (Fig. 1.1). Land use acts as an interface between natural systems and socioeconomic systems.

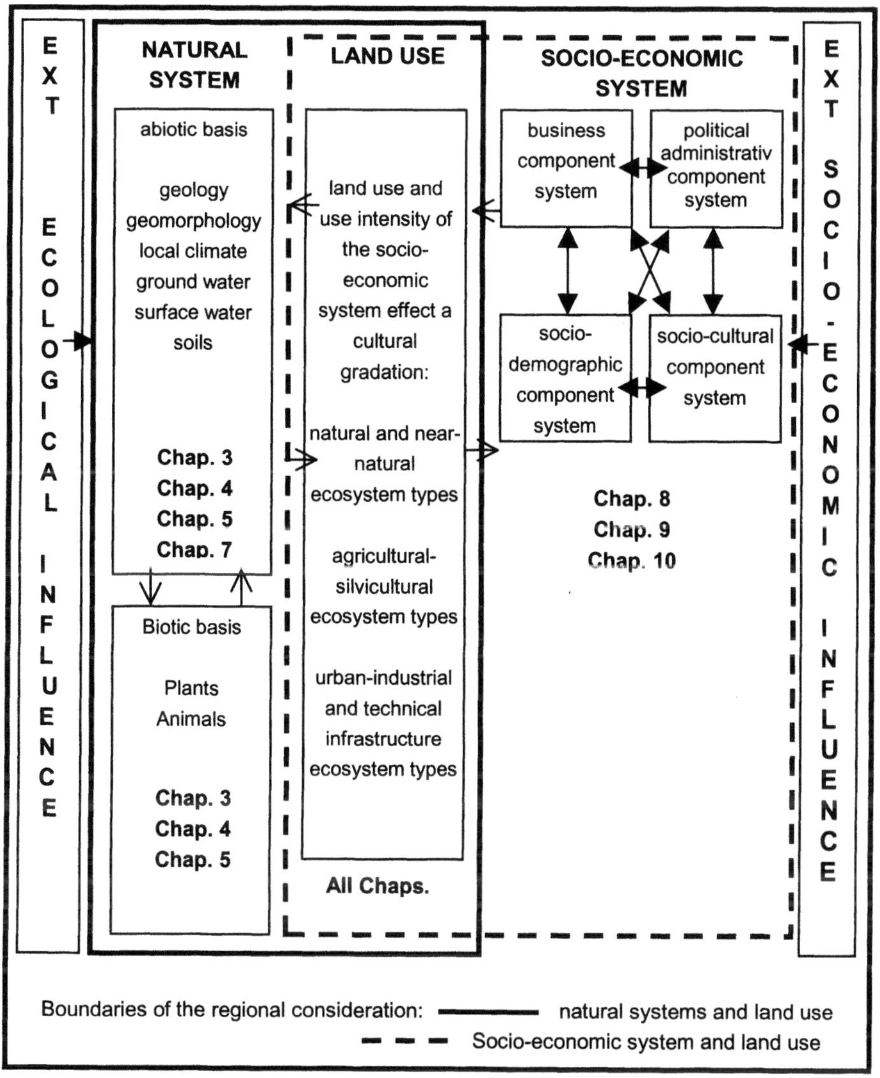


**Fig. 1.1:** Schematic presentation of a regional socioeconomic ecological system (Messerli and Messerli 1979), with an allocation of the chapters in the book dealing more or less with the thematic categories mentioned. Chap. 5, which deals with scales and dimensions in landscape ecology, is not mentioned here separately, since it is actually related to all these categories.

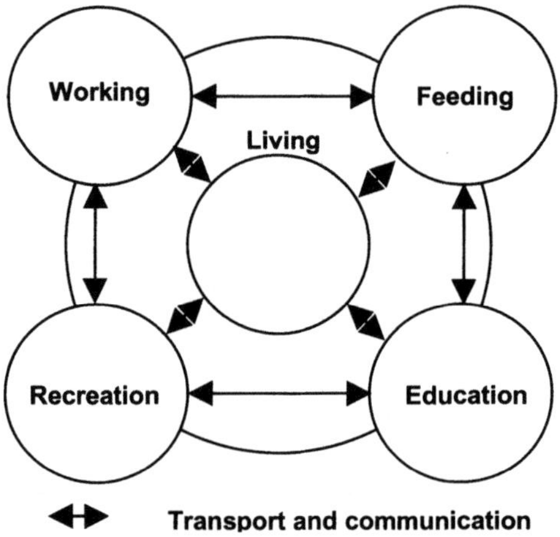


**Fig. 1.2:** Basic human needs: fundamental human activities that are immanent in all social ranks and that can be measured temporally and spatially. The number of basic human needs depends on the cultural group as well as the epoch. In our zone the basic human needs are living, working, feeding, recreation and education within communities. Transport and communication are not considered basic human needs, although they are essential activities for their realization.

Landscapes are subject to permanent changes and developments due to the natural processes taking place in them and their anthropogenic use. The anthropogenic use of landscapes results from the working and living activity of mankind to satisfy his needs (Fig. 1.2). The overlay of societal demands resulting from mankind's aspiring to well-being and complex socioeconomic interactions. As a result the frequently production-oriented use of landscapes leads or contributes to very different environmental stresses (the greenhouse effect and the depletion of the ozone layer, eutrophication, acidification, toxic contamination, the loss of biological variety and biodiversity, the pollution and consumption of soil, water, forest and fish resources, waste dumping, the consumption and destruction of land, the decrease in environmental quality in urban areas stemming from air, water and soil pollution, noise, and the sealing of land).

The change of both land use and cultivation practices (i.e. ploughing, melioration, fertilization, draining, sealing of soil etc.) is one of the most visible features of landscape change with its related far-reaching ecological consequences. Due to natural changes and in view of the history of mankind and his impact on the environment, landscape changes can have time scales ranging from thousands of years (e.g. climatic change since the last ice age), centuries (e.g. the cultivation of arable land, settlement, etc.), decades (change of agricultural cultivation practices, suburbanization, opencast mining, changes of the weather sequences and water balance, etc.) and years (e.g. crop rotations) to single years (e.g. seasons, phenology and land cover), or even individual (short-term) events (volcanic eruption, earthquakes, flooding). Thus, landscapes have a history (genesis), a current condition or state, and a tendency of development, as well as a potential natural condition or state (as an abstraction of the current or real landscape). They include renewable and unrenewable natural resources and potentials. Against this complex background, mankind aspires to particular conditions of the landscape corresponding to his system of values and demands - but landscapes, like other systems, are characterized by a re-oscillating to the initial state, or after interference to a condition similar to the initial state (resilience). Anthropogenic objects and influences (activities) also have a more or less capacious potential of persistence against change. Thus, cultural landscapes develop within an interplay between persistence and dynamics (change).

Landscapes fulfil different functions. On the one hand, there are landscape functions concerning the protection of the functioning of the landscape itself and its related geoecosystems: the regulation functions ('internal functions'). On the other hand, such functions occur that are important for the immediate use of society and man: production, carrier and information functions ('external functions'). The permanent fulfilment of these external functions can only be reached by the preservation, improvement and protection of the internal functions (regulation functions). For instance, nature and landscape protection serve as an important instrument for the preservation and improvement of the regulation functions - which also includes the elimination of critical environmental stresses.

Any landscape fulfils several functions of any functional class simultaneously, albeit with different weightings. The weighting of landscape functions depends on the human value system and the land use type. Land use optimization tries to determine, assess and weight land use functions in order to take into account different ecological-socioeconomic demands regarding environmental and natural resource protection towards the aim of sustainable development. However, before carrying out any kind of landscape-related analysis, we have to decide what reference areas or assessment units have to be used for the investigations. Some remarks on this topic are contained in the following chapter.

## 1.3 Nature areas, landscape units or hydrological response units?

All the disciplines of the earth sciences have developed their own classification of area or reference units, and this is especially true for landscape ecology. German landscape-ecological approaches differentiate between natural areas and landscape areas. Additionally, watersheds are used to balance fluxes of water and material within landscapes. All these units are organized hierarchically (Meyer, Krönert & Steinhardt 1999, Steinhardt & Volk 2000).

"The natural area *(Naturraum)* is an area of land (a section of the earth's land crust) characterized by a uniform structure determined by natural laws and by a monoform complex of abiotic and biotic components; it represents the relationship (in terms of functions) between the geosphere and biosphere." (Freistaat Sachsen) The geosphere quite clearly represents the abiotic part of the natural complex. However, it is not quite clear from the above definition exactly what the biosphere is meant to represent. Whenever landscape-ecological research into natural areas does in fact take the biosphere into consideration, it usually involves an examination of the existing weed vegetation in arable land, meadow or woodland vegetation communities with regard to their indicator values and some kind of declaration of the natural potential vegetation. The reference to the natural potential vegetation (i.e. the vegetation that would appear if all human activities ceased) is purely hypothetical in nature; we cannot expect the woodland communities that preceded man to reappear on fertilized or built-up land, and it is unclear how plant species would migrate or compete today, or how neophytes would behave. In research, natural areas are extrapolated and kept separate from actual land uses - including agriculture and forestry. Early approaches (Billwitz 1968; Haase 1968; Krönert 1968), which sought to integrate agricultural land use into landscape-ecological research into natural areas, have never been implemented on a widespread or systematic basis. We can conclude that natural areas no longer exist in most parts of the world today. They can merely serve as 'guide constructions' in approaching our complex environment.

Regarding the hierarchical organization of nature, natural areas are hierarchical in nature too. Mannsfeld (1997) differentiates between the following landscape-ecological spatial units in terms of topological and chorological dimensions (Table 1.1).

The importance of being able to research, identify and draw conclusions from natural areas is obvious when we consider how we might otherwise evaluate the following:

**Table 1.1:** Spatial dimensions and the indicator characteristics used to differentiate between reference areas (Mannsfeld 1997).

| Spatial dimension | Delimiting characteristics |
|---|---|
| Top | geomorphological-energetic-material associations and lateral and vertical processes |
| Nanochore | Top mosaics defined by<br>▪ ecological similarities<br>▪ ongoing dynamic processes<br>▪ historical dynamic processes |
| Microchore | Landscape genetic "petromorphological" associations (structures, substrates, relief, drainage); differences and similarities in structure |
| Mesochore (lower order) | Landscape genetic similarities (orographical structure and situation; meso- and macroclimatic associations) |
| Mesochore (upper order) | Landscape genetic diversity; orographical and hydrological associations; internal and external site connectivity |

- Natural area potentials (e.g. biotic yield potential, building potential)
- Natural area functions (particularly regulation functions)
- Naturalness
- Pressures on the natural system (e.g. with regard to groundwater contamination) and its capacity to tolerate such pressures (e.g. susceptibility to erosion)

Of course, these evaluations rely on hypothetical/potential natural conditions. Real land use is, of course, also mapped in order to be able to evaluate naturalness, the pressure on natural areas and land use impacts.

After many decades of detailed work, an appropriate methodology has finally been developed by the *Naturhaushalt und Gebietscharakter* research group of the Saxon Academy of Sciences. The group is now mapping natural areas throughout Saxony at the microgeochore level and at a scale of 1:50,000 using numerous template mappings (Freistaat Sachsen 1997).

Besides 'natural area', the term 'landscape' is also used: "The term landscape describes the contents and essence of an area (which can be considered as a section of the earth's crust or 'landscape area') whose underlying characteristics are dictated by natural conditions, but which has been influenced and formed by society. Landscape is a space-time construct which is determined by the flows of matter and energy between man and nature." (Haase1991)

Like natural areas, landscape also has a hierarchical structure and includes:

- *landscape elements,*
- *landscape units, and*
- *landscape regions,*

each of which in turn consists of a hierarchy of different levels. Land use is the principle element used to define the structure of cultural landscapes.

Landscape elements include linear elements, such as hedges and individual trees, as well as parcels of land sharing a common use. Landscape elements can also be described using various hierarchical keys. "Landscape units are sections of landscape with different dimensions and chorological structure. Each landscape unit can be distinguished by its own, relatively stable set of natural and anthropogenic factors, and its functional expression is characterized by a specific complex of landscape elements." (Niemann (1982) This complex is the crucial factor behind the performance of a landscape unit in terms of use. The elements making up this complex are "the smallest area of a landscape which can be distinguished according to actual land use, structural properties and social influences (broader uses, technological measures, management treatments, decisions) (Niemann (1982), slightly modified by Grabaum (1996)). "Components of the natural area and real land use are given equal weighting when defining landscape units." (Freistaat Sachsen (1997) We would argue, however, that the actual land use must be the guiding criterion used when identifying landscape elements and landscape units.

Like natural areas (see Table 1.1), landscape units also need to be categorized according to their spatial characteristics. Table 1.2 lists the relevant spatial dimensions, spatial characteristics and criteria used to distinguish between different units. These definitions are based on German experience with landscape ecology and do not match the equivalent landscape hierarchy proposed by Forman (1995). The characteristics Forman used to delimit landscape units are based on North American experience and cannot be applied to central European conditions. They are also more appropriate for structuring natural areas. Indeed, the classification of spatial dimensions given by Forman (1995) can at best be described as a mixture between a structural classification for landscapes and one for natural areas.

Table 1.2 calls for some remarks. Firstly, there are obvious problems when dealing with settlements and urban areas: should they be included or excluded in landscape classification? Often, settlements are not considered as a part of the landscape. Secondly, this spatial classification of landscape units follows the above-mentioned hierarchy of natural areas (top - chore - region etc.). As landscape units are now also classified in terms of these dimensions, a second terminology should be avoided. Thirdly, land use or land cover is not the routing criterion in landscape classifications over all scale levels.

**Table 1.2:** Spatial dimensions, spatial characteristics and characteristics used to delimit landscape units (Meyer, Krönert & Steinhardt 2000).

| Spatial dimension | Spatial characteristics | Delimiting characteristics |
|---|---|---|
| Landscape element | Homogeneous, usually clearly delimited land-use unit; heterogeneous, small-scale complexes (defined by use) in settlements | Land use, land cover |
| Landscape unit 1st order | Heterogeneous land-use mosaics, usually dominated by one of these land uses and including villages | Land use mosaics; clearly defined borders based on microchores; functional areas in cities, or towns with up to 5,000 inhabitants |
| Landscape unit 2nd order | Heterogeneous land-use mosaics including small towns of up to 5,000 inhabitants | Land use mosaics; less precise border definition based on lower order mesochores; combinations of functional areas in cities, or towns with between 5,000 and 20,000 inhabitants |
| Landscape unit 3rd order | Heterogeneous land-use mosaics including small towns of up to 20,000 inhabitants | Land use mosaics; less precise border definition based on higher order mesochores; combinations of functional areas in cities, or towns/cities with between 20,000 and 200,000 inhabitants |
| Landscape region | very heterogeneous land-use mosaics | Land use mosaics; less precise border definition based on macrochores; large conurbations and municipal regions |

Microscale classifications such as the "Mapping units of the biotope type and land use mapping of Saxony" (Luftbild Brandenburg 1994) on a scale of 1:10,000 (topological and chorological dimension) often follow land cover classification rules. In contrast, macroscale classifications like "Ecoregions of the Continents and Oceans" (Bailey 1996) neglect land use. The delineation of units is only based on natural components. Special problems occur when dealing with landscape units on the mesoscale. For instance, Krönert & Erfurth (1994) provided a "Landscape Typification for the Agglomeration Area Leipzig-Halle", which initially distinguished between urban areas, mining areas, agricultural areas and forest areas, as well as several combinations of the last two types. All these types are guided by land use. But in addition there is also a landscape type "floodplains and valleys" - guided by a geomorphological criterion. This shows the difficulties in dealing with landscapes in the mesoscale.

Natural areas as well as landscape units are delimitated by more or less structural parameters. This approach leads to unsatisfactory results if processes (fluxes of matter, energy and information) are investigated. Process-oriented research needs process-oriented reference units. Thus watersheds - often considered as 'quasi-closed systems' - have to be used as investigation units if the studies take a more process-oriented viewpoint (water balance, water-coupled fluxes of material). They allow cycles of water and matter to be modelled on the landscape scale. Their natural boundaries and hierarchical organizational form an appropriate structure for environmental impact analysis. In this context hydrological system analysis and the HRU concept should be mentioned. HRUs (hydrological response units) can be defined as follows (Flügel 1996): "Hydrological Response Units are distributed, heterogeneously structured entities having a common climate, land use and underlaying pedo-topo-geological associations controlling their hydrological transport dynamics." The HRU concept uses three basic assumptions:

1. Each land use class on a specific pedo-topo-geological association is characterized by a homogeneous set of hydrological process dynamics.
2. These dynamics are controlled by the land use management (type of vegetation, management practices) and the physical properties of the respective pedo-topo-geological association.
3. The variation of the hydrological process dynamics within the HRUs is small compared with the dynamics in a different HRU.

The HRUs serve as modelling entities that can be derived from GIS analyses of the soil-vegetation-atmosphere (SVAT) interface. The HRU concept is capable of preserving the heterogeneity of the three-dimensional physiographical properties of the drainage basin and can therefore be used for spatial scale transfer in regional hydrological modelling. The insights gained from a thorough, detailed hydrological system analysis are essential for defining the criteria to be common to each HRU such as land use and the underlying pedo-topo-geological association.

With respect to the discussed "Framework for community action in the field of water policy" (Directive 2000), both the truth and the practical pertinence of this approach are self-evident. In contrast to previous specific planning activities, no administrative units are used. However, planning in the field of water-related problems (water quality, water management, etc.) must overcome this obsolete position. Water-bound fluxes does not stop at administrative borders.

## 1.4 Data base and indicators: About difficult dependencies and relationships

For integrated landscape ecological investigations, gathering and providing the basic data is essential. At present, between half and three-quarters of the whole time spent on investigations is devoted to data procurement and preparation. Data should always be gathered in relation to the research aims, and their accuracy must be verified. Time-related series of data (e.g. climate data) have to be complete, representative and should cover a couple of years. One important aspect is that the data have to be suitable for the scale of investigation in terms of their spatio-temporal resolution as otherwise results (e.g. the derivation of indicators) may be produced which are incorrect, especially if several different data layers are used for integrated analysis. Moreover, sometimes the form of the required data and their availability is often insufficient, and the user has to process them. Volk & Steinhardt (1998) pointed out these problems and suggested methods for standardizing comprehensible methods for aggregating and generalizing the data.

The basic data have to enable the derivation of area-related spatial information, or to be transferable to areas. The linkage of the data to derive integrative information about landscapes is very important for landscape ecology. This is true for cartographic representations, but also for the modeling of interactions and processes within landscape systems, and the calculation of conceptual indices. The linkage of the data is not at all trivial, because it has to be gathered for and transferred to discreet points, raster data, polygons and administrative units, which can all be managed in databases. Thus, analogue data must first be transformed into a digital format. One problem is that the data mostly originate from different sources, and so information has to be processed in order to make it homogeneous. Afterwards, it must then be adapted to the structure of the database. The data sources which can be used include field measurements, data from governmental or other measuring networks, or existing digitized information ("Fachinformations-systeme": specific information systems, e.g. soil information systems). Examples of digitized information include topographic and thematic maps, aerial photographs, satellite data, information from government authorities, and environmental and geological surveys, as well as data from agricultural and forest managers. In connection with this, problems arise especially with study areas crossing administrative borders, because the different authorities responsible often use different methods and techniques to gather and manage the data. This is especially the case for data gathered at the level of federal states. Thus, data catalogues serve as important manuals for checking existing information for investigations (Bastian & Schreiber 1999, Zepp & Müller 1999). One important step towards resolving these problems would be to improve cooperation between government authorities and surveys, planning agencies and research institutes. Additionally, responsibilities should be divided among the parties involved to ensure, for instance, the continuous management of information and its updates.

The environmental conditions of natural areas, landscapes or watersheds cannot be characterized in their totality. Therefore, information must be reduced so that useful indicators can be derived. Indicators should supply assertions about characteristics, processes or structures of objects, which are in our case landscapes or the environment. The conclusion and provision of indicators should be modest, with a high degree of explanation. Indicators are thus used to characterize very complex objects by the application of just a few parameters. E. Neef, for instance, introduced as fundamental ecological features soil type, vegetation society and the soil moisture regime into the landscape-ecological complex analysis of ecotopes, in the sense of indicators linked in various ways. Nevertheless, the definitions of these indicators for the meso- and macroscale still need some work. To characterize environmental stresses, the pressure-state-response approach of the OECD (1994) is often chosen used (c.p. Walz et al. 1997), in which pressure, state and response are characterized with indicators. This approach is important for characterizing the environmental conditions, their main causes (driving forces) and their reaction to the conditions and changes. This approach also supplies important indications on how to rectify environmental damage. However, this method is not sufficient for the characterization of ecosystems and interactions within landscapes and watersheds This requires the investigation of processes and balances (i.e. water balance, water-coupled fluxes of material and nutrients, surpluses of nutrients within a balance) with measuring and modelling which allow the identification and localization of the causes and consequences of land use changes. On this basis, recommendations can be derived that have positive effects on the landscape balance. Processes and balances can partly also be characterized by indicators as a result of model calculations. For instance the level of mineral nitrogen in the soil after harvesting can provide an indicator for sustainable agricultural management.

## 1.5 Scales and dimensions in landscape ecology: Transformation, aggregation and disaggregation of landscape information

Several studies in the environmental sciences deal with the hierarchical organization of ecosystems (O'Neill et al. 1986, Burns et al. 1991, and several others). Klijn (1995) gives an overview of hierarchies in landscape ecology. Unfortunately, in landscape ecology, treatments of hierarchies were mostly limited to the consideration of the different spatio-temporal resolution of the basic data. In fact, the problem of scales and transformation with narrow limits is related to the compilation of the data and the choice of the indicators. The scale problem results from the transfer from one hierarchical level to another, which is true for both the "top down" and the "bottom up" approach. Both approaches have to consider generalizations suitable for the scale, and changes to the criteria of homogeneity. According to the diagram by Herz (1973) dealing with this problem, homogeneity can be achieved at each level of consideration by agglomeration or generalization. An increasing hierarchical order is often accompanied by an increase in heterogeneity. Thus, during studies it should be made clear whether mean values or data of dominant conditions and processes have been used, or whether features of heterogeneity (frequency, minima and maxima, variance, etc.) are also considered.

The crucial point is to define scale-specific processes: what processes act at which scale? If this is known, we can derive the data necessary to describe these processes. Of course, the problem still remains of whether these data are actually available.

The choice of hierarchical level depends on the scientific question formulation. At any hierarchical level, the arrangement and classification of the landscape can be made using its components like soil, climate or relief. This method can include the whole nature complex abstracting from land use within the order of natural areas or classification. Beside the nature complex, land use is considered as the feature in landscape classifications or the delineation of landscape units. These spatial units are organized in spatiotemporal hierarchies, which can be approached at micro-, meso- and macroscale levels (Steinhardt 1999, Steinhardt & Volk 2000). The main problem here results from the transformation and transfer of information from one spatiotemporal scale level with less complexity to one with a higher complexity. Thus, difficulties arise with the fact that no area congruency exists and the spatiotemporal structuring of the objects is unsuitable (for instance combining the results of the (East) German mesoscale agricultural site mapping with forest site mapping is problematic and has to be processed and verified critically in order to obtain a consistent database covering the investigation area as a whole). One provisional solution often proposed is the intersection of the information and the derivation of the smallest common topologies. This method is unsatisfactory because existing interactions within a system as well as between systems or the unsteadiness of the transition zones between two ecosystems (for instance, ecotones, edges, etc.) are neglected. Additionally, the derivation of smallest common topologies produces enormous data files, often exceeding computer capacity.

## 1.6 Integrated landscape analysis and assessment as a base for sustainable landscape development: Theory and spatial (and economic) planning practice

Nature and landscape protection as well as landscape architecture and landscape development should be oriented towards guidelines, environmental quality targets and environmental quality standards (SRU 1994). The concept of sustainable development can be considered as an overall guideline in the sense of a vision which includes ecological, economic and societal components. The starting-point of this concept is based on the definition of sustainability by the World Commission on Environment and Development (WECD 1987). This definition includes a durable environmentally sound development incorporating the principle of precaution.

The following rules obtain as guidelines:

- The consumption of renewable resources must not exceed their rate of regeneration.
- The consumption of non-renewable resources must not exceed their rate of substitution.
- Residual emissions must not exceed the rate of assimilation.
- Environmental functions (regulation functions) must be preserved and improved.
- Environmental conditions must ensure human health and well-being .
- Environmental targets and standards must be oriented towards critical consumption rates of resources and the supply of resources, as well as critical values of stresses and concentrations, material and nutrient inputs, critical changes to structures (e.g. spatial structures), and critical stresses and health risks for mankind.

Unfortunately, most of the above critical values are only partly defined. Forman (1995) put forward a proposal over which features should be applied for sustainable landscapes: "... a sustainable environment is an area in which ecological integrity and basic human needs are concurrently maintained over generations." Ecological integrity includes the productivity and biodiversity of a landscape, as well as the preservation of soil functions (for instance resistance to soil erosion) and the preservation of water quality and quantity. "Basic human needs are considered to be adequate food, water, health, housing, energy, and cultural cohesion." (Forman 1995) Sustainable landscapes have an adaptability against disturbances, and "a mosaic stability, where interactions among neighbouring elements dampen fluctuations from disturbances." (Forman 1995) Since only a few parts of the world fulfil this condition, ways have to be shown how to reach this condition.

The assessment of landscapes must therefore be based on the theories and knowledge about the capacity of natural areas, as well as about the condition, use and development of landscapes. This calls for a few remarks. Potentials of natural areas describe the capacity of natural areas to fulfil the functions of natural areas. They are mainly used to identify the margin for the demands of land use between usability and resilience under special consideration of multiple land use (Haase & Mannsfeld 1999). The potentials of natural areas have been defined with reference to the societal demands on the natural area. It is not possible to assess them as an integral potential; instead they have to be determined in accordance with individual functions of natural areas.

As potentials of nature areas, Haase (1978) differentiates between:

- biotic yield potential
- potential of water availability
- potensial of (waste) disposal
- biotic regulation potential
- geoenergetic potential
- potential of mineral resources
- building potential
- recreation potential

Land use is an indicator of the intensity for the usage of natural resources, the use of sites, and also the spatial structure of landscapes (Haase & Lüdemann 1972). An integrated consideration of land use and land use intensity, and of course of the potential natural conditions and potentials of natural areas, enables information to be derived about the degree of naturalness or hemeroby and the conditions of cultural landscapes. Problems arise here with correlation to the reference point (what is "natural"?). Sukopp & Zerbe (1998) suggest determining the degree of naturalness at a primary natural reference condition (which is mostly very difficult or impossible), and expressing the intensity of the changes on a scale of hemeroby (taking into consideration irreversible anthropogenic changes).

On the other hand, Haber (1972) developed the concept of differentiated land use and subsequently continued the related theories (Haber 1998). In contrast to all technical systems, highly developed natural systems are maintenance-free due to self-regulating mechanisms. Hence all natural systems have features which are desirable for our environment: the (relative) stability of environmental conditions that can't be guaranteed by high-end technologies. The European agricultural landscape remained stable for centuries as the first farming activities and the first settlements increased diversity, causing the European cultural landscape to develop. However, the techniques later introduced led to a demixing of land use forms. Original natural regulation processes had to be replaced by artificial ones. And therefore we have to focus on the re-implementation of land use forms that are more varied. The concept of differentiated land use includes some ideas from Odum (1971). It is also based on the theory of segregating the landscape into protective and productive parts, and for linkage to actual land use (which stands in contrast to other concepts aimed at the protection and use of a whole area, the "integration concept").

The application of the concept of differentiated land use requires observance of the following rules:

1. Within a spatial unit, pollutant and intensive land use should not take up 100% of the total area. At least 10-15% (on average) of the area must be available or reserved for relieving and buffering types of land use.
2. Each dominant land use has to be diversified to prevent large uniformed areas.
3. Within spatial units characterized by intensive types of land use, at least 10% (on average) of the total area must be available or be reserved for nature conservation, preferably in network distributions.

(The protection areas described in 1 and 3 may be identical or overlap each other).

Although this concept is certainly widely accepted by the representatives of nature conservation, who are often guided by a biological viewpoint and a structural orientation (towards the protection of species), it seems to be inadequate for the protection of the abiotic components of nature. Areas for immissions are permitted indirectly: for instance, atmospheric nitrogen inputs, or by fertilization. In view of the demands for a 50% reduction in nitrogen inputs into seas and inland waters, this concept is insufficient.

Conceptions for an area-wide nature conservation refer to the German Nature Conservation Act (*Bundesnaturschutzgesetz*). Under Section 1, the German Nature Conservation Act requires not only the protection of the biotic components, but also the protection of a functioning nature balance. The landscape water balance is one of the most important components of the nature balance owing to its role as a link between the climate, soil and land use - especially in our temperate climates. Additionally, water is the main transport medium for all material fluxes within the landscape. Ideally, the landscape water balance should be differentiated into its drain components surface runoff, interflow and base flow, although this is increasingly difficult the bigger the study area.

Knowledge about the landscape water balance is the main key for understanding the outwash processes for nitrogen and other nutrients and pesticides, as well as for the soil erosion processes responsible for the soil particle-coupled outwash of phosphorus. The diminution of both the eutrophication of water and soil erosion entails improving the retention capability of the landscape (retention of water, soil substrate and nutrients). It is clear that intense outputs of material have to be influenced and coupled with a decrease incritical material input, for instance by means of fertilization restrictions or cutting aerial pollutants. In contrast to former segregative approaches, Jedicke (1995) suggested a concept for an integrative, area-wide nature conservation. He defined nature conservation as an equitable fulfilment of the four interpenetrative responsibilities protection of species, protection of biotopes, protection of the (abiotic) resources and protection of processes. In addition, a fifth responsibility about the protection of the aesthetics of landscapes (i.e. diversity, peculiarity and beauty) is conceivable. Jedicke (1995) includes soil, water and air in the protection of the (abiotic) resources.

For the protection of processes, he suggests the following division:

- The protection of parts of natural landscapes: the long-term preservation of natural abiotic and biotic processes, without human intervention or impact.
- The protection of cultural landscapes: the preservation of anthropogenic land use processes in the sense of a sustainable practiced land use with the careful handling of our natural resources, including positive side-effects for nature conservation (protection through land use).

Nature conservation oriented towards development guidelines and environmental quality targets which embrace the German Nature Conservation Act (and also European strategies for environmental protection) requires the combination of segregative area protection with different land use intensities and area-wide protections. This request stems from the concept of differentiated land use according to Haber (1972) and the demands of Jedicke (1995). The background for this purpose is provided by several empirical studies on regionally differentiated land use, its ecological impacts, as well as the driving forces of land use in combination with the investigation of the main features of the landscape balance. In this connection, the work of our Department of Applied Landscape Ecology is carried out mainly on mesoscale levels, with analysis on the microscale level. Thus, our studies aim at both the improvement of the basis of spatial and environmental planning and the progress of theories of landscape ecology. From a theoretical viewpoint we are close to the approaches of the IGBP/HDP-project "Land Use and Land Cover Change" (IGBP/IHDP 1999).

Landscape assessment has to decide what mosaics of land use best fulfil the desired landscape functions (Forman 1995). Besides the spatial distribution of land use types, the overlay of both land use types and societal demands must also be considered, namely:

- The main land use type (e.g. arable land, forest, urban areas, etc.)
- Land use types regulated by law (e.g. biosphere reserves, conservation areas for drinking water, etc.)
- "Spontaneous" uses (e.g. sinks and sources of fluxes of material, etc.)

It is apparent that approaches to sustainable landscape development have to consider and weight all these different functions and demands simultaneously. Despite the establishment of the landscape function concept in planning, this is frequently also a hindrance for its application in planning practice because of its complexity, lack of time and finances, and the structure of the relevant authorities, the weight and pressure of political and economic interests, etc. (Petry 2001, Auhagen 1999).

On the other hand, it also seems evident that the theoretical and scientific basis for realizing these approaches already exists to a large extent. The increasing number of multiple criteria decision-making methods (e.g. Climaco 1994, Haimes & Steuer 1998) applied in landscape-related branches may serve here as one of several examples of this tendency (Grabaum 1996, Drechsler 1999). During the last 10-15 years, integrated assessment approaches have increasingly shifted away from the main consideration of the capability of the landscape balance to ecological-socioeconomical approaches (Dabbert et al. 1999), which can be also observed at the UFZ Centre for Environmental Research Leipzig-Halle (Horsch & Ring 1999). Thus increasing acceptance of such often very complex methods can be expected in the many branches of spatial, environmental and economic planning. Nevertheless, prospective environmental planning for the preservation and development of natural resources is one of the main tasks of political precaution for existence and life (Bastian & Schreiber 1999). This overall goal can only be achieved by improved communication and cooperation among governmental authorities and surveyors, planning agencies, research institutes, the business community and the general public.

## 1.7 Final comments about general trends of German landscape ecology

The aim of this chapter was to introduce the topics of the book "Landscape balance and landscape assessment", considering all its related fields and problems within landscape ecology. In connection with this, it is also designed to improve knowledge and integration of German (including our) approaches within the international scientific community of landscape ecology. The long history of German landscape ecology has produced a great potential of theories, methods and suggestions for its application in planning. Unfortunately, during the past 10-20 years only a few contributions by German landscape ecologists have appeared in international journals dealing with landscape-ecological topics. Besides the fact that many German landscape ecologists have published much of their work only in German, this is also accounted for by a number of tendencies that can be observed in the development of the scientific structure. In Germany, a large part of landscape ecology is moulded by geography. During the past few years, several classical branches of geography, and partly also biology, have broken away to form their 'own' disciplines with their own faculties at universities. Examples include geoecology, geobotanics, landscape ecology, biogeography, remote sensing, environmental protection and planning, etc. Whether this breakaway trend really supports the development of common strategies of landscape ecology is a moot point.

On the other hand, the foundation of a German regionalchapter of the International Association of Landscape Ecology (IALE) in May 1999 gives reason to hope for the future development of common strategies of German landscape ecology and an opportunity to contribute to several open questions within international landscape ecology.

In our opinion, topics which must in future be tackled by landscape ecology include the following :

- Establishing standardized databases and frameworks for integrated analysis and assessment methods for different spatiotemporal scales as bases for landscape models
- The development and application of comprehensive (digital) landscape models
- Improving knowledge about the interactions of landscape structure and processes within the landscape
- Improving the integration of remote sensing methods and process-structure-oriented methods to develop knowledge-based systems.
- The development of methods for pattern recognition in landscapes.
- Improving knowledge about the self-organization and adaptability of landscape systems against natural and anthropogenic disturbances (resilience and resistance).
- The further development of multicriteria decision systems considering both ecological and socioeconomic assessment approaches (based on a comprehensive value system).
- The development of solution strategies for sustainable land use and land management systems (transferability/applicability of methods into/within spatial planning).

## 1.8 References


Auhagen A (1999) Verwendung von Bewertungsverfahren in der Landschaftsplanung. In: Bastian, O & K-F Schreiber (eds) (1999): Analyse und ökologische Bewertung der Landschaft. Springer, Berlin Heidelberg, S 394-402

Bastian O, Schreiber K (1999) Analyse und ökologische Bewertung der Landschaft. Spektrum, Heidelberg-Berlin

Bayley RG (1996) Ecosystem Geography. Springer, New York

Billwitz, K (1968) Die Physiotope des Lößgebietes östlich Grimma und seines nördlichen Vorlandes in ihren Beziehungen zur Bodennutzung. Dissertation, UniversitätLeipzig

Burns TP, Pattan BC, Higashi H (1991) Hierarchical Evolution in Ecological Networks. In: Higashi H, Burns TP (eds) Theoretical Studies of Ecosystems: The Network Perspective. New York.

Climaco J (1994) Multicriteria Analysis. Proceedings of the XIth International Conference on MCDM, 1-6 August 1994, Coimbra, Portugal - Springer

Dabbert S, Hermann S, Kaule G, Sommer M (eds) (1999) Landschaftsmodellierung für die Umweltplanung. Springer, Berlin-Heidelberg-New York

Directive 2000//EC of the European Parliament and of the Council of establishing a framework for Community action in the field of water policy (http://www.umweltdaten.de/down-d/reswfd.pdf; 20.11.2000)

Drechsler M (1999) Multikriterielle Analyse von Szenarien zur Landnutzungsänderung. In: Horsch H, Klauer B, Ring I, Gericke H-J, Herzog F (eds) (1999): Nachhaltige Wasserbewirtschaftung und Landnutzung: Methoden und Instrumente zur Entscheidungsfindung und -umsetzung. UFZ-Bericht 24/2000, Leipzig, pp 47-49

Finke L (1986): Landschaftsökologie. Das Geographische Seminar, Westermann, Braunschweig

Flügel WA (1996) Hydrological Response Units (HRU's) as modelling hydrological river basin simulation and their methodological potential for modelling complex environmental processes. Results from the Sieg catchment. Die Erde 127:43-62

Forman RTT (1995) Land Mosaics. The ecology of landscapes and regions. University Press, Cambridge

Freistaat Sachsen (ed) (1997) Naturräume und Naturraumpotentiale des Freistaates Sachsen. Materialien zur Landesentwicklung 2/1997, Dresden

Grabaum R (1996) Verfahren der polyfunktionalen Bewertung von Landschaftselementen einer Landschaftseinheit mit anschließender "Multicriteria Optimization" zur Generierung vielfältiger Landnutzungsoptionen. Shaker, Aachen

Haase G (1968) Inhalt und Methodik einer umfassenden landwirtschaftlichen Standortkartierung auf der Grundlage landschaftsökologischer Erkundung. Wiss. Veröff. Dt. Inst. f. Länderkunde N.F. 25/26:309-349

Haase G (1978) Zur Ableitung und Kennzeichnung von Naturpotentialen. Petermanns Geographische Mitteilungen 12:113-125

Haase G (1991) Gesellschaftliche und volkswirtschaftliche Anforderungen an den Naturraum. In: Haase G (ed) Naturraumerkundung und Landnutzung. Geochorologische Verfahren zur Analyse, Kartierung und Bewertung von Naturräumen, Beiträge zur Geographie 34, Berlin pp 26-28

Haase G unter Mitwirkung von Barsch H, Hubrich H, Mannsfeld K, Schmidt R (1991) Naturraumerkundung und Landnutzung. Geochorologische Verfahren zur Analyse, Kartierung und Bewertung von Naturräumen. Beiträge zur Geographie 34, BerlinText- und Beilagenband

Haase G, Lüdemann H (1972) Flächennutzung und Territorialforschung - Gedanken zu einem Querschnittsproblem bei der Analyse und Prognose territorialer Strukturen. Geographische Berichte 17:13-25

Haase G, Mannsfeld K (1999) Ansätze und Verfahren der Landschaftsdiagnose. In: Haase G (ed) Beiträge zur Landschaftsanalyse und Landschaftsdiagnose. Abhandlungen der Sächsischen

Akademie der Wissenschaften zu Leipzig. Mathematisch-naturwissenschaftliche Klasse, Band 59, Heft 1, pp 7-17
Haber W (1972) Grundzüge einer ökologischen Theorie der Landnutzungsplanung. Innere Kolonisatio 21:294-298
Haber W (1998): Das Konzept der differenzierten Landnutzung - Grundlage für Naturschutz und nachhaltige Naturnutzung. In: BMU, Ziele des Naturschutzes und einer nachhaltigen Naturnutzung in Deutschland. Tagungsband zum Fachgespräch, 24. u. 25. März 1998, Bonn, pp 57-64
Haimes YY, Steuer RE (1998) Research and Practice in Multiple Criteria Decision Making. Proceedings of the XIVth International Conference on Multiple Criteria Decision Making (MCDM) Charlottesville, Virginia, USA, June 8-12, 1998 (Lecture Notes in Economics and Mathematical Systems Vol. 487), Springer, New York
Herz K (1973) Beitrag zur Theorie der landschaftsanalytischen Maßstabsbereiche. Petermanns Geogr Mitt 117: 1-96
Horsch H, Ring I (eds) (1999) Naturressourcenschutz und wirtschaftliche Entwicklung. Nachhaltige Wasserbewirtschaftung und Landnutzung im Elbeeinzugsgebiet. UFZ-Bericht 16/1999, Leipzig
IGBP Report 48/ IHDP Report (1999) Land-Use and Land-Cover Change. Implementation Strategy. Stockholm
Jedicke E (1995) Ressourcenschutz und Prozeßschutz. Diskussion notwendiger Ansätze zu einem ganzheitlichen Naturschutz. Naturschutz und Landschaftsplanung 27:125-133
Klijn JA (1995) Hierarchical concepts in landscape ecology and its underlying disciplines. DLO Winand Staring Center Report 100, Wageningen
Krönert R (1968) Über die Anwendung landschaftsökologischer Untersuchungen in der Landwirtschaft. Wiss. Veröff. Dt. Inst. f. Länderkunde, N.F. 25/26, pp 181-308
Krönert R, Erfurth S (1994) Landnutzung und Landschaftsverbrauch im mitteldeutschen Ballungsgebiet. Geographie und Schule.16:18-24
Leser H (1976, 1991) Landschaftsökologie. Ansatz, Modelle, Methodik, Anwendung. Mit einem Beitrag zum Prozeß-Korrelations-Systemmodell von T Mosimann, Ulmer, Stuttgart
Leser H (1999) Das landschaftsökologische Konzept als interdisziplinärer Ansatz - Überlegungen zum Standort der Landschaftsökologie. In: Mannsfeld K, Neumeister H (eds) Ernst Neefs Landschaftslehre heute. Petermanns Geographische Mitteilungen Ergänzungsheft 294, Gotha, pp 65- 88
Luftbild Brandenburg (1994) Einheiten der Biotoptypen- und Landnutzungskartierung im Freistaat Sachsen. Kartierschlüssel, (unv. Manuskript)
Marks R, Müller MJ, Leser H, Klink HJ (eds) (1989) Anleitung zur Bewertung des Leistungsvermögens des Landschaftshaushaltes (BA LVL). Forschungen zur Deutschen Landeskunde Bd 229, Trier
Mannsfeld K (1997) Etappen und Ergebnisse landschaftsökologischer Forschung in Sachsen. In Dresdener Geographische Beiträge 1, pp. 3-21
Messerli B, Messerli P (1979) Wirtschaftliche Entwicklung und ökologische Belastbarkeit im Berggebiet (MAB Schweiz), Geogr. Helvetica 33: 203-210
Meyer BC, Krönert R, Steinhardt U (2000) Reference areas and dimensions in landscape ecology and application of evaluation functions. In: Mander Ü, Jongmann RHG (eds) Consequences of Land Use Changes, Advances in Ecological Sciences 5, Southampton, Boston, pp 119-146
Meynen E, Schmithüsen J (eds) (1953-1962) Handbuch der naturräumlichen Gliederung Deutschlands. Erste bis bis Fünfte Lieferung, Remagen
Naveh Z, Lieberman A (1994) Landscape Ecology. Springer, New York-Berlin
Neef E, Schmidt G, Lauckner M (1961) Landschaftsökologische Untersuchungen an verschiedenen Physiotopen in Nordwestsachsen. Abhandlungen d. Sächs. Akad. D. Wiss. zu Leipzig, Math.-nat. Kl., Bd. 47, H. 1, Berlin
Niemann E (1982) Methodik zur Bestimmung der Eignung, Leistung und Belastbarkeit von Landschaftselementen und Landschaftseinheiten. Wissenschaftliche Mitteilungen 12, IGG Leipzig
Odum E P (1971) Fundamentals of Ecology. W.B. Saunders, Philadelphia

O'Neill RV et al (1986) A hierarchical concept of Ecosystems. Princeton University Press, Princeton, N.J.

Petry D (2001) Landschaftsfunktionen und planerische Umweltvorsorge auf regionaler Ebene: Entwicklung eines landschaftsökologischen Verfahrens am Beipiel des Regierungsbezirks Dessau. PhD-thesis. Universität Halle.

Schmithüsen J (1953) Einleitung. Grundsätzliches und Methodisches. In: Handbuch der naturräumlichen Gliederung Deutschlands, Erste Lieferung, Remagen, pp 1-44

SRU: Der Rat der Sachverständigen für Umweltfragen (1994) Umweltgutachten 1994. Für eine dauerhaft -umweltgerechte Entwicklung., Bonn

Steinhardt U (1999) Die Theorie der geograohischen Dimensionen in der angewandten Landschaftsökologie. In: Schneider-Sliwa R, Schaub D, Gerold G (eds) Angewandte Landschaftsökologie. Grundlagen und Methoden. Springer, Berlin-Heidelberg-New York, pp 47-64

Steinhardt U, Volk M (2000) Von der Makropore zum Flußeinzugsgebiet. Hierarchische Ansätze zum Verständnis des landschaftlichen Wasser- und Stoffhaushaltes. Petermanns Geographische Mitteilungen 144:80-91

Sukopp H (1997) Indikatoren für Naturnähe. In: BMU (eds) Ökologie. Grundlagen einer nachhaltigen Entwicklung in Deutschland. Tagungsband zum Fachgespräch, Bonn - Bad Godesberg, pp 71-84

Troll C (1939) Luftbildplan und ökologische Bodenforschung. Zeitschr. d. Gesellsch. f. Erdkunde Berlin 7/8: 241-298

Volk M, Steinhardt U (1998) Integration unterschiedlich erhobener Datenebenen in GIS für landschaftsökologische Bewertungen im mitteldeutschen Raum. Photogrammetrie, Fernerkundung, Geoinformation 6:349-362

Walz R u. a. (1997) Grundlagen für ein nationales Umweltindikatorensystem - Weiterentwicklung von Indikatorensystemen für die Umweltberichterstattung. Umweltbundesamt, Texte 37, Berlin

WCED (1987) Our Common Future (The Brundtland Report), World Commission on Environment and Development, Oxford

Zepp H, Müller MJ (eds) (1999) Landschaftsökologische Erfassungsstandards. Ein Methodenbuch.- (= Forschungen zur Deutschen Landeskunde, 244) Deutsche Akademie für Landeskunde, Flensburg

# 2 History of landscape assessment

Eckhard Müller, Martin Volk

## 2.1 Introduction

Human society has constantly undergone changes - not only with respect to quantitative aspects, such as population and its distribution, but also standards and judgements about life's values and its relationship with the environment. This connection between controlling and protecting the environment can be traced back to sources in the Old Testament in Christian culture: Genesis 1.28 "Be fruitful and multiply; fill the earth, and subdue it.:.." and Genesis 2.15.: "And the Lord God took the man, and put him into the garden of Eden to work it and to take care of it." We need look no further than here for the roots of an anthropocentric view of the world.

A change in the landscape is effected by mankind and his visible demands on the landscape. Table 2.1 shows the main aspects of the changes between society and landscape.

The authors justify the question mark hovering above the stage of landscape development in the post-industrial phase since not enough time has passed to be able to judge this development. There appears to be a growing intensity in the Mankind-Nature metabolism, with the formerly local effects now being seen and felt both regionally and globally. Priorities are set and decisions are made within the demands on nature by humans. It is thus important to be able to analyze objects, statuses and processes and, depending on their suitability, to be able to evaluate the consequences of their use or indeed their exploitation and their need for protection and organization.

## 2.2 The basis of landscape-ecological assessment

'Landscape' is one of the basic geographical terms whose meaning has been the subject of controversy for decades. This discussion influenced developments in the field of landscape ecology but also had the effect of making advances in landscape assessment a central point in landscape ecology hypotheses. Therefore the main terms will be discussed in the following sections.

### 2.2.1 The term "landscape"

The German word *Landschaft* or 'landscape' is based on the old German language (Old and Middle High German) and is "a collective term for land that belongs together with respect to its qualities." As a colloquial term it is used in several different senses (Schmithüsen 1964):

1. Pictorial representation of a section of the earth in art
2. A sensory impression of the earthly environment
3. Appearance of a part of the earth
4. Natural qualities of an area
5. Cultural qualities of an area
6. General character of an area of the earth
7. Restricted region of the earth
8. Political-legal society or organization
9. Area or expansion area of a certain category of objects

| | |
|---|---|
| Pre-neolithic phase | S Hunter and gatherer<br>L: Natural landscape |
| Neolithic revolution | |
| Pre-industrial phase | S: Agricultural society<br>L: Natural landscape→semi-natural cultural landscape |
| First industrial revolution (2nd half of the 19th century) | |
| Industrial phase | S: Industrial society<br>L: Cultural landscape dissociated from nature with relics of the semi-natural cultural landscape. Endangerment of the basics of life- water, air etc. |
| Second industrial revolution (2nd half of the 20th century) | |
| Post-industrial Phase | S: "Post-industrial society"<br>L: ?? |

S = Status of the development of society
L = Status of the development of the landscape

**Fig. 2.1:** Phases of the changes between society and landscape in central Europe (Buchwald, Engelhardt 1980)

Alexander von Humboldt is seen as the founder of scientific geography and it was he who described "the typical character of an area" as landscape. In the above list, definition 6 comes closest to this description. According to a study by Hard (1970, 1970a,b) the term 'total character' attributed to Humboldt in the literature up until recently (e.g. Naveh & Lieberman 1994) did not exist in his work

The question of whether landscape deals with an individual or a type is an important issue in the discussion of the term landscape (Paffen 1953, Schmithüsen 1964). In landscape physiology (Passarge 1912) the landscape is composed of a synthesis of a variety of single elements. This idea was employed again in the structuring of natural areas. The synthesis should thus take on central significance for landscape ecology.

Another important factor is dealt with in the discussion about the term landscape: the dimension. Troll (1950) defined small landscapes as typical area arrangements of physiotopes and ecotopes into a mosaic in which, in this definition, the smallest area components (physiotope and ecotope) are not taken into consideration. Neef (1967a,b) argues that the size and thus the related exclusion of total units cannot be a defining characteristic of a landscape (see also Carol 1957).

By landscape Neef (1967) understands "a concrete part of the surface of the earth that is characterized by its uniform structure and similar structure of effects."

Using this definition, landscape can be seen as a structurally differentiated part of the earth's surface as well as an area with a specific process-dynamic condition (Aurada et al. 1989). Most authors use these terms as they agree with the main landscape-ecological viewpoint.

It must be emphasized that this discussion over the most central term within the field of geography may have led to more confusion than clarity in neighbouring disciplines (Finke 1994). For this reason the question of the general validity, temporal dependency and the demand for detail of assessment should correspond to the theoretical statements by Neef (1967): "Geographical landscape is a compound, a material system resulting from various construction processes and different causal connections"; "all geographical systems exist in space and time"; "All geographical boundaries are boundaries in continuum"; "... they are lines or seams of a change in form."

## 2.2.2 The term (landscape) 'assessment'

Similarly to the term landscape, assessment is also used in a variety of ways in the literature. According to Wiegleb (1997) the following meanings can be distinguished:

1. Analysis ("evaluation" of data, real "assessment")
2. Judgement
3. Sequencing (relative comparison)
4. Current/potential status comparison (assessment in more precise sense)

Bastian (1997) used these meanings to develop his multi-step assessment model (Fig. 2.2).

Assessment requires the existence of a subject (to carry out the assessment) and an object (which is to be valued) that are related to each other (in this case the changing effect between society and nature). The assessment object (in this case landscape) is normally not recorded as a whole but instead through the use of a model. This reduction in complexity can be justified by keeping the time and effort spent by the surveyor manageable and the limited knowledge of the subject. Depending on the assessment goal, the same assessment object can have different meanings, but an agreement on the area relationship (scale) is necessary.

Until the 1970s landscape assessment normally had economic goals of land use based on increasing yields and the effective quarrying of resources, etc., which led to an intensification of land use. Little consideration was paid to the negative effects these one-sided uses of large areas of landscape had on the functioning of the ecosystems and on the countryside.

Since the early 1970s there has been an increase in the development of methodologies for ecological assessment. This enables the assessment of areal structures, functions and potentials with respect to the capabilities of the natural balance of the landscape. Kias and Trachsler (1985) mention that ecological assessment methods are always open to attack "since the real inter-relationships in a system are much more complex than their depiction in an assessment method." Thus the problem of landscape assessment becomes clear as it is always connected with anthropogenic interference in the natural balance. These interventions compel the assessment of ecosystems, which, in the natural sense, are actually free from values. Assessments arise to a certain extent out of certain social environments and are thus subject to changes. This can be clarified through the 'assessment' of a forest, whose nutritional and timber capabilities were the main concerns up until the 19th century (Mantel 1990). Until the mid-20th century the wake theory ("Kielwasser theory") represented the view that the protection and recreation function of a forest could fulfilled following in timber production's wake (Arbeitskreis Forstliche Landespflege 1994).

## 2.3 Use-related assessment of landscape components

### 2.3.1 Relief

The relief can be seen as a boundary area between the atmosphere/hydrosphere and the pedosphere/lithosphere (Dikau & Schmidt 1999). Due to its presence as a regulating factor in the landscape, it is rarely evaluated in isolation and usually incorporated into complex assessments, for example in processes such as water erosion or in the visual landscape quality. The assessment of relief should be related to characteristics of sculpture, structure and dynamics.

Many methods were developed between the 1960s and the 1980s to assess relief - an important landscape component due to its characteristic steering force. These methods enabled the geomorphological characteristics of areas and a geomorphological structuring of areas to be carried out. They include the work by Kugler (1977) on the geomorphological relief characteristics and the work of the DFG research programme "Geomorphological map 1:25,000 of Germany" (see Barsch et al. 1978). These two methods determine relevant information for geo-ecological questions from the genetic and sculptural survey of the geo-relief,

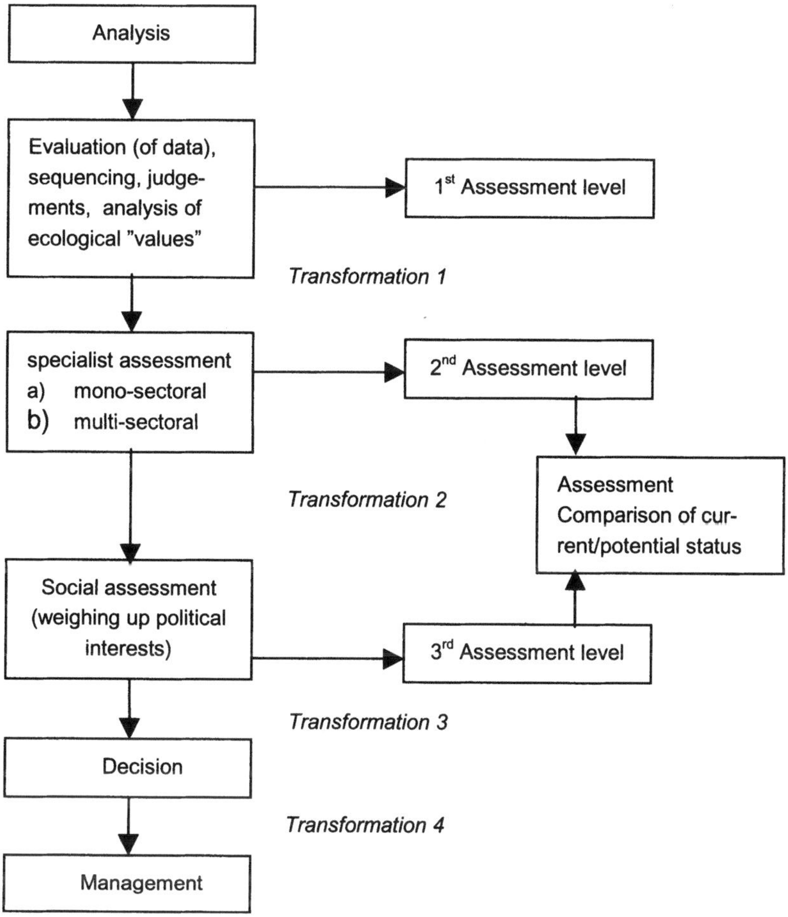


**Fig. 2.2:** Multi-step assessment model (Bastian 1997)

which can be used for practical applications. Examples include the assessment of the use of modern, technical equipment in agriculture and forestry, susceptibility to soil erosion, the building potential and the dynamics of drainage. They also include assessment for suitability for recreation with respect to psychological and physical attractions through the geo-relief. Despite the simplification and systemizing of the methods mentioned and the support of Kugler (1977) for a "geo-ecological geomorphology", the authors' expectations concerning the increased use of geo-morphological maps in practice has not materialized. This may be due to the very specialized point of view and the complexity of the maps (see also Finke 1980).

### 2.3.2 Soil

The main points regarding the assessment of soil stem from the functions of the soil such as:

- Comprising the basis for life and a habitat for people, animals, plants and soil organisms
- Forming part of the natural system (water, nutrient cycles)
- Its role in breakdown, compensation/balancing and production for material effects due to its filtering, buffering and material change characteristics, especially for the protection of groundwater
- A site for agriculture and forestry
- A location for other economic and public uses, e.g. production, waste disposal, transport
- An area for building development and recreation
- A 'stockyard' for raw materials
- An archive of natural and cultural history

The soil (as a component of the landscape) has a central position in the landscape balance due to its manifold functions. Neef, Schmidt & Lauckner (1961) state the soil type, in addition to the soil moisture regime and the vegetation, as being one of main ecological characteristics. Due to soil's economic importance, it is no surprise that an assessment methodology was developed quite early on - for example the soil value estimation 1935-45 (*Reichsbodenschätzung*). This standardardized method enabled a comprehensive analysis and assessment of the natural yield potential of the entire German Reich to be carried out. Finke (1994) emphasized that the results of this method can also be used today to provide important information about the soil and the general ecological situation of an area of landscape (see also Arens 1960, Mertens 1964). Work by, for example, Heinecke et al. (1986) shows that these data are also being increasingly used and evaluated in digital map work to solve ecological problems. However Bastian and Schreiber (1999) demonstrate the problem of the use of soil estimation since it does not relate to the current situation of the soil. In their opinion, the lack of continuing soil assessment in eastern Germany has prevented the knowledge base from growing. Thus the measures carried out later (such as soil, hydro and relief melioration) are not taken into consideration. However, local additions to the soil assessment were carried out in the 1970s and 1980s, including site surveys (Kasch 1967). They included the production of a soil structure map (interpretation of the substrate relationships, new survey of soil-water relationships, determination of the stone content of the soil) as well as a relief map (1:10,000), the latter providing almost blanket coverage for the whole of East Germany. In some districts, widespread local additions to the soil assessment were carried out (e.g. Frankfurt/Oder, Potsdam, Cottbus) whereas elsewhere (such as in Leipzig, Halle and Dresden) work was not even started (Lieberoth 1982).

In the states of western Germany, assessment of the soil was requested by the financial department and has been continued and updated ever since. In East Germany the mesoscale agricultural maps (*Mittelmaßstäbige Landwirtschaftliche*

*Standortkartierung* or MMK, scale 1:100,000) were drawn up between 1976 and 1982 to provide an evaluation basis for a land use appropriate to the whole agriculturally used area. The results of the MMK enabled the heterogeneous local units to be characterized by the soil substrate, soil profile and soil water relationships, the structure of the soil surface and the relief (Schmidt, Diemann 1981). It was possible to carry out assessments for planning industrial scale plant production using this information (Lieberoth 1980).

In 1952 the forestry mapping of East Germany began on a scale of 1:10,000 (Kopp 1975, Schwanecke 1983). In this mapping the local soil forms were determined from the links between the types of soil, soil water, relief and climate as well as the relationship with the vegetation type. Important values and relationships of these components are demonstrated in the explanations. In the widest sense the goal of local forestry mapping is to provide an assessment basis for optimum forestry planning (choice of tree species depending on afforestation methods, fertilizer recommendations, melioration, etc.), i.e. mainly for the biotic yield potential. Unfortunately local forestry mapping has been developed independently in the individual states and thus "despite mapping recommendations from the working group 'Local Mapping', it remains a very individual, state-related tool." (Bastian & Schreiber 1999).

Soil and rock are particularly important due to their buffering functions. Regarding these functions, assessment methods are based mainly on the issues of mechanical pollution, nitrates, heavy metals, organic pollutants and the groundwater protection functions as well as the choice of waste tip locations. The scientific discussions and debates about the reduction and recycling of waste and the choice of suitable locations together with the increasing public awareness led to the development of the technical directive (*Technische Anleitung/TA*) on waste (01.04.1991) and the *TA* on settlement waste (14.05.1993). These directives give strict criteria for the construction of surface tips.

Another issue ripe for assessment is the mechanical resistance of the soil to soil erosion by water and wind as well as to soil compaction. Soil erosion processes can be measured, mapped and modelled. A lot of mainly natural-science based methods are available. when evaluating the triggering moments and taking measures for soil protection, interdisciplinary methods must be applied together with practical users (Herweg 1999).

### 2.3.3 Water

The main fluxes of material and energy transport within the landscape ecosystems are carried by water (on the surface as well as in the unsaturated and saturated zone, e.g. Hermann 1977) - at least in temperate climates. The ecological importance of surface water means that plenty of methods and parameters exist for its characterization and assessment (water availability, flood dangers, self-cleansing abilities, retention of nutrients through protective strips on riverbanks etc.); these are described by Bastian and Schreiber (1999). These methods and parameters are also used to estimate the water quality and the degree of stress or the stress limit and are partly integrated into the methods of watercourse assessment or the assessment of ecological water quality.

General methods used for watercourses comprise species and diversity indices, lack of species, the halobien index, hemeroby and others. The best known and typical parameter for flowing water - besides the value from individual organisms and the macro index - is the saprobiotic index. Assessment tables can be used to judge the eco-morphological state of the naturalness of a watercourse (see Werth 1987). Giesshübel (1993) described a system based on nature conservation for survey and assessing watercourse structures using aerial photographs. The assessment of a lake or other large standing water bodies can be carried out using criteria such as the oxygen content, nutrient conditions and the organisms living in the water. The choice of further technical literature for the characterization and assessment of surface watercourses has become almost overwhelming. Several are named to provide an overview: Succow (1991), Friedrich & Lacombe (1992), Klapper (1992) and Schönborn (1992).

The groundwater fulfils the most important ecological function of a regulator (e.g. transport of material) due to its seasonal variability. For this reason numerous methods for the documentation or assessment of the available field capacity (soil water storage available for plant uptake), ecological moisture level, groundwater replenishment, water penetrability of the soil etc. have been developed (AG Bodenkunde 1996, Thomas-Lauckner & Haase 1967). Depending on the parameters of the soil, condition values are allocated (Marks et al. 1992). Methods have also been developed which allow the conclusion of important ecological soil parameters through vegetation surveys, or the indication of soil characteristics through ecological indicator plants (Ellenberg 1991, Kunzmann 1989).

### 2.3.4 Climate

This component of the landscape provides a category for assessment due to the mixing of global, regional and local influences of human beings in the interaction between the atmosphere and the earth's surface. Although rules (such as air quality regulations) and environmental laws are usually specific to one state or country, climatic influences do not stop at state boundaries. Thus there is a problem of assessment in border areas. There are no legal frameworks for many meteorological elements.

The main points for assessment are in the field of:

- Air quality, movement of air and heat stress
- Assessment of growth climate structures
- Areas endangered by late frosts
- Bio-climatic effects
- Climatic compensation in urban landscapes
- Emissions and noise protection through vegetation structures.

In addition to the purely climatic questions, landscape-ecological studies try to clarify and evaluate the interaction of the climate with other parts of the landscape as a whole. In bio-ecological (Ellenberg 1982, Schmithüsen 1968), agricultural (Uhlig 1954 ) and urban-climatic studies (e.g. Jacobeit 1996) and within other planning problems (e.g. suitability for recreation, Knoch 1963), climatic elements are evaluated with respect to their effects on the suitability for use and/or their ecological effects. Unfortunately information on the local climate is usually not available for planning; apart from in the urban region of the Ruhr (see Klink 1990, KVR 1990).

### 2.3.5 Flora

Assessment of flora is almost focussed on:

- The naturalness of vegetation
- The ability for regeneration and replaceability
- Diversity
- Regional viewpoints
- Rarity/endangerment of species and ecosystems
- Complex habitat value
- Indirect assessment criteria for the functions of habitats

The work by Schlüter (1985, 1991) is cited as an example of the assessment of vegetation as a landscape component. Since the naturalness of the vegetation depends to a large extent on the intensity of the use and stress on an area, which is shown by the level of human influence on an ecosystem, it is an important aspect for the ecological assessment of habitats or landscapes. The 9-point scale for the naturalness of vegetation used by Schlüter (1991) enables an ecological vegetation assessment to be carried out of the effects of all types of use of differing intensity on the vegetation for areas of similar use, and their mapping on a large scale. The availability of land use maps, aerial photos and satellite images makes the surveying and mapping of the current ecological state of the vegetation for large areas possible with justifiable output.

### 2.3.6 Fauna

According to the Environmental Impact Assessment (EIA; *UVP/Umweltverträglichkeitsprüfung*) in German conservation law, animals should be integrated more closely into landscape ecology studies as indicators of certain environmental conditions. In addition to their mobility, their hidden ways of life and their often unknown ecological requirements compared to vegetation, however, it is "due to the large number of animal species as well as financial, personnel and time difficulties ... largely impossible to obtain a comprehensive overview of the animal world, even for small areas. Thus representative species or groups of animals must be selected." (Bastian & Schreiber 1999, Zucchi 1990) Trautner (1992) and Reinke (1993) give comprehensive reports of the methodical standards for surveying animal species groups in ecological studies. Leser (1991a) criticizes the assessment methods, which are used selectively and are thus are unable to survey or portray the functions of the entire area. Finke (1994) points out that this is "expressly aimed at" by the evaluators, since only the aspects relevant to the question are surveyed, evaluated, weighted and aggregated to provide a result. He also points out that improved knowledge about the entire interactions of the system is needed to select the convenient indicators and to link them to a judgement about suitability, pollution effects, endangerment, the need for protection etc.

## 2.4 Integrated assessment of the potential and functions of landscapes

In connection with the criticism by Leser (1991b) (i.e. on the surveying and assessment of the function of the entire area), the integration of ecological conditions in assessment and planning with respect to the priorities for the development of areas was strongly supported back in the early 1970s (Buchwald & Engelhardt 1980).

### 2.4.1 Potential

Differentiation of specific capabilities of the natural area into the so-called partial natural area potential (biotic yield potential, water availability potential, waste removal potential, biotic regulation potential, geo-energetic potential, raw material potential, building and recreation potential) was carried out by Haase (1978) based on the use of the term 'potential' (landscape potential) by Neef (1966, 1969). Representative terms are coined for many exemplary studies, e.g. building potential (Hrabowski 1978), waste disposal potential (Wedde 1983), water availability potential (Müller 1984).

Mannsfeld (1983) suggested an algorithm for the determination of natural area potential units of landscape areas. This algorithm covers the selection of indicators, assessment, and the determination of suitability ranks of natural area units for a partial natural area potential. Comparisons of potentials (Barsch & Schönfelder 1983) provide indications concerning the types of multiple use and possible conflicts that may arise.

### 2.4.2 Functions of landscapes

The 'natural' or 'environmental functions' can also provide a basis for the assessment of landscapes. De Groot (1992) defines them as the capacity of natural processes and components to make goods and functions available which fulfil human needs directly or indirectly. The natural functions are then divided into four categories: regulation function, carrying function, production function and information function. These categories are verified using examples with several functions.

According to Bastian and Schreiber (1999) "only integration (of nature area and landscape points of view) leads to natural areas being described as landscape, due to the use of the area by society down the ages, and to the use of landscape units as synthetic units for the determination of the functional capability." The term *landscape function* evolves from the assessment of the landscape structure. Bastian (1991) distinguishes between the following groups of functions: production functions (economic functions), regulation functions (ecological functions) and habitat functions (social functions), which are subdivided so that the "individual effect and interactive relationships between the user's requirements on the one hand and the landscape structure on the other become recognizable."

According to Marks et al. (1992) the capabilities of the landscape stem from the sum of the functions and potentials. Thus they suggest that the term 'function' should be used as a basis for the tasks and capacity of the landscape (soil/relief, water, climate/air, biotic functions, recreation). By contrast, the term 'potential' remains limited to the natural area potential and real objects (e.g. the relief of the landscape) or there is a viewpoint that it should no longer be used due to its orientation towards resources (Daniels & Lüttich 1982) and the confusion caused by this at ecological assessment and planning (Bastian & Röder 1996). Although there is general agreement surrounding the recognition of the groups of functions, no uniform methodology can be recognized with regard to the further differentiation, the selection of determining characteristics, the related scale of the data sources and the use of the results of the interpretations and assessments (Bastian & Schreiber 1999).

### 2.4.3 Landscape elements, landscape units

The typification, characterization and categorization of landscapes, taking into account the characteristics of nature areas and land use, provide the basis for the sustainable development of landscape. Thus landscape is hierarchically viewed for the categorization and the comparison of study areas. This structuring can occur using natural area characteristics but is also possible using land use. Thus *landscape units* can be defined by relatively stable nature area and anthropogenic factors and are definable through a specific gradient of landscape elements of marked landscape sectors of various dimensions (Niemann 1982). Landscape elements are the smallest structural units that can be defined using current land use, structural characteristics and social effects, although a hierarchical subdivision into homogeneous and heterogeneous landscape elements can be considered. In contrast to De Groot (1992), who sees the functions as having a global character, Niemann (1982) visualizes the landscape-ecological functions as being identical with the landscape-ecological effects in connection with structure and site characteristics - thus having a concrete relationship to the landscape.

The development of the *Multiverfahren* (multiple methods) began in the mid-1960s. Within this approach, several forms of land use are evaluated, and ideas of environmental protection gradually found a way onto the assessment scene (see Brahe 1972, Bischoff, Weller & Gekle 1974/75, Bechmann 1977, Marks 1979).

Towards the end of the 1970s, Niemann began developing the *polyfunctional assessment system.* This work is based on landscape units or landscape elements and thus reached a congruency between management and assessment units. Despite using cost-benefit analysis, he tried to integrate numerous ecological situations. Although the benefit here is also mainly defined by means of economic points of view, the resulting integration of ecological situations (e.g. regulation function of the landscape) with regard to land use should be seen in a positive light. Ecological assessment methods thus became increasingly important (e.g. Buchwald &Engelhardt 1980). At the same time as these authors were using a point system for their assessment (which included cost-benefit analysis methods), Sporbeck (1979) continued to use the strict form of cost-benefit analysis. He analysed and evaluated the physical potential use of two study areas affected by coal mining - not only the landscape before the mining of coal, but also after recultivation e.g. for agriculture, forestry or recreation. The basis of the assessment results (areas with differing grades of suitability) and a potential comparison between the two conditions made it possible to obtain increases or decreases in value for potential use and to differentiate these into complexes of soil, relief and local climate. It was thus clear from this study that even when using pure cost-benefit analysis, surveying and assessing the natural potential cannot be covered by one scientific discipline alone, thus demonstrating the need for an interdisciplinary taskforce.

Zangemeister (1973) and Bechmann (1978) developed the mathematical assessment models of cost-benefit analysis of the first and second generation. These models are designed to enable quantification of the assessment for recreation, so that it can be set up more objectively. According to Finke (1994) "there is an increase in criticism about the cost-benefit analysis (developed in economy), which should not be used in a strictly methodical manner in the context of ecosystems." Thus it was that a move away from the pure use of the cost-benefit analysis ensued in the following years (Leser & Schmidt 1981).

### 2.4.4 GIS-aided methods

With the rapid development and use of computers and software, in particular of geographical information systems and their linkage to modern landscape assessment methods, there is further hope that more complex landscape-ecological topics can be developed. Such methods appear to enable the management requirement for the improvement of the ecological situation (including the economic interests and requirements) in landscape units to be worked out.

Investigations have already been conducted into the possibility of simplifying the analysis of the influence and function of the partial complex 'relief' through digital terrain models (DTM). Computers are used since investigations of large areas usually involve time-consuming mapping and field work. Strobl (1988) demonstrates the possible use of DTM from the "geographic-landscape-ecological point of view" with geomorphologic, climatic, hydrolic and landscape-ecological aspects. Despite the available software, the production, availability and precision of DTMs remains a problem, which also limits their incorporation into landscape ecology and landscape planning (Volk & Steinhardt 1996). Köthe & Lehmeier (1993) developed a system for automatic relief analysis (SARA) for the DTM, which has found limited access in practice. Ideally digital terrain models (integrated into a GIS) allow - apart from the visualization of landscape structures (Suter et al. 1996) - the analysis and simulation of relief characteristics as an important steering force in the landscape. Unfortunately, trends seem to show that usually only visualization is used in practice. Besides the computer-aided methods for the analysis and assessment of individual partial complexes, GISs also offer the potential for carrying out complex, integrated assessment tasks.

The 'integrated approach' is oriented towards the regulation and modelling concepts of the geo(eco)system (Mosimann 1985; Klug & Lang 1983). The aim is to improve the link between the availability of means of environmental conservation and recreation with the concurrent supply of consumables and environmental goods in regional approaches (SRU 1996). Barsch & Saupe (1994) see the modular integration of landscape-ecological and socio-economic data in spatial planning as a way of reducing and resolving user conflicts between ecological and economic interests. As an example of this complex integrated approach (including functions and potentials) they employed assessment methods in a GIS (ArcInfo) together with digital image analysis using the Erdas Imagine image processing software. In connection with this a 'fuzzy set data model' was defined, for both the integration of measured and estimated values and the adaptation of various methods to the assessment scale while taking account of the sites' existing heterogeneity. This allowed an assessment to be carried out even when uncertain or incomplete data sets are used.

Grabaum (1996) developed a computer-aided method in which landscape elements of a landscape unit can be altered via the optimum consideration of various goals and the weighing-up of ecological and economic interests. This method enables for example the land use of an intensively used agricultural area (mainly cultivated land) to be optimized using the optimum compromise between the four goals of reduction of erosion, retention of drainage regulation, improvement of groundwater production and retention of the production, as well as from an ecological point of view (Meyer & Grabaum 1997). This method has been tried out on a topological level and requires numerous, relevant information levels of various types, which are only available in the rarest of cases for large study areas, and so usually need to be created in a labour-intensive process.

## 2.5 Systemizing different landscape assessment methods

Marks et al. (1992) show that all ecological assessment methods developed so far can be traced back to four "basic methodological patterns". The structure of these methods and the basic pattern of the different landscape assessment methods (Fig. 2.3) are systematically drawn together by Hase (1996).

### 2.5.1 Unsolved problems in assessment methods and future perspectives

When reviewing the development and current status of landscape-ecological research, many problems come to light in the methods of landscape assessment. One of the main problems is that despite the development of a multitude of methodologies, there is not yet an area or a landscape assessment system that can be used generally (see Leser 1991a). The surveying, assessment and depiction of ecological phenomena can only be incomplete since these phenomena are too complex and so far not enough is known about the ecological structures. Thus "the aggregation of several ecological parameters in single processes is always afflicted with uncertainty, since the interactions are unknown" (see Marks et al. 1992). Richter et al. (1997) mention the fact that when carrying out surveys of geo-referenced processes on different scales, it becomes clear that not all processes relevant to landscape ecology are precisely quantifiable. Even the modelling of well known physical processes (e.g. water and material transport in the soil) is beset by uncertainty over the model's parameters on the scale of the landscape.

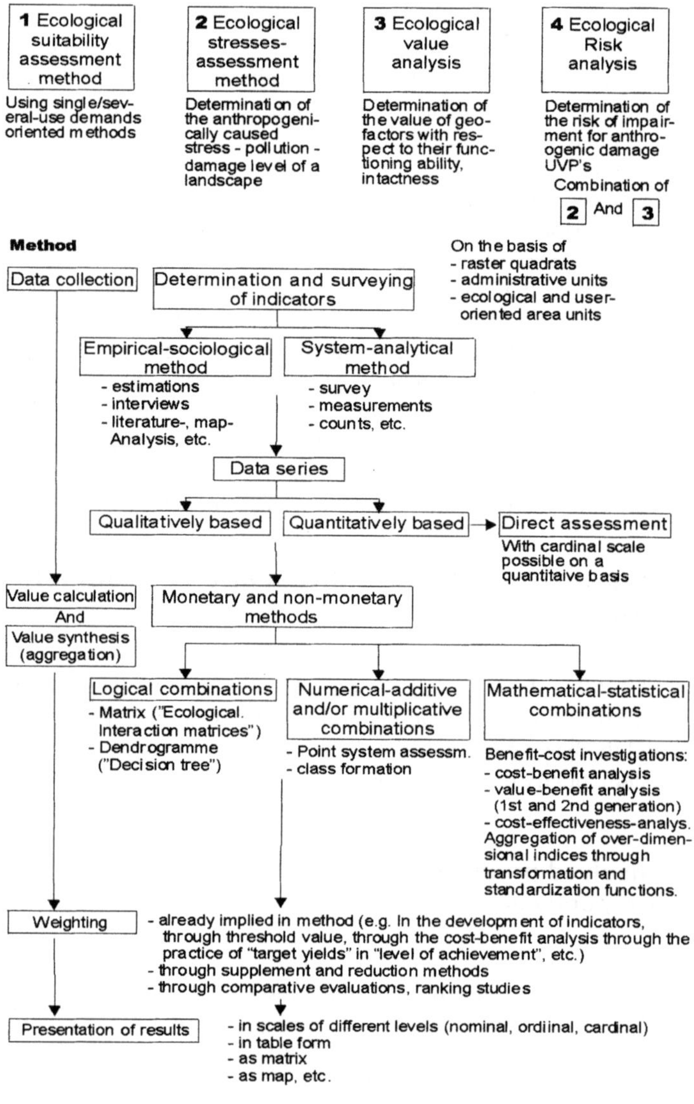


**Fig. 2.3:** Various basic pattern of ecological assessment methods

In relation to the review and ease of transfer of assessment methods to different scales, which play a major role not only in landscape ecology but also in spatial planning practice, it can be said that there is no generally valid theory "that allows the derivation of rules for, for instance, regionalization." (Richter et al. 1997) At present there are two different basic points of view: on the one hand efforts are being made to find a model that allows the transfer of methods valid in a small area to larger areas using suitable indicators or transfer functions ("upscaling").

For this to work, new transfer functions have to be developed for each new area of use. In terms of soil science, this problem appears to have been solved successfully through "pedotransfer functions" (Tietje & Tapkenhinrichs 1993). The other opinion is that specific working methods are required on each survey level. Future research should be carried out here in order to answer the question about connection points between the two points of view. Without looking at this problem it would certainly be advantageous for regionalization methods if the use of uncertain data at various scales were enabled, since the spatial and/or temporal resolution of available data for the study areas in different project themes is insufficient (Moevius & Murschel 1997). The use of 'fuzzy-set data models' (Steinhardt 1998) is plausible here; their use is discussed in this volume.

Another problem exists in the usability of methods developed in research institutes without explicit practical application. Finke (1994) emphasized that Haase's landscape ecology spatial structuring (Haase 1968) is one of the few geographical methods geared towards practical implementation in agricultural planning. In contrast to the pure, scientifically based landscape-ecological basic research, which currently endeavours to do things in a 'holistic' way (i.e. to consider as many ecological and economic parameters or their interactions within and among landscape ecosystems as possible), the practice-oriented work already consciously selects the data required. Due to financial and temporal limits, this is reduced to the main aspects, which means that ecological points are often neglected. Kleyer et al. (1992) note that landscape planning is thus often subject to juggling with other user interests, because the proof of damage of environmental conditions can be neither carried out precisely nor undertaken for the whole area or include the most important interactions. According to Kiemstedt (1979) this indicates that an ecological orientation of the spatial planning in the sense of a comprehensive steering of the whole ecological system oversteps the working framework of our sociopolitical relationships. Thus for example integrated methods in environmental planning are defined as neither classical spatial planning nor specialist planning: according to Peters (1996, in Durwen 1997) they are based much more on interrelationships, evaluated and defined environmental standards. Within the planning balance an optimization rule is formulated in which locational and regional environmental plans are to be developed. In practice it must be underlined that assessment methods "must be as simple and comprehensible as possible, so that the interested citizens are able to reconstruct the results and thus understand the justified decisions." (Finke 1994) From the above contrasting requirements it is clear that a compromise is required between the "scientific exact, holistic" research method and the "clear, economic and comprehensible" method for the practical user. This problem can only be solved through dialogue between researchers and planning practice.

Over the past few years Environmental Information Systems have been developed by many authorities to provide bases for more complex landscape-ecological assessment (Senstadtum 1990, Anders 1997). It must be noted that their practical use is limited due to the lack of sufficiently qualified personnel (who must be able to manage often highly complicated systems). Moreover, there is often a lack of basic data, or the available data are heterogeneous because they stem from different requirements over the course of time that were or are dependant on each other. Unfortunately many of these data sets have not been continued or updated, turning them into individual records of a single time period. It is hoped that in the coming years the data base will be improved and that generally valid integrated assessment methods will be installed.

This presupposes the recognizable trend for the development of 'user-friendly' systems, which is apparent from the use of ArcView, "a little sister of the highly complex system ArcInfo" (Kratz & Suhling 1997). Two fundamental requirements are an effort to improve cooperation between local government departments, agencies and research institutions, and also the legal simplification of the necessary exchange of data (DFG 1996). It must be emphasized here that the extent of the implementation of assessment methods and development scenarios in practice often has less to do with the quality of the results than with economic requirements and political structures (Roweck 1995).

## 2.6 Assessment focal points in German ecosystem research centres

Since it is impossible here to include all research and planning institutions in Germany that deal with landscape assessment, a representative selection is given along with examples of their research field :

- UFZ Centre for Environmental Research Leipzig-Halle

This is a recognized centre of expertise on the restoration and renaturation of polluted landscapes, as well as the conservation of semi-natural landscapes. Its principal research topics are:

- Rehabilitation and renaturation strategies for highly polluted landscapes
- Inland water: development and règeneration
- Urban ecology and urban development
- Biodiversity, land use and resource protection
- Development and application of biotechnological processes and methods

These tasks are carried out in twelve departments, four interdisciplinary projects and two scientific groups within the institution. The research carried out by the Department of Applied Landscape Ecology contributes to sustainable landscape development. The focus is on the ecological consequences of land use and changes in land use. The department studies structures, functions and balances within landscapes at different spatial dimensions. The information generated is used to evaluate the functional performance of landscapes, with particular emphasis placed on multi-functionality. While selecting and developing scale-specific assessments, problems of data aggregation, standardization, regionalization and upscaling data are taken into consideration. The assessment results are aimed at users in planning practice. (http://www.ufz.de 2000)

- Ecology Center of the University of Kiel (ÖZK)

The centre maintains a research department specializing in environmental assessment and planning. Its work includes a management-oriented analysis of the results of ecosystem research focusing on the use of assessment criteria for the development of landscape conservation, renaturation and management strategies. An integrated approach is taken including the following contributions:

- Community ecology contributions (vegetation, ornithological; terrestrial and limnological invertebrate research)
- Contributions of ecotone research
- Hydro-chemical contributions
- Contributions from the research field "movement of materials"
- Ecotoxological contributions.

Study areas are situated in the state of Schleswig-Holstein, in particular the region of the Bornhöveder lakes. (http://www.pz-oekosys.uni-kiel.de 1999)

- Centre for Agricultural Landscape and Land Use Research at Müncheberg (ZALF)

The primary scientific objective of the center is to analyze, evaluate and predict processes (including their interactions) in agricultural landscapes of the northeast German lowlands. The properties and potentials of landscapes are evaluated with emphasis on ecological and economic aspects.

For example the Institute of Soil Landscape Research has carried out studies on the:

- Analysis, assessment and description of soil functions as a means of controlling landscape water balance (field scale to regional scale)
- Development and validation of a soil indication and assessment system for the derivation of a decision support system for strategies of soil protection
- The socioeconomic assessment of substance transfer in landscapes through erosion
  (http://www.zalf.de 1999)

- Institute of Landscape Planning and Ecology of the University of Stuttgart (ILPÖ)

Its main research topics are in the following areas:

- Species and habitat conservation research (nature conservation, ecosystem research and population biology)
- Protection of the capability of natural systems (landscape analysis, landscape assessment, modelling, GIS, data preparation)
- Practical implementation in spatial planning (landscape planning, state planning, environmental impact assessment, agricultural planning, regionalization)

The second field is seen not as a biological or technical problem, but rather as a prerequisite for qualified planning. The goal is thus to analyze and evaluate the effects of changes of use on the natural system in planning on all levels and in all

specialist areas. The dimensions of the research range from large-scale urban planning to the development of guidelines to support sustainable agricultural at the level of the European Union.
(http://www. uni-stuttgart.de/forschung 1999)

- Research Center on Forest-Ecosystems of the University of Göttingen (FZW)

The main focus is on the development of a model for multifunctional forestry use as well as the assessment of habitat, regulation, production and social functions of the forest. Other tasks relate to biodiversity and regionalization as well as the emergence and whereabouts of climatically relevant atmospheric gases. Long-term study facilities are maintained in Solling, Harz, Göttingen Forest, Lüneberger Heide and near Kassel, which range in size from several hundred square metres to 76 ha. The extensive information system is to be specified at the level of forestry organizations and the landscape.
(http://www.gwdg.de/~fzw 1999).

- Bayreuth Institute for Terrestrial Ecosystem Research (BITÖK)

A central research theme of the institute is to determine how altered boundary conditions (e.g. climate, nitrogen and sulphur deposition) affect forest ecosystems in the long term. An estimation of the stability and load-bearing capacity of the forest ecosystem under anthropogenic environmental pollution and an assessment of the sustainability of the use of forest functions follow on from the main study. An assessment of the forestry techniques is being undertaken using simulation models of the growth of the forest. The largest units studied are small forested watersheds, e.g. Lehstenbach in the northern Fichtelberg (Fichtel mountains), which measures 4 $km^2$. Selected forest stands are also being used as intensive sites for investigations of processes and experiments under controlled laboratory conditions. (PGM-Statistik 2000, http://www.uni-bayreuth.de 2000)

- Munich Research Alliance for Agroecosystems at Freising-Weihenstephan (FAM)

The main research topics are the control, quantification and assessment of interventions in land use on oxygen, carbonates, water, material, organisms and economic balances. The aim is to investigate the effects of environmentally beneficial agricultural systems on the quality of resources and on the steering processes in a typical agricultural central European landscape. Representative indicators are used to evaluate the effects, and forecasting models are to be developed on planning relevant scales. The study area is the monastery in Scheyern, north of Munich. Studies are being carried out on two agricultural systems measuring 114ha and on study parcels (38ha). (PGM-Statistik 2000, http://fam.weihenstephan.de 1999)

## 2.7 Outlook/ Perspectives

An attempt has been made in this chapter to crystallize the development and status of landscape assessment from the almost overwhelming number of methods that exist. The advantages of the currently favoured methods geared towards a 'holistic' process should not be ignored despite the numerous problems which beset their use. However the development of generally valid methods and definitions should be the main goal of future research. Neef's conclusion that the method of a truly scientific synthesis still has to be developed remains valid (Neef 1967b). Integrated methods endeavour to include as many ecological and economic parameters or their interactions with and among landscape ecosystems as possible and thus try to come as close as possible to the complex reality.

Since quantification in relation to the above methods is not completely possible and also from "the experimental side of modelling often only incomplete and/or imprecise data are available" (Richter et al. 1997) and the fact that sustainable landscape development requires compromise-oriented decisions, new ways have to be found and adopted. The development and inclusion of mathematical methods in landscape assessment, with respect to the use of uncertain data sets, and linking ecological models to geographic information systems (see Johnston 1993) for future decision making is of vital importance, especially with respect to the transferability of assessment methods and models to smaller scales (regionalization). The inclusion of economic viewpoints (as in Dabbert, Herrmann et al. 1999) is becoming increasingly important with regard to ecological and economic assessment in various scenarios as well as the modelling and assessment of landscape-ecological topics such as nitrogen leaching or changes in vegetation through eutrophication.

## 2.8 References

AG Bodenkunde (1996) Bodenkundliche Kartieranleitung. Hannover

Anders V (1997) Das Umweltinformationssystem des Kreises Merseburg-Querfurt. Workshop "Digitale Geowissenschaftliche Daten - Bedarf, Nutzung, administrative Regelungen", Halle Tagungsband, pp 13-14

Arbeitskreis Forstliche Landespflege (1994) Waldlandschaftspflege. Ecomed, Landsberg/Lech

Arens H (1960) Die Bodenkarte 1:5000 auf der Grundlage der Bodenschätzung, ihre Herstellung und ihre Verwendungsmöglichkeiten. - Fortschritte in der Geologie von Rheinland und Westfalen 8

Aurada D, Haase G, Neumeister H (1989) Funktionsweise und Leistungsfähigkeit der Landschaft. In: V. Geogr. Kongress d. DDR, Potsdam, pp 7-8

Barsch D et al (1978) Das GMK Musterblatt für das Schwerpunktprogramm geomorphologische Detaikartierung in der Bundesrepublik Deutschland. Berliner Geogr. Abh. 30:7-19

Barsch H, Saupe G (Hrsg) (1994) Bewertung und Gestaltung der naturnahen Landschaft in Schutzgebieten, Erholungs- und Freizeitgebieten. Potsdamer Geogr. Schriften 8

Barsch H, Schönfelder G (1983) Landscape Diagosis as a Geographical Contribution to Landscape Management. GeoJournal 7: 2-

Bastian O (1991) Biotische Komponenten in der Landschaftsforschung und Planung. Probleme ihrer Erfassung und Bewertung. - Habil.-Schr., Matin-Luther-Univ. Halle-Wittenberg

Bastian O (1997) Gedanken zur Bewertung von Landschaftsfunktionen - unter besonderer Berücksichtigung der Habitatfunktion. NNA-Berichte 10:106-125

Bastian O, Schreiber K-F (Hrsg) (1999) Analyse und ökologische Bewertung der Landschaft. - Umweltforschung, Spektrum, Heidelberg

Bastian O, Röder M (1996) Beurteilung von Landschaftsveränderungen anhand von Landschaftsfunktionen. Naturschutz und Landschaftsplanung 28:302-312

Bechmann A (1977) Ökologische Bewertungsverfahren in der Landschaftsplanung. Landschaft + Stadt 9: 170-182

Bechmann A (1978) Nutzwertanalyse, Bewertungstheorie und Planung. Haupt Bern, Stuttgart.

Bischof T et al (1974/75) Verbesserte Beurteilung landwirtschaftlicher Flächen in der Agrar- und Landschaftsplanung. Berichte über die Landwirtschaft 52. 547-557

Brahe P (1972) Matrix der natürlichen Nutzungseignung einer Landschaft bei der Auswertung landschaftsökologischer Karten für die Planung. Landschaft + Stadt 4. 133-141

Buchwald K, Engelhardt W (Hrsg) (1980) Handbuch für Planung, Gestaltung und Schutz der Umwelt. Bd.: 3 Die Bewertung und Planung der Umwelt. BLV Verlagsgesellschaft, München

Carol H. (1957) Grundsätzliches zum Landschaftsbegriff. Petermann's Geogr. Mitt. 101: 93-97

Dabbert S, Herrmann, S. et al. (1999) Landschaftsmodellierung für die Umweltplanung. Springer, Berlin-Heidelberg-New York

Daniels Cv, Lüttich W (1982) Geowissenschaftliche Karten des Naturraumpotentials als Unterlagen für Raumordnung und Landesplanung. Graz

Deutsche Forschungsgemeinschaft (DFG 1996) Forschungsfreiheit - ein Plädoyer der DFG für bessere Rahmenbedingungen der Forschung in Deutschland. Bonn

de Groot RS (1992) Functions of Nature. Wolters-Noordhoff, Groningen

Dikau R, Schmidt J (1999) Georeliefklassifikation. In: Schneider-Sliwa,R. Schaub, D., Gerold, G. (Hrsg) Angewandte Landschaftsökologie. Grundlagen und Methoden. Springer, Berlin-Heidelberg-New York

Durwen K-J (1997) Landschaftsökologisches Informations-System und digitaler CD-Atlas BADEN-Württemberg: Nutzung für großflächige Schutzkonzepte. In: Kratz, R. & F. Suhling (Hrsg): GIS im Naturschutz: Forschung, Planung, Praxis, pp 223-233

Ellenberg H (1982) Vegetation Mitteleuropas mit den Alpen in ökologischer Sicht. Stuttgart

Ellenberg H jun (1991) Ökologische Veränderungen in Biozönosen durch Stickstoffeintrag. Arten- und Biotopschutzforschung in Deutschland. Ber. ökol. Forsch. 4: 75-90

Finke L (1980) Anforderungen aus der Planungspraxis an ein geomorphologisches Kartenwerk. Berliner Geogr. Abh. 31: 83-90

Finke L (1994) Landschaftsökologie. - Das Geographische Seminar, Westermann, Braunschweig

Friedrich G, Lacombe J (Hrsg) (1992) Ökologische Bewertung von Fließgewässern. Limnologie aktuell 3

Giessübel J (1993) Erfassung und Bewertung von Fließgewässern durch Luftbildauswertung. Schriftenreihe für Landschaftspflege und Naturschutz, pp 37- 77

Grabaum R (1996) Verfahren der polyfunktionalen Bewertung von Landschaftselementen einer Landschaftseinheit mit anschließender "Multicriteria Optimization" zur Generierung vielfältiger Landnutzungsoptionen. Shaker-Verlag, Aachen

Haase G (1968) Inhalt und Methodik einer umfassenden landwirtschaftlichen Standortkartierung auf der Grundlage landschaftsökologischer Erkundung. Wiss. Veröff. Dt. Inst. F. Länderkunde, N. F. 25/26, pp 309-349

Haase G (1978) Zur Ableitung und Kennzeichnung von Naturpotentialen. Petermann's Geogr. Mitt. 122: 113-125

Hard G (1970) Die "Landschaft" der Sprache und die "Landschaft" der Geographie. Semantische und forschungslogische Studien zu einigen zentralen Denkfiguren in der deutschen geographischen Literatur. - Colloquium Geographicum II

Hard G (1970a) Noch einmal: "Landschaft als objektivierter Geist". Zur Herkunft und zur forschungslogischen Analyse eines Gedankens. Die Erde 101: 171-197

Hard G (1970b) Der “Totalcharakter der Landschaft”. Re-Interpretation einiger Textstellen bei Alexander von Humboldt. - In: Alexander von Humboldt. Eigene und neue Wertungen der Reisen, Arbeit und Gedankenwelt. Geogr. Z., Beih.: 49-73.

Hase E (1996) Grundlagen, Problemfelder und Konsequenzen von Landschaftsbewertungsverfahren. In: Preu, C. & P. Leinweber (Hrsg): Landschaftsökologische Raumbewertung: Konzeptionen - Methoden - Anwendungen. Vechtaer Studien zur Angewandten Geographie und Regionalwissenschaft. Bd. 16, pp 23 - 32

Heineke H-J et al. (1986) Automatisierte Aufarbeitung der Daten der Bodenschätzung zur Unterstützung der bodenkundlichen Landesaufnahme und zur Ableitung planungsrelevanter bodenökologischer Kriterien. Kieler Geogr. Schriften 64:19-29

Herrmann R (1977) Einführung in die Hydrologie. Teubner, Stuttgart

Herweg K (1999) Von der Bodenerosionsforschung zum angewandten Bodenschutz. In: Schneider-Sliwa R, Schaub D, Gerold G (Hrsg) Angewandte Landschaftsökologie. Grundlagen und Methoden. Springer, Berlin-Heidelberg-New York

Hrabowski K (1978) Die ökonomische Bewertung des naturräumlichen Bebauungspotentials. - In: Beiträge zur planmäßigen Gestaltung der Landschaft. Wiss. Abh. Geogr. Ges. der DDR 14.

http://www.fam. weihenstephan.de (1999)

http://www. gwdg.de/~fzw (1999)

http://www. pz-oekosys. uni-kiel.de (1999): Umweltbewertung und -planung

http://www.ufz.dc(2000): Umweltforschungszentrum Leipzig-Halle GmbH. Forschungsprogramm, Organisation

http://www. uni-bayreuth.de (2000): Forschungsbericht

http://www. uni-stuttgart.de/forschung (1999): Forschungsbericht

http://www. zalf.de (1999): Zielsetzung und Aufgabengebiet des ZALF

Jacobeit J (1996) Bewertungsverfahren in der Stadtklimatologie. Vechtaer Studien zur Angewandten Geographie und Regionalwissenschaft 16, pp 33-42

Johnston CA (1993) Introduction to quantitative methods and modeling in community, population and landscape ecology. - In: Goodchild MF, Parks BO, Steyaert LT: Environmental modeling with GIS, pp 276-283

Kasch W (1967) Arbeitsrichtlinie zur Durchführung der Standortkundlichen Ergänzung der Bodenschätzung. Akademie der Landwirschaftswissenschaften zu Berlin

Kias U, Trachsler H (1985) Methodische Ansätze ökologischer Planung. - Schriftenreihe zur ORL 34, pp 53-77

Kiemstedt H (1979) Methodischer Stand und Durchsetzungsprobleme ökologischer Planung. - FuS 131: 46-62

Klapper H (1992) Eutrophierung und Gewässerschutz. G. Fischer, Stuttgart

Kleyer M et al (1992) Landschaftsbezogene Ökosystemforschung für die Umwelt- und Landschaftsplaunng. Z. Ökologie u. Landschaftsplanung 1: 35-50

Klink H (1990) Ergebnisse siedlungsökologischer Untersuchungen im Ruhrgebiet. BDL 64/2: 299-344

Klug H, Lang R (1983) Einführung in die Geosystemlehre. Die Geographie. Einführungen in Gegenstand, Methode und Ergebnisse ihrer Teilgebiete und Nachbarwissenschaften. Wissenschaftliche Buchgesellschaft, Darmstadt

Knoch K (1963) Die Landesklimaaufnahme. Wesen und Methodik. Berichte des Deutschen Wetterdienstes 85, Bd.12, Offenbach

Kopp D (1975) Kartierung von Naturraumtypen auf der Grundlage der forstlichen Standortserkundung. Petermann’s Geogr. Mitt. 119: 96-114

Köthe R, Lehmeier F (1993) SARA - Ein Programmsystem zur Automatischen Relief-Analyse. Z. f. Angewandte Geographie 4. 11-21

Kratz R, Suhling F (Hrsg) (1997) GIS im Naturschutz: Forschung, Planung, Praxis. - Westarp-Wissenschaften

Kugler H (1977) Die geomorphologische Reliefcharakteristik im Atlas DDR. Geogr. Ber. 22: 187-197

Kunzmann G (1989) Der ökologische Feuchtegrad als Kriterium zur Beurteilung von Grünlandstandorten, ein Vergleich bodenkundlicher und vegetationskundlicher Standortmerkmale. - Diss. Bot. 134

KVR (Kommunalverband Ruhrgebiet) (Hrsg) (1990) Datenanalyse Ermscher-Stadtökologie. Bestandsaufnahme und Defizitanalyse ökologischer Daten im Planungsbereich der IBA. - Essen.

Leser H (1991a) Landschaftsökologie. UTB 521, Stuttgart

Leser H (1991b) Ökologie wozu? Der graue Regenbogen oder Ökologie ohne Natur. - Berlin.

Leser H, Schmidt R-G (1981) Die Naherholungsgebiete im Schweizerischen Umland der Stadt Basel: Bestandsaufnahme der Typen und Möglichkeiten für die Planung. - Materialien zur Physiogeographie 2, Basel

Lieberoth I (1980) Auswertung der MMK für die Planung der industriemäßigen Pflanzenproduktion. Arch. Acker- u. Pflanzenbau u. Bodenkunde 11, Berlin

Lieberoth I.(1982) Bodenkunde. Akademie, Berlin

Mannsfeld K (1983) Landschaftsanalyse und Ableitung von Naturraumpotentialen. Abh. d. Sächs. Akad. d. Wiss. zu Leipzig, Math.-naturwiss. Kl. 55

Mantel K (1990) Wald und Forst in der Geschichte. Schaper, Alfeld-Hannover

Marks R (1979) Ökologische Landschaftsanalyse und Landschaftsbewertung als Aufgabe der Angewandten Physischen Geographie. Bochumer Geogr. Arb., Materialien z. Raumordnung 21

Marks R et al. (1992): Anleitung zur Bewertung des Leistungsvermögens des Landschaftshaushaltes (BA LVL). Forschungen z. Deut. Landeskunde 229, Trier

Mertens H (1964) Über die Verwertbarkeit der Bodenschätzungsergebnisse für die bodenkundliche Kartierung. Forsch. und Beratung B 10: 21-34

Meyer B, Grabaum R (1996) Szenarien zur Einschätzung der Bodenerosionsgefährdung durch Wasser mit GIS (ARC/INFO) am Beispiel des Untersuchungsgebietes Jesewitz/Sachsen. Geoökodynamik 17: 45-67

Moevius R, Murschel B (1997) Methoden zur Regionalisierung von Standorteigenschaften. - In: Kratz, R. & F. Suhling (Hrsg) (1997) GIS im Naturschutz: Forschung, Planung, Praxis, pp 155-164

Mosimann T (1985) Untersuchungen zur Funktion subarktischer und alpiner Geoökosysteme, Finnmark (Norwegen) und Schweizer Alpen. Physiogeographica, Basler Beitr. z. Physiogeographie 7, Basel

Müller E (1984) Zur Bestimmung von Aspekten des Wasserpotentials aus Ergebnissen der topischen und chorischen Naturraumerkundung. Dissertation A, Univers. Leipzig

Naveh Z, Lieberman A (1994) Landscape Ecology. Springer, New York- Berlin

Neef E (1966) Zur Frage des gebietswirtschaftlichen Potentials. Forschungen und Fortschritte 40: 65 - 70

Neef E (1967a) Die theoretischen Grundlagen der Landschaftslehre. Haack, Gotha

Neef E (1967b) Entwicklung und Stand der landschaftsökologischen Forschung in der DDR. Wiss. Abh. Geogr. Ges. DDR 5, pp 22-34

Neef E (1969) Der Stoffwechsel zwischen Gesellschaft und Natur als geographisches Problem. Geographische Rundschau 21: 453 - 459

Neef E, Schmidt G, Lauckner M (1961) Landschaftsökologische Untersuchungen an verschiedenen Physiotopen in Nordwestsachsen. Abh. Sächs. Akademie Wiss. zu Leipzig Math.-Nat. Kl. 47

Niemann E (1982) Methodik zur Bestimmung der Eignung, Leistung und Belastbarkeit von Landschaftselementen und Landschaftseinheiten. IGG Leipzig, Wiss. Mitt. Sonderheft 2.

Passarge S (1912) Über die Herausgabe eines physiologisch-morphologischen Atlas. In: Verh. 18. Dt. Geogr. Tages zu Innsbruck, Berlin, pp. 236-247

Peters H-J (1996) Die Einführung einer Umweltleitplanung durch das Umweltgesetzbuch. Unveröff. Manusk., 7 S.

Paffen K-H (1953) Die natürliche Landschaft und ihre räumliche Gliederung. Eine methodische Untersuchung am Beispiel der Mittel- und Niederrheinlande. Forschungen zur deutschen Landeskunde, Trier

PGM-Statistik (2000): Zentren der Ökosystemforschung in Deutschland. Petermanns Geographische Mitteilungen 144:38 - 41

Reinke E (1993) Verfahrensansatz zur Berücksichtigung zoologischer Informationen bei der UVP. Naturschutz und Landschaftsplanung 25: 5-10

Richter O et al (1997) Koppelung Geographischer Informationssysteme (GIS) mit ökologischen Modellen im Naturschutzmanagement. In: Kratz, R. & F. Suhling (Hrsg) GIS im Naturschutz: Forschung, Planung, Praxis: 5-29

Roweck H (1995) Landschaftsentwicklung über Leitbilder? LÖBF-Mitt. 4/95: 25-34

Schlüter H (1985) Kartographische Darstellung und Interpretation des Natürlichkeitsgrades der Vegetation in verschiedenen Maßstabsbereichen. Wiss. Abh. d. Geogr. Ges. d. DDR 18

Schlüter H (1991): Ökologische Gliederung und Bewertung des Bezirkes Leipzig nach Vegetationsmerkmalen. In: Neumeister H (Hrsg) Ausgewählte geoökologische Entwicklungsbedingungen Nordwestsachsens. Leipzig

Schmidt R, Diemann R (Hrsg) (1981) Erläuterungen zur Mittelmäßigen Landwirtschaftlichen Standortkartierung. Eberswalde

Schmithüsen J (1964) Was ist eine Landschaft? Erdkundl. Wissen 9:1-24

Schmithüsen J (1968) Allgemeine Vegetationsgeographie. Erdkundliches Wissen 9

Schönborn W (1992) Fließgewässerbiologie. G. Fischer Verlag, Stuttgart

Schwanecke W (1983) Zur Rolle der Waldvegetation bei der Kennzeichnung von Naturraumeinheiten auf der Grundlage der forstlichen Standortserkundung im Mittelgebierge/Hügelland der DDR. Petermann's Geogr. Mitt. 127: 245-254

SENSTADTUM (Senatsverwaltung für Stadtentwicklung und Umweltschutz) (Hrsg) (1990): Ökologisches Planungsinstrument Naturhaushalt/Umwelt. Berlin

Sporbeck O (1979) Bergbaubedingte Veränderungen des physischen Nutzungspotentials. Bochumer Geogr. Arb. 37

SRU (Sachverständigenrat für Umweltfragen 1996) Konzepte einer dauerhaften umweltgerechten Nutzung ländlicher Räume. Sondergutachten AGRA Europe 12

Steinhardt U (1998) Applying the fuzzy set theory for medium and small scale landscape assessment. Landscape and Urban Planning 41: 203- 208

Strobl J (1988) Reliefanalyse mit dem Computer. Anwendungsmöglichkeiten digitaler Geländemodelle in der Physischen Geographie. Geogr. Rundschau 40: 38-43

Succow M (1991) Seen als chorische Naturraumeinheiten. Beiträge zur Geographie 34, pp 100-114

Suter M et al. (1996) Virtuelle Simulation realer Landschaften als Basis für ein virtuelles GIS. Salzburger Geogr. Materialien 24

Thomas-Lauckner M, Haase G (1967) Versuch einer Klassifikation von Bodenfeuchteregime-Typen. Albrecht-Thaer-Archiv 11:003-1020.

Tietje O, Tapkenhinrichs M (1993) Assessment of Pedo-Transfer-Functions. Soil Sci. Soc. Am. J. 57: 1088-1095

Trautner J (Hrsg) (1992) Methodische Satandards zur Erfassung von Tierartengruppen. Arten und Biotopschutz in der Planung, 254 S.

Troll C (1950): Die geographische Landschaft und ihre Erforschung. Studium Generale 3: 163-181

Uhlig H (1954) Beispiel einer kleinklimatologischen Geländeuntersuchung. Zeitschr. Meteorologie 8: 66-75

Volk M, Steinhardt U (1996): Erstellung und Anwendung digitaler Geländemodelle mit ERDAS Imagine und ARC/INFO. GUGM'96 - GEOSYSTEMS User Group Meeting, 8.-10. Oktober: Tagungsband

Wedde D (1983) Zur Kennzeichnung des räumlich differenzierten Entsorgungspotentials im glazial bestimmten Tiefland der DDR. Wiss. Z. d. Päd. Hochschule Potsdam 27, 3

Werth W (1987) Ökomorphologische Gewässerbewertung in Oberösterreich (Gewässerzustandkartierung). Österr. Wasserwirtschaft 39: 122-128

Wiegleb G (1997) Leitbildmethode und naturschutzfachliche Bewertung. Zeitschrift für Ökologie und Naturschutz 6. 43 - 62

Zangenmeister C (1973) Nutzwertanalyse in der Systemtechnik. München

Zucchi H (1990) Gedanken zur Erstellung faunistisch-ökologischer Gutachten. Mitt. d. Landesanst. f. Ökologie, Landschaftsentw. u. Forstplanung Nordrhein-Westfalen (LÖLF) 15:13-21

# 3 Databases, data organization and data processing

Annegret Kindler, Ellen Banzhaf

## 3.1 Introduction

In the age of transition towards the information society, many different kinds of information and data are being continuously produced and distributed all over the world. Faster computers, the internet, earth observation satellites, and the development of new software and new technologies for various disciplines shape the production, transfer and exchange of information in all spheres of life. Moreover, many questions cannot be solved on a local, regional or national level because they have an international or even global nature. Considering the sheer volume and diversity of information, ever greater demands are being made on the collection, processing, structuring, standardization, sources, accuracy, updating and presentation of data. One general rule is that data collection is very time-consuming and expensive.

This chapter takes a look at selected aspects concerning the spatial data needed in the field of landscape-ecological investigations.

## 3.2 The problems faced

Landscape-ecological research deals with the study of various landscapes. Each landscape is the result of its natural and anthropogenic development over a long time and represents a certain part of geographical space. Landscapes consist of abiotic, biotic and anthropogenic components and are subject to permanent changes and developments. They are areas where humans, animals and plants live and act, and reflect the interactions between mankind and the environment.

Various sciences such as geography, landscape ecology, landscape planning, biology and environmental science investigate landscapes from their own specific angle. In addition to natural components, man-made influences and their consequences are studied. Apart from the capture and analysis of the state landscape structures are in and how they are changing, the processes taking place (e.g. fluxes of matter and energy) are becoming increasingly important for the latest landscape-ecological studies. Yet besides analysis, the ecological assessment of landscapes is also playing an growing role within landscape-ecological investigations, with selected small parts of landscapes being investigated and assessed. Furthermore, growing attention is being directed to statements and predictions about future landscape development, especially for larger coherent geographical spaces. Some aspects of the history of landscape assessment can be found in the article by Müller & Volk in Chap. 2 of this book.

Landscape-ecological investigations take place on different spatial scales (micro-, meso-, macroscale), in different dimensions (e.g. topological, chorological, regional, geospherical dimensions), and in different periods. This complexity needs to be considered, as it influences the availability, selection and use of information necessary for such research, its validity, and the transfer of results to other landscapes or to other parts of a landscape.

Such complex, diverse investigations call for a large and varied quantity of data. In fact without different kinds of data on different scales and dimensions, landscape-ecological investigations are simply not possible. The permanently increasing quantity of data makes high demands on methods and procedures of its collection, processing, analysis, standardization, visualization and storage in databases. In this field geographical information systems (GISs) and digital image processing have become indispensable and efficient tools for landscape-ecological investigations. But nevertheless some problems arise between the demands on landscape-ecological data on the one hand and reality on the other.

Regarding the increasing role of data, some aspects concerning their need, their sources, methods of capture and processing, data quality, updating, storage and data organization, the problems of standardization, and the possibilities and boundaries of their use are to be investigated and discussed in this chapter.

## 3.3 The need for data resulting from tasks, objectives and specific features of landscape ecological investigations

Landscape-ecological investigations have always needed plenty of different data. Depending on changing and increasing tasks and objectives, the need for data has grown with respect to type, relevance to the present, accuracy and availability. Besides the introduction and usage of new and better information technologies, demands have increased such that the data have to be available to many users, not only in an analogue form but also in a digital, standardized form.

Data needs result from the tasks and objectives, as well as from certain specific features of landscape-ecological investigations.

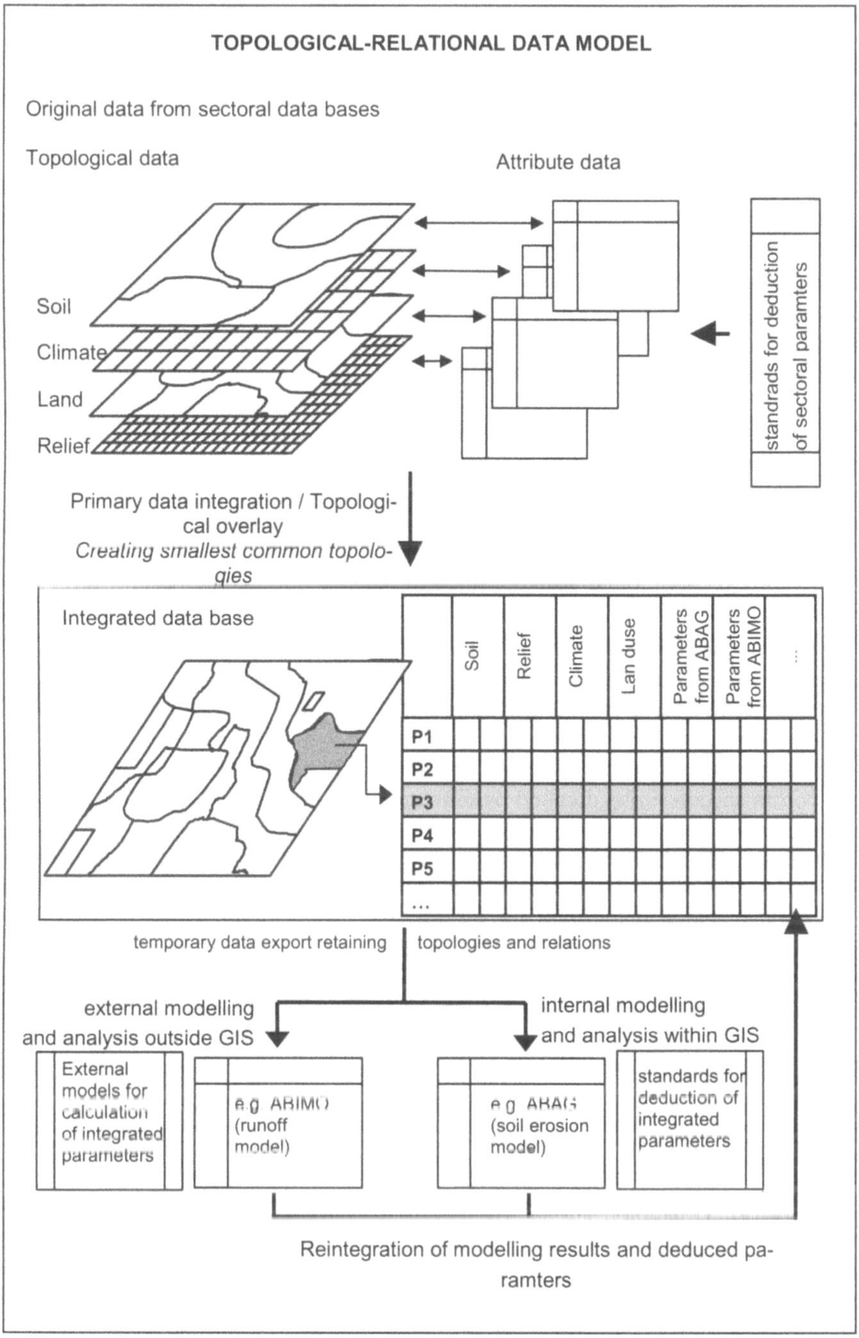


**Fig. 3.1:** Example for a topological-relational data model (Petry 2001)

### 3.3.1 The specific of landscape ecological investigations

Questions arising landscape-ecological research are of enormous breadth and diversity. According to Leser (1997), landscape ecology deals with interacting factors of a landscape ecosystem which is functionally and visually represented in a landscape. In reality, a landscape ecosystem is a very complex structure of natural, biotic and anthropogenic activities. By means of direct and indirect relations, these factors form a higher functional coherence which is spatially represented in a landscape. Just mentioning that landscapes represent the living space per se.

According to Leser (1997), some axiomatic basic rules apply to landscape-ecological systems regardless of the specific viewpoint of disciplines' broad research nature. These axioms are as follows:

- Landscape-ecological systems are everywhere.
- Landscape-ecological systems appear as geographical spaces.
- Landscape-ecological systems function all the time.
- Landscape-ecological systems function as three-dimensional structures of activities.
- Landscape-ecological systems function locally, i.e. in the topological dimension
- Landscape ecological systems always function in laws of nature even if they have been changed or they have been changed anthropogenically
- Landscape ecological systems are structures of activities with a basic dependence on biotic factors and abiotic factors
- Landscape-ecological systems function in a variable manner, albeit generally within certain limits dictated by climate.

Assuming that landscape-ecological systems or landscapes always function on the basis of natural laws, regardless of anthropogenic influence, landscape-ecological research deals with the abiotic and biotic components of landscapes, their appearances, their mutually relative nature, their mutual positive or negative influences, their anthropogenic influences and their consequences for a certain time, vertical and lateral fluxes of matter and energy, and so on.

### 3.3.2 Tasks and objectives of landscape ecological work

Landscapes exist in space and time, they are highly complex, and they cannot be measured. Therefore parts of landscape ecosystems, e.g. soils, water or vegetation, individual processes within their balances (water, fluxes and air), and individual functions are investigated in order to characterize a landscape.

Mosimann (1999) describes the central themes of landscape-ecological work as follows:

- Land use and the protection of resources
- Natural structural elements of landscape and their importance for flora, fauna, and landscape balance
- Estimation and assessment of landscape-ecological functions
- Ecological space division
- Ecological risk analysis
- Environmental monitoring and assessment.

Yet apart from analysing individual landscape elements and their mutual influence, landscape-ecological work also involves other tasks. Meyer presents several procedures for landscape-ecological assessment in Chap. 8. One main task consists in landscape-ecological process research, i.e. determining and understanding the vertical and lateral processes which take place in landscapes (or parts thereof). Furthermore, forecasts for the future development of landscapes are required - an important aspect for landscape planning and practice. Steinhardt (1999) emphasizes that after decades of mainly location-related research, a modern landscape ecology has to provide both information for geographical spaces whose dimensions are relevant for planning processes, and the required methodological equipment. Observation in space and time and the description of processes taking place in large landscapes are of fundamental interest but are difficult to organize because of the natural variability of real processes and the heterogeneity of spaces of reference.

Landscapes have to simultaneously fulfil several functions: production functions, regulation functions, capability functions and information functions. One main aim of landscape-ecological research comprises preserving and re-establishing the multiple use of landscapes by avoiding and/or minimizing use conflicts. Therefore investigating landscape balance is necessary to derive strategies for future landscape developments.

### 3.3.3 The role of geographical dimensions and scales

The assumption that nature is structured hierarchically and based on the holistic axiom - that the whole is more than the sum of its individual components - was introduced into ecology back in the 1940s (Steinhardt 1999). In this sense an ordered whole consists of a hierarchy of multilayered systems, where each higher level is formed by systems of a lower level. The change or transformation process between different levels is always connected with qualitative leaps and bounds. This theoretical concept can be applied to geoecological systems, too. According to Steinhardt (1999), investigating geosystems is tied to hierarchical principles: besides capturing their components (elements, relations, structures), separation must be performed vis-à-vis the surroundings, which always involves a certain range.

Landscape-ecological data are directly related to different geographical dimensions, a term which was first defined by Ernst Neef in the early 1960s (Neef 1963). He described scale zones with the same contents as a dimension. If a change in scale opens a new level of geographical facts and changed statements are possible, a change in the dimension of the geographical observation will take place. Based on the theory of geographical dimensions, which is important for geographical space divisions, a distinction is drawn between a topological, chorological, regional and geospherical dimension. Each dimension corresponds to a certain degree of generalization and reflects a particular degree of homogeneity and heterogeneity.

In the topological dimension, the 'tops' represent the smallest landscape-ecological base units. The tops should be of an approximately homogeneous content, i.e. the area of a top should have the same characteristics, the same structure and the same structure of activities, a unique mechanism of matter and energy balance, and exhibit the same ecological behaviour pattern. However, the homogeneity of the tops is relative and not necessary for all their components. Selected reference parameters and certain intervals are determined within which variation is allowed and neglected. Although tops are homogeneous, they can be distinguished into smaller parts of a heterogeneous character. Steinhardt (1999) wrote that the degree of heterogeneity between units of the same dimension level must differ significantly from the degree of heterogeneity within this unit. Therefore the heterogeneity of the unit at the lower level of dimension is a standardized measurement to signify separation from the higher dimension level.

In the chorological dimension, the chores are the landscape-ecological base units and represent a certain spatial structure. They have a heterogeneous character and consist of several homogeneous topological spatial units. Their determination takes place according to the landscape-ecological base units, landscape-genetic parameters of the chores, their spatial context and present dynamic characteristics which cover the whole unit.

The next higher level after the chorological dimension is the regional one. It represents large spatial extents of the geosphere. Therefore they cannot be derived from topological units. They are only partly based on chorological units. They form homogeneous areas on a very high level of integration.

The geospherical dimension serves for a landscape-ecological description of phenomena important for the whole geosphere. They can be landscape zones, continents or the earth. They are determined by planetary processes which are represented by a unique structure of a certain landscape zone and visually and/or ecofunctionally differ from neighbouring geo-regions (Leser 1997). Steinhardt (1999) discusses the present developments concerning the theory of geographical dimensions and the links between homogeneity and heterogeneity in different geographical dimensions and contributes to this topic in Chap. 6.

The specific features of each geographical dimension can only be represented by an adequate dimensional framework and on a certain (map) scale. However, strict limits do not exist between the scales. Instead, the transitions are fluid between the scales during a change from a lower to a higher dimensional stage. Steinhardt (1999) gives an overview of the different geoscientific levels of hierarchy exemplified by hydrological processes. An attempt to assign map scales to different geographical dimension stages can be found in the "Brockhaus abc Kartenkunde" (Ogrissek 1983). Topological units such as climatopes, physiotopes and ecotopes are represented cartographically at large scales between 1:5,000 and 1:25,000, which reflect the microscale of the geographical dimension theory. The chorological dimension includes three-dimensional stages, the microchores at a scale of 1:25,000 up to 1:100,000, the meso chores at a scale of 1:100,000 up to

1:1,500,000, and the macro chores at a scale of 1:1,500,000 up to 1:5,000,000. The microregion, mesoregion and macroregion belong to the levels of the regional dimension. They can be represented in maps of a small scale which can start from 1:750,000 (1:500,000) and reach up to 1:10 million or even smaller. Both the chorological dimension and the regional dimension represent the mesocale of the geographical dimension theory. The geospherical dimension can be represented by landscape zones or landscape belts at very small scales, e.g. 1:10 million or even smaller. They can be assigned to the macroscale.

### 3.3.4 The need for data

Knowing very well that it will never be possible to recognize and understand all the processes taking place in landscapes, it will nevertheless be possible to improve our knowledge about the function of landscapes step by step. The more information we have, the better the possibilities and chances to understand individual landscape-ecological processes and landscapes as a whole. The availability of complex landscape-ecological knowledge is a basic condition for keeping, protecting, restoring and developing diverse landscapes with different structures.

Because of the high demands on landscape-ecological investigations, many different sorts of data are necessary. Information about soil, water, vegetation, morphology, climate, land use, land cover, biotope types, landscape units, weather, landscape and natural conservation, administrative divisions and many more characteristics exist for different geographical spaces and at different scales. This is exactly what explains one major difference from non-spatial data.

Different tasks and objectives of landscape-ecological investigations essentially determine the data needed to represent a certain part of a landscape in a particular geographical dimension and at an adequate scale.

## 3.4 Types of data, data sources and methods of data acquisition

From analysis to assessment all landscape ecological investigations represent an integrative field of research and require a solid and diverse data base. The data can be differentiated concerning their topological or theme, the geographical space, the geographical dimension and the scale, the method of registration, the source, and the time (point in time or period).

With regard to the topological it is possible to divide data in two main parts, topographic data and thematic data.

As follows an overview shall be given over kinds of data nessecary for landscape ecological investigations, their sources and methods of acquisition.

### 3.4.1 Topographic information

Topographic data provide fundamental information about objects and conditions, their kind, geographical location, size, form, arrangement, and neighbourhood which allow an orientation in a certain terrain. These include information about settlements, industrial and other technical building complexes, traffic routes, stretches of running and standing water, relief, soil conditions, land cover and land use and others. Topographic information are mainly represented in official topographic maps on different scales, which are produced, updated and offered by national ordnance surveys. In the following the specific situation concerning the availability of topographic information in the Federal Republic of Germany shall be described.

In Germany the national ordnance survey of each federal state is responsible for the supply of topographic information for the whole particular federal state for different users in different spheres like ordnance survey and cartography, army, economy, planning, sciences, tourism and private persons. The national ordnance surveys are the only authorities which are allowed to produce and offer topographic information. They also have to update them in certain periodes. This is a sovereign task and of national importance.

### 3.4.1.1 Topographic maps

Topographic information has long been represented in topographic maps on different scales. Generally speaking, we differentiate between large-scale, medium-scale and small-scale official topographic maps. However, the boundaries between the different scales are not hard and fast, and may be very small depending on the viewpoint. Among the large-scale topographic maps, a difference is found between the staes in western Germany and those in the east. This different approach is explained by the historical development of Germany as two separate states from 1949 until 1990. The states in western Germany use a topographic map for the whole country with a scale of 1:5,000, the so-called German basic map. By contrast, the basic scale for topographic maps in eastern Germany is 1:10,000. Both topographic maps have been the basis for all other topographic maps on smaller scales and have been updated in certain periods. Since unification they continue to be maintained as the two basic scales for topographic maps. Because of the very high costs and the time needed time, integrating the two basic scales and creating one single topographic basic map for the whole country with a single scale has not yet been considered.

In the field of medium and small scales, there exists a unique order of scales in both western and eastern Germany. Topographic maps on the scales of 1:25,000, 1:50,000 and 1:100,000 are available in medium scales. Topographic maps on scales less than 1:100,000 are among the small-scale maps. In this field topographic maps on the scale of 1:200,000 exist for the whole country. Topographic maps on scales smaller than 1:100,000, the so-called topographic overview maps, on scales of 1:200,000, 1:500,000, 1:1 million (international world map), 1:1.5 million and 1:2.5 million are produced, updated and supplied by the Federal Agency for Cartography and Geodesy based in Frankfurt am Main, Leipzig and Berlin.

Besides these topographic maps mentioned, the national ordnance survey of each federal state offers specific topographic maps such as topographic regional maps and topographic special maps on different scales and maps of administrative divisions into municipalities, districts, administrative districts and federal states.

Maps for certain surroundings, selected administrative units like districts, administrative districts, individual federal states or for the whole country belong to the topographic regional maps. Topographic special maps are topographic maps combined with specific thematic information, such as maps of natural parks, maps for hiking and recreation, and "leisure" maps.

Besides the present topographic maps, historical topographic maps on the scale of 1:25,000 (*Messtischblätter*) exist in printed form for the federal states for different points in time. They can be used to study the development and change of the topographic situation. In the field of landscape-ecological investigations, historical topographic maps serve for example to record the historical landscape-ecological situation, past land cover and land use, landscape structuring elements, relief and others, and their changes in time. Yet there are also other historical topographic maps on different scales, for instance the "Map of the German Reich" on a scale of 1:100,000.

All these maps are available and can be bought in the form of multicoloured or sometimes printed maps which can be coloured in by the user.

#### 3.4.1.2 Topographic raster data

Parallel to the development and introduction of new information technologies into many spheres during the last 10 years, the need for topographic information in digital form has also arisen. Therefore the federal ordnance surveys have begun transferring their information from analogue to digital form. They now offer digital topographic information in two new products - raster data and the ATKIS data.

Topographic raster data are easily produced. As topographic maps are printed in single colours, the folios for printing each colour are scanned at a certain resolution and transferred to digital form. "Digital raster data" are the result of this process. Every map sheet consists of four data layers, representing map elements in one color. The ground plan (map frame, boundaries, settlements, transport routes, map names in black) consists of all the map elements which are printed in black. The second layer contains all the map elements printed in blue, i.e. the contours of water, water areas and map names in blue. Relief with the contour lines and map names in brown represent the third layer. The fourth layer comprises the vegetation, printed in green. Small scales also contain a layer for roads and built-up areas. In addition to these layers, every map sheet has a header file with the coordinates and pass points. Some federal ordnance surveys offer more than the four mentioned layers for selected scales, but this varies from one federal state to the next. In these cases, the map names in black are separated from the other elements of the ground plan, and the contours of water and map names in blue are separated from the water areas. Sometimes the administrative boundaries are shown in an independent layer. These raster data can be bought on the scales 1:10,000, 1:25,000, 1:50,000, 1:100,000 and 1:200,000. They are very useful for simply augmenting thematic maps with topographic elements. However, before they can be used within a geographical information system (GIS), they must first be georeferenced. This means the pixels have to be given specific coordinates delivered by the federal surveys. This step could be saved for many users if federal surveys offered the data in a georeferenced form from the start.

One disadvantage of such raster data is that only complete individual layers can be combined with other data. The selection of individual topographic elements from one layer, for example a certain water area, is not possible. Different numbers of layers in individual federal states cause problems if topographic raster data is to be used for a geographical region which is located in two or more federal states. By an agreeing on a unique number of layers on different scales in each federal state, standardization could be achieved in the field of topographic raster data.

### 3.4.1.3 ATKIS data

Unlike topographic raster data, ATKIS data are digital topographic vector data. Since the end of the 1980s, the national ordnance surveys and the land registry authorities of all the federal states in Germany have been working on a unique basic information system for topographic information. The acronym ATKIS stands for Official Topographic/Cartographic Information System. The development of ATKIS results from the increasing need for digital topographic vector data among science, trade and industry, planning, etc. The advantages of such a system include a unique spatial basis for the whole country, a unique data structure, a unique interface in the form of the unique database interface (EDBS), the independence of the map sheet vector data, and a larger share of permanently updated content.

ATKIS is being developed in the form of a digital landscape model (DLM) for three scale ranges. Digital landscape models have been created as DLM25 for scales from 1:10,000 up to 1:50,000, as DLM200 for scales from 1:50,000 to 1:500,000, and as DLM1000 for scales from 1:500,000 to 1:1 million (Landesamt für Landesvermessung und Datenverarbeitung Sachsen-Anhalt 1999). DLM25 has been built up by the ordnance survey authorities of the federal states. The Federal Agency for Cartography and Geodesy is responsible for developing DLM200 and DLM1000 for the whole of Germany.

ATKIS is structured in different components, the digital landscape model and digital administrative boundaries (DVG). The digital landscape model consists of two parts, a digital situation model (DSM) on a scale of 1:25,000 and a digital terrain model (DGM). Two-dimensional vector data structured on objects and the objects describing attribute data belong to the contents of the digital situation model. This information is independent of scale. On the recommendation of the ordnance surveys, it can be used on scales of 1:10,000 to 1:50,000. The contents of the digital situation model mainly correspond to the topographic map on the scale of 1:25.000. Individual objects are registered in the "ATKIS-OK", a catalogue of object types created and published by a working group of the ordnance survey authorities of the German federal states in 1998 (Arbeitsgemeinschaft der Vermessungsverwaltungen der Länder der Bundesrepublik Deutschland [AdV] 1998). The aim of this catalogue is to structure landscapes by topographic points of view, classify topographic phenomena in landscape objects, determine the contents of the digital landscape model, and provide instructions necessary for modelling (AdV 1998). The catalogue is structured on an attribute-oriented basis. Landscapes can be roughly structured in types of objects, and with the help of attributes they can be structured in detail. Thus the catalogue enables a free selection of topographic phenomena, and - if already integrated - of specialized facts. It is an open catalogue which can be augmented with further types of objects, topographic and specialized attributes.

ATKIS data are registered and offered for sale by each federal ordnance survey only for its respective territory. Although data covering a geographical region within two or more federal states currently have to be ordered in two or more federal states, the newly founded Geo-data Centre of the Federal Agency for Cartography and Geodesy will eventually offer all ATKIS data for the whole country.

Besides the digital situation model, the digital terrain model (DGM) belongs to the digital landscape model. Whereas digital situation models completely covering each federal state already exist with small differences from one federal state to the next, digital terrain models are not yet available for the whole area of each federal state. Digital terrain models contain elevation information in the form of a regular point raster with different resolutions. In comparison to two-dimensional data, ATKIS data digital elevation models provide three-dimensional terrain data. Each registered point of the surface (of the earth) is represented by both an x and y coordinate, and by a z value for the elevation. The distance between the elevation points, i.e. the raster size, varies among 10 metres, 20 metres, 25 metres, 40 metres, 50 metres, 80 metres and 200 metres in different federal states. The elevation accuracy is determined by data sources, aerial photographs and topographic maps on different scales, and by raster size. In Saxony-Anhalt, for example, the elevation lines of the topographic map on the scale of 1:10,000 have been digitized to derive the digital elevation model. Each federal state uses its own interpolation algorithm without any form of coordination with the other federal states. The federal ordnance surveys claim accuracy of about ± 2 metres for high-resolution digital terrain models. DGMs with a lower resolution of 25 metres up to 50 metres have an accuracy of between ± 2 meters and ± 5 meters. As a result, problems can occur if data of digital terrain models of different federal states are to be integrated.

The derivation of three-dimensional digital terrain models from the DGMs provides basic information for landscape-ecological investigations because relief is one of the most important landscape-ecological regulation parameters concerning the spatial structure of landscapes, fluxes of matter and energy, vegetation, land use and so on.

Depending on a certain landscape-ecological task, the spatial resolution of a digital elevation model and its use for a certain scale have to be considered before it is used. Digital administrative boundaries represent the second main part of ATKIS. The federal surveys recommend using these data on a scale of 1:50,000 up to 1:1 million. The following table gives an overview of the types of topographic information available in Germany.

**Table 3.1:** Overview of the available topographic information in the Federal Republic of Germany

| **Type of Topographic Information** | **Scale** | **Spatial availability** | **Producer and provider** |
|---|---|---|---|
| Topographic maps | 1:5,000 (Deutsche Grundkarte) | Old federal states only: Bavaria, Baden-Württemberg, Saarland, North Rhine Westphalia, Hesse, Rhineland-Palatinate, Schleswig-Holstein, Lower Saxony, Bremen, Bremerhaven, Hamburg, Berlin | Ordnance Survey of each federal state |
| | 1:10,000 | New federal states only: Saxony, Saxony-Anhalt, Thuringia, Brandenburg, Mecklenburg-Vorpommern, Berlin | Ordnance Survey of each federal state |
| | 1:25,000 | Old and new federal states | Ordnance Survey of each federal state |
| | 1:50,000 | Old and new federal states | Ordnance Survey of each federal state |
| | 1:100,000 | Old and new federal states | Ordnance Survey of each federal state |
| | 1:200,000 | Old and new federal states | Ordnance Survey of each federal state; Federal Agency for Cartography and Geodesy |
| | 1:500,000 | Old and new federal states | Federal Agency for Cartography and Geodesy |
| | 1:1 million and smaller | Old and new federal states | Federal Agency for Cartography and Geodesy |
| Topographic special maps | different scales | Old and new federal states or selected areas | Ordnance Survey of each federal state |

continued

| | | | |
|---|---|---|---|
| Historical topographic maps | 1:25,000 (Messtischblätter) | Old and new federal states | Ordnance Survey of each federal state |
| | 1:100,000 (Karte des Deutschen Reiches) | Old and new federal states | Ordnance Survey of each federal state; Federal Agency for Cartography and Geodesy |
| Digital raster data | 1:10,000 | New federal states | Ordnance Survey of each federal state |
| | 1:25,000 | Old and new federal states | Ordnance Survey of each federal state |
| | 1:50,000 | Old and new federal states | Ordnance Survey of each federal state |
| | 1:100,000 | Old and new federal states | Ordnance Survey of each federal state |
| | 1:200,000 | Old and new federal states | Ordnance Survey of each federal state |
| Digital ATKIS-data | | | |
| 1. Digital landscape model (DLM) | | Old and new federal states | Ordnance Survey of each federal state |
| DLM25 | usable on 1:10,000-1:50,000 | | |
| DLM200 | usable on 1:50,000-1:500,000 | | Federal Agency for Cartography and Geodesy |
| DLM1,000 | usable on 1:500,000-1:1million | | |
| 1.1 Digital situation model (DSM) | 1:25,000 (usable on 1:10,000-1:50,000) | | Ordnance Survey of each federal state |
| 1.2 Digital terrain model (DGM) | | | Ordnance Survey of each federal state |
| 2. Digital administrative boundaries (DVG) | usable on 1:50,000-1:1 million | | Ordnance Survey of each federal state |

Topographic information has a great advantage in contrary to other information necessary for landscape-ecological investigations. It is generated on the basis of unique viewpoints, with sufficient accuracy over a certain range of different scales ensured by federal ordnance surveys. Topographic information is available for the whole country and is updated at certain periods. It is offered for sale by the federal ordnance surveys in both analogue form and digital form. When acquiring topographic information in the form of maps or digital data, users have to comply with strict rules imposed by the federal ordnance surveys concerning the duplication and publication of topographic data, and they have to pay for each data representation.

ATKIS with its components provides updated, more or less unique topographic information which can be used to build up spatially orientated professional information systems. ATKIS has not yet been completed. At present, individual parts of ATKIS are available at the different levels of their development in each federal state. For the whole of Germany, the first level of ATKIS - the creation of the geodatabase - was finished at the end of the 1990s. Despite a few problems, the development of ATKIS is an example for the standardization of topographic information for the Federal Republic of Germany as a whole.

### 3.4.2 Thematic information

Thematic information plays the more important role for landscape-ecological investigations. In comparison to topographic information, the situation in this field is quite different. Due to the diverse tasks and objectives of landscape-ecological work, much thematic information already exists or has been produced. However, it is much more heterogeneous then topographic information. Thematic information is only available for selected geographical spaces, in a certain geographical dimension and scale, with a certain accuracy, for a specific point in time. In a few cases it covers the area of the whole country. Mostly, however, the information has been created for small, limited geographical spaces. Thematic information has different sources. On the one hand it can be of an original nature. This means it is the result of terrain investigations or terrain measurements. Such investigations are expensive and call for huge volumes of time and resources. Therefore they can only be carried out for small geographical spaces. On the other hand, thematic information mainly results from secondary sources. The latter may be thematic maps, thematic atlases, descriptions and others. The information is available for larger geographical spaces.

The information is scattered throughout various scientific institutions, official authorities and agencies, and environmental and planning bureaus. As nobody knows who has what kind of information in an analogue and/or digital form, inquiries about data are arduous and sometimes unsuccessful. The fact is that each regional landscape-ecological study entails specific data inquiries.

The specific situation in the field of thematic information for landscape-ecological investigations in Germany is considered below. Important sources of thematic information are mentioned. The basic thematic information available and the differences stem from different historical periods in the former "German Reich", the existence of two German states over a period of 40 years, and the present Federal Republic of Germany. The information is very heterogeneous regarding its accuracy and sharpness of content, scale and availability.

In view of the great variety of thematic information, only selected aspects concerning type, availability, sources and importance for landscape-ecological investigations can be discussed.

### 3.4.2.1 Information of classification of nature areas

Natural areas are parts of the geosphere which can be characterized by certain abiotic and biotic geocomponents and geoprocesses determined by natural laws and anthropogenic influences. They have a unique structure and a unique structure of activities. Natural areas can be divided with the help of two methods, the natural spatial order and the classification of natural areas. Both methods represent different geographical dimensions. Depending on the extent of the geographical space investigated, natural area units of different geographical dimensions are distinguished due to the theory of geographical dimensions. The method of the natural spatial order enables natural area units of a homogeneous character to be extracted, characterizing homogeneous ecological units of a certain function with a specific flux and energy balance. The method of classifying natural areas serves the extraction of landscape spaces in the form of types and in a certain hierarchical order.

The results of such divisions of natural areas are represented in 'natural area maps', landscape-ecological maps or maps of classification of natural areas on different scales. In such complex or synthetic maps, natural area units of the topological, chorological or regional dimension with a unique landscape-ecological character determined by climate, water, soil, substratum, relief and vegetation give an overview of the classification of natural areas of a certain geographical space.

One example of such a map is that of the natural area types on a scale of 1:750,000 in the "Atlas Deutsche Demokratische Republik" (Akademie der Wissenschaften der Deutschen Demokratischen Republik 1981). In East Germany, the plan was to draw up a map of the natural spatial types on scales of 1:50,000 and 1:200,000 for the whole country. However, this project was not completed; only individual map sheets were made on both scales, and so there is no map of the natural areas on a larger scale then in the national atlas for the territory of East Germany.

An updated map of the natural areas and natural spatial potentils has since been created covering the whole of Saxony at a scale of 1:50,000.

The "Digitaler Landschaftsökologischer Atlas Baden-Württemberg" (Institut für Angewandte Forschung "Landschaftsentwicklung & Landschaftsinformatik" [IAF] der Fachhochschule Nürtingen 1996), the digital landscape-ecological atlas of Baden-Württemberg (at a scale of 1:200,000), consists of 37 thematic maps for the whole territory of this federal state. One basic spatial division is a map of natural areas structured in so-called complexes of sites, partial landscapes and large landscapes representing different geographical dimensions.

An updated map of the division of the whole country into natural areas does not exist. Therefore landscape ecologists use the historical map of the classification of natural areas of Germany, the "Naturräumliche Gliederung Deutschlands mit

Höhenschichten", on a scale of 1:1 million created and described by Meynen and Schmithüsen in their "Handbuch der naturräumlichen Gliederung Deutschlands" (1952-1963), a handbook for the classification of natural areas of Germany in the 1950s.

These are just a few examples of the great number of maps of the natural spatial order and the classification of natural areas in Germany.

### 3.4.2.2 Soil information

Thematic information about the soil, which is one of the most important landscape-ecological parameters, exists in different kinds and on different scales.

One of the oldest sources of thematic information covering the whole country is the "Reichsbodenschätzung". Between 1935 and 1956, agricultural soils (subdivided into arable land and grassland) were taxed depending on their natural conditions and fertility. According to Bastian and Schreiber (1999, p. 484), characteristics such as soil substrates, different state levels and their modes of origin were used for the taxation of arable land. Grassland was taxed with the help of soil types, different state levels, climate and water. The original data were mapped at large scales (1:500, 1:1,000, 1:2,000, 1:2,500 or 1:2,730). Later on, the results were presented in thematic maps. The "Deutsche Grundkarte - Böden" on a scale of 1:5,000 was developed on the basis of the results of the "Reichsbodenschätzung" for the old Federal Republic of Germany. The results were shown for the area of East Germany on the scale of 1:10,000. More information about the "Reichsbodenschätzung" is contained in Chap. 2 of this book. Two other available sources for soil information, albeit only for the whole of East Germany, are the "Mittelmaßstäbige landwirtschaftliche Standortkartierung (MMK)" and the "Forstliche Standortkartierung (FSK)". The first one, made between 1976 and 1982, provides basic information for the land use of agricultural areas depending on their location. The MMK can be used to characterize heterogenic location units by substrate, soil and soil water conditions, the structure of the soil cover and the relief conditions. The results of the MMK are represented on two scales, 1:25,000 and 1:100,000.

The FSK only contains soil information for areas used by forestry. It provides information about the soils, soil water, relief and climate and their vegetation. The results were shown in maps on a scale of 1:10,000.

The MMK and FSK complement each other, making soil information of areas used for agriculture and by forestry available for the territory of eastern Germany. However, they do not contain any information about areas which are not used by agriculture or forestry, such as built-up areas or opencast mining districts. Both maps have been transferred to a digital form, and are available from various environmental authorities in eastern Germany. The combination of both maps reveals differences regarding the spatial separation of areas used by agriculture and by forestry. In some cases agriculturally used areas in the MMK differ from the areas shown with the same use in the FSK. The same situation is found in the areas used by forestry. Therefore, joint use entails fitting both maps to each other. They are one of the most important sources for obtaining detailed information about soil conditions in the topological and chorological dimension.

The Federal Institute of Geosciences and Natural Resources (BGR) sells soil maps on different scales covering the whole of Germany - both as printed colour maps and in digital form.

The first one is the "Soil Map of the Federal Republic of Germany" (BÜK1000) on a scale of 1:1 million. It gives an overview of the distribution of dominant soil types and parent material throughout Germany. This map is the result of unifying the digital soil map of western Germany on a scale of 1:1 million and the soil map of eastern Germany on a scale of 1:500,000 (Bundesanstalt für Geowissenschaften und Rohstoffe, Niedersächsisches Landesamt für Bodenforschung, Geowissenschaftliche Gemeinschaftsaufgaben 1999). At first "it was necessary to match the soil systems used in East and West Germany and to develop standardized descriptions of soil units." (Bundesanstalt für Geowissenschaften und Rohstoffe 1995, p. 4) "The map shows 72 soil mapping units, described in the legend on the basis of the German and FAO soil system. Each soil unit has been assigned a characteristic soil profile as an aid to map interpretation. For the first time the subdivision of the country into 12 soil regions has been represented on the map.... It is an important part of the spatial database integrated in the Soil Information System currently being established at the Federal Institute of Geosciences and Natural Resources (FISBo BGR). It can be used together with the characteristic soil profiles to derive thematic maps related to nationwide soil protection. The scale of the BÜK 1000 makes it especially suitable for small-scale evaluations at the federal or EU level." (Bundesanstalt für Geowissenschaften und Rohstoffe 1995, p. 4)

Based on the BÜK 1000, a second soil map on a scale of 1:2 million (BÜK 2000) was derived covering the whole country. Through an insignificant generalization of the 72 soil mapping units, the map now shows 61 soil mapping units. The third soil map offered by the BGR is a soil map on the scale 1:5 million (BÜK 5000). This map is the result of a significant generalization of the BÜK 2000, with the original 61 soil mapping units being generalized and summarized to form 19 soil mapping units.

In addition to these soil maps, the BGR has started creating a large-scale map of soil landscapes for the whole country. This map is based on the BÜK 1000 map and represents 38 mapping units with 329 areas. Moreover, on the same basis a map of the distribution of soils at the level of soil regions with 12 mapping units with 173 individual areas exists. These soil regions differ with respect to orography and lithography.

In connection with the geographical dimensions, the soil maps mentioned above can be used for investigations at the regional dimension.

Furthermore, a soil map with a scale of 1:200,000 is being developed. The map will ultimately consist of 56 map sheets with between and 100 mapping units. At present only individual map sheets are available, e.g. CC7934 Munich, CC3926 Braunschweig and CC4734 Leipzig. Other map sheets are in preparation. Because of its larger scale, this soil map can be used for investigations in the chorological dimension.

Information about the maps and data produced and available from the Federal Institute of Geosciences and Natural Resources can be found on the internet under http://www.bgr.de. In addition, important descriptive information about each product is contained in the metadata catalogue (Bundesanstalt für Geowissenschaften und Rohstoffe, Niedersächsisches Landesamt für Bodenforschung, Geowissenschaftliche Gemeinschaftsaufgaben 1999).

Apart from soil data and maps on small scales covering Germany as a whole, each federal state has soil maps on larger scales at its disposal. In western Germany, soil maps on different scales exist for the whole federal state or parts thereof. For example, in Bavaria and North Rhine Westphalia soil maps on a scale of 1:50,000 are available. In Lower Saxony soil maps on a scale of 1:200,000 and 1:25,000 have been developed. After German unification in 1990 and the administrative division of East Germany into five federal states, there was a great demand for (not to mention a severe shortage of) various up-to-date thematic maps for the territories of these new administrative units. The Federal Authorities of the Environment and Geology have tackled the the development of new maps and digital data for each federal state. In Saxony and Thuringia, for example, a soil map (BÜK400) has been developed on a scale of 1:400,000 covering the whole territory of each federal state. The soil map of Saxony-Anhalt was developed on the scale 1:500,000. A comparison of these maps along the boundaries shows that the maps were made separately without any adjustment between the federal states. This is an avoidable disadvantage if a user needs soil information crossing the boundaries of federal states.

In addition to the soil map on the scale of 1:400,000 in Saxony, a new soil map on the scale of 1:50,000 has been prepared covering the whole territory. The first three map sheets have already been published. This map can be used for landscape-ecological investigations in the microchorological and mesochorological dimension.

#### 3.4.2.3 Geological and hydrogeological information

The main sources of geological and hydrogeological information are geological and hydrogeological maps on different scales. Geological maps show the spatial distribution, stratification, age and other characteristics of rock. As a rule the beds underground below the soil are shown.

Older and younger geological maps on the scales 1:25,000, 1:50,000, 1:100,000 and 1:200,000 exist for the whole of the Federal Republic of Germany. In East Germany, two specific geological maps on the scale 1:50,000, the lithofacies map and the hydrogeological map, were produced for the whole territory.

In addition, the above-mentioned authorities produced new geological maps for the territory of their federal states. At first in Saxony and in Thuringia they made maps on the scale of 1:400,000 which gave an overview of the geological situation. Whereas the soil maps are island maps, the geological maps are framework maps. This means that the problem of adjustment does not arise in the same manner if boundaries are crossed during investigations.

The Saxony Federal Authority of the Environment and Geology has developed a new geological map of the glacial covered areas of Saxony on a scale of 1:50,000. This map consists of 20 map sheets. Each map sheet represents the solid rock and the loose rock of the terrain surface. The map is designed to act as a basis for the assessment of soil formation, water flow, the distribution of usable solid and loose rock, possible reaches of environmental damage, and the possibilities of building landfills, etc.

The location, size, depth and quality of groundwater occurrences, possibilities of extracting groundwater and the water yield of different rock beds are shown in hydrogeological maps. Moreover, they contain hints about filtering ability, the chemical conditions and the risk of groundwater contamination. Hydrogeological maps on scales of 1:25,000 and 1:50,000 exist for western Germany. In East Germany, the hydrogeological map on a scale of 1:50,000 was developed for the whole territory. Each map sheet consisted of between three and ten map sheets of different topics. They are the Hydrogeological Basic Map - Quaternary Aquifers, the Map of Hydrological Characteristic Values, the Map of Groundwater Contours, and the Map of Groundwater Risks.

#### 3.4.2.4 Relief information

Relief is one of the most important regulation parameters of landscape-ecological processes. The main sources of relief information are the topographic maps on various scales and the digital elevation models. These maps and digital elevation models can be used to determine the elevation, the slope and the relief intensity. A detailed description of available topographic maps and digital elevation models is contained in Chap. 3.4.1.

Specific relief maps provide information and a spatial impression of the relief with the help of shading. One example is the relief map in the "Atlas Deutsche Demokratische Republik" (Akademie der Wissenschaften der Deutschen Demokratischen Republik 1981) on a scale of 1:750,000. In this atlas, a hypsometric map gives an overview of the spatial distribution of the elevation conditions. Another map shows the georelief and local processes influencing relief on the same scale.

#### 3.4.2.5 Information about watercourses

The system of stretches of surface and underground flowing and standing water is one of the most important components of the landscape balance and the living space of the bios. Water is a very important medium of transport for fluxes of matter, both underground and on the surface. Information about the surface water, the surface runoff, its quality and quantity have been measured and collected. By comparison, there is a lack of information about subsurface runoff. The vertical and lateral transport processes and the spatial distribution of fluxes with the help of water have not yet been sufficiently investigated, and there is a great need for data to ascertain and understand the role of water within the landscape and the landscape balance.

Basic information about the location, width and length, size, depth, natural and artificial water conduits and the direction and rapidity of flow, the water level of surface flowing and standing water, watersheds, underground water conduits, springs and others aspects is contained in topographic maps on various scales. Information about the groundwater isohypses are part of hydrogeological maps.

Water grade maps show the quality of important running water with the help of different classes. Each class provides information about the degree of load and the degree of water pollution. These maps give initial information about the polluted and unpolluted parts of rivers and streams. An example of such a map is the "Gewässergütekarte 1991" (Sächsisches Landesamt für Umwelt und Geologie 1992), a water grade map of Saxony on a scale of 1:400,000.

Information about the water flow and the rate of flow provide flood measurement points.

Further information is contained in individual thematic maps and regional thematic atlases. However, some parameters still have to be especially measured.

For example, the "Atlas Deutsche Demokratische Republik" (Akademie der Wissenschaften der Deutschen Demokratischen Republik 1981) contains a map with a hydrographic overview of the stretches of water and their density on a scale of 1:750,000. This atlas also contains a hydrological map with information about the average amount of runoff and the runoff of the surface waters.

### 3.4.2.6 Climatic information

The climate reflects the summary of atmospheric conditions over a long period for a certain geographical space, for example a certain location, a landscape, a region or a country. It is one of the factors of the landscape balance. The main climatic elements are the radiation, air pressure, air humidity, temperature, wind, evaporation, rainfall and cloudiness. These climatic elements determine in their combination with climatic factors such as geographical latitude, altitude, exposition, land cover and density of settlements the climate in a certain geographical space. Climatic elements and their changes in time are permanently measured with the help of climatic and meteorological stations distributed at different distances and densities throughout the whole country. The Deutsche Wetterdienst (DWD) - the German Meteorological Service - collects, examines, processes, analyses and assesses all the data and their changes over time. The DWD is the most important provider of climatic data in Germany. It runs and maintains about 4,000 weather, climate, rainfall, sunshine and wind stations in Germany. Depending on the spatial density of the measuring stations, the data are interpolated for a raster of 1km x 1km and are offered for sale in digital form to various users. The accuracy of the data differs from parameter to parameter depending on the spatial density of measuring stations. Besides providing climatic information, the DWD develops specific climate maps, for example wind maps and radiation maps on the basis of rasters with different resolutions (1km x 1km or 200m x 200m) for the whole country and for individual federal states or for a certain location using different models. In addition, the results of climatic measurements are shown and described in various thematic atlases, especially in climate atlases or wind atlases, for example in the "Klimaatlas von Bayern" (Bayrischer Klimaforschungsverbund (BayFORKLIM) 1996) or in the "Bayrischer Solar- und Windatlas" (Bayrisches Staatsministerium für Wirtschaft, Verkehr und Technologie 1997). More information about the tasks, data, products and publications of the DWD can be found on the internet (http://www.dwd.de).

### 3.4.2.7 Vegetation information

The vegetation is one of the main landscape-ecological characteristics and represents the summary of all plants and societies of plants and their spatial distribution in a certain geographical space or in a landscape. It is determined by and developed under specific conditions such as location, soil, geology, relief, climate, water, land cover and land use. The vegetation is the result of the development of landscapes. The existence, development, rarity and/or loss of a certain vegetation provides information about the structure and the balance of a landscape. The vegetation reacts to changes of landscape balance resulting from anthropogenic influence faster then the soil.

Updated vegetation maps are very rare because the vegetation cover changes. Updated information about the vegetation can be obtained with the help of vegetational mapping in the terrain. As this ties up considerable time and resources, such mapping can only be conducted for selected small geographical spaces.

The biotope maps of the federal states are another source of certain updated vegetation information (cp. 3.4.2.8).

The distribution of species of plants is shown in so-called floristic raster mappings for the whole of Germany. Vegetation is mapped on the basis of topographic maps on a scale of 1:25,000. The results are shown in small scales.

One result of this work is the "Verbreitungsatlas der Farn- und Blütenpflanzen Ostdeutschlands" (Benkert, Fukarek, Korsch 1998), an atlas of the distribution of ferns and phanerogams in eastern Germany on a scale of 1:4 million.

Besides the present vegetation, the potential natural vegetation - the vegetation which would have been developed under natural conditions without any anthropogenic influence - is shown in specific thematic maps. One example is the map of natural vegetation in the "Atlas Deutsche Demokratische Republik" (Akademie der Wissenschaften der Deutschen Demokratischen Republik 1981) on a scale of 1:750,000, which also contains some maps of the spatial distribution of selected plants and plant societies.

### 3.4.2.8 Land use information

Land use reflects the social use of the limited natural resource area. Natural conditions in different landscapes and social demands determine land use and cause conflicts. Land use influences and/or changes the natural conditions in the landscapes. This means land use is the key and the regulatory instrument for the preservation of these natural systems and re-establishing the multiple use of landscapes. In comparison with the geological or hydrogeological situation, land use is a parameter which is subject to permanent change. Therefore there is a great demand for updated land use information. Given this, more effective methods for capturing land use and land use changes have to be developed in order to derive updated land use information and land use maps.

Presently aerial photographs and satellite images are the main sources for gaining updated information about land use and its changes over time. More in formation about their role for landscape-ecological investigations is contained in part 3.5.4 of this chapter.

The Statistical Federal Authority recorded land cover information for the whole country during the European project CORINE Land Cover in the 1990s. On the basis of satellite images, using topographic maps and additional sources, the land cover was captured on a scale of 1.100,000 for the whole of Germany for the first time. This information about the main types of land cover is available in digital form. The following land cover types were registered: built-up areas, agriculturally used areas, woodlands and quasi-natural areas, moist areas and water areas. These five land cover types are divided into 15 subtypes - which in turn are subdivided into many more subtypes. Depending on the objective of investigation, detailed subtypes can be aggregated in subtypes and/or main types.

Some federal states have biotope-mapping at their disposal. In Saxony, for example, the State Ministry of the Environment and Agriculture (the former State Ministry of the Environment and Regional Development) of Saxony and the Federal Authority for the Environment and Geology has performed CIR biotope and land use mapping for the whole state. Colour-infrared (CIR) aerial photographs from 1992/93 were used to document the state of natural conditions and land use in Saxony for the first time after German reunification and the formation of the new federal states in eastern Germany. More than 16,000 aerial photographs were interpreted and cartographically prepared on a scale of 1:10,000, and the information was also made available in digital form.

This biotope and land use mapping is the most important data basis for various scientific institutions, official authorities and planning agencies. It provides information about the state of the natural environment and landscapes, and provides data for scientific and planning purposes on a local up to a nationwide level (Frietsch 1997, p.10). The biotopes were initially divided into the following eight main groups: 1. stretches of water; 2. moors and swamps; 3. meadows and ruderal fields; 4. rocks; 5. groups of trees, hedges and bushes; 6. wood and forests; 7. arable field and specific sites; 8. built-up areas, infrastructure and green spaces. These 8 main groups were subdivided into subgroups. The next level is the biotope followed by form, use, secondary use and specific use. Each area has a key-code corresponding to its use. The key-code enables a certain area to be characterized regarding the different subgroups and main groups. This means that depending on a specific objective, an aggregation and generalization of the land use and biotope types is simple with the help of the key-code. Moreover, it means that the CIR biotope and land use mapping of Saxony is a very good example of how to structure data for multiple use in different geographical dimensions. The very detailed mapping allows investigations into the topological, chorological and regional dimension on the basis of the same source and with the same accuracy. This is a great advantage over other thematic data available at different scales with different accuracies and based on different sources.

The other states in eastern Germany have also carried out such biotope and land use mapping. The main disadvantage is that each federal state made its own map without any coordination or adjustment concerning the types and division of biotopes between the neighbouring federal states. As a result, the biotopes along borders do not match. This is a serious problem if borders are crossed during a certain investigation, entailing much work to fit the different areas together.

Besides CIR biotope and land use mapping, the Saxony Federal Authority for the Environment and Geology has also produced a land use map known as the "Landnutzungskarte des Freistaates Sachsen" (Sächsisches Landesamt für Umwelt und Geologie 1996) on a scale of 1:100,000 for the whole of Saxony. Using basis satellite images from 1992/1993, 17 land use types were automatically classified and represented in 15 map sheets based on pixels with a size of 25m x 25m. This map gives an overview of the spatial distribution and size of main land use types. Such a map needs to be permanently updated in order to have up-to-date land use information in Saxony and as a source for the conclusion of main land use changes.

#### 3.4.2.9 Information about landscape protection

Information about protected parts of landscapes or landscapes which need to be protected in the form of nature reserves, landscape reserves and water reserves are an important factor of landscape-ecological investigations. They mainly result from such complex investigations. Natural and landscape protection serves the preservation of landscapes and parts thereof, including all the components necessary to preserve their ecological capability, variability, biodiversity and attractiveness. The protection of landscapes or parts thereof is a consequence of the anthropogenic influence over a certain period. Selected land uses influence landscapes or the landscape balance in a negative way, and so types of land use have to be restricted or even prohibited. Therefore land uses disturbing or changing the natural balance are forbidden in nature reserves, landscape reserves and water reserves.

The spatial distribution of such protected areas is represented in specific thematic maps on different scales. Two examples are the maps "Naturschutz im Land Sachsen-Anhalt. Karte der Schutzgebiete" (Landesamt für Umweltschutz Sachsen-Anhalt 1996) and the map "Naturschutz. Schutzgebiete in Sachsen" (Sächsisches Staatsministerium für Umwelt und Landesentwicklung 1996) on a scale of 1:200,000. A combination of nature reserves and landscape reserves, national parks, biosphere reserves, natural parks and bird reserves in different levels of protection is represented.

#### 3.4.2.10 Field information

Field information is an essential source to augment the other data sources for landscape-ecological investigations, especially in the topological dimension. Mainly in the field of the mesoscale and macroscale level, i.e. in the chorological and regional dimension, data are available in most cases. However, there is a lack at the microscale level. This gap can only be closed by detailed field work in a certain terrain. Terrain measurements and mapping over a certain period provide realistic (original) information about selected landscape-ecological parameters. Knowledge of the details contributes to the cognition and understanding of processes taking place within the landscape, the landscape balance and the landscape as a whole. Field information also plays an important role for validating models used for landscape-ecological investigations.

Standards for landscape-ecological capture were developed by Zepp and Müller (1999). In their book they describe principles and methods for surveying and processing landscape-ecological aspects in the terrain which are necessary for the characterization and spatial differentiation of landscape ecosystems. These standards can only be regarded as proposals because universal norms do not exist for characterizing ecosystems in their complexity. The idea of this book is to give an orientation about essential aspects for the characterization of landscape ecosystems from a scientific point of view. Demands and possibilities are to be represented for a balanced consideration of ecological facts composed of the compartments atmosphere, relief, soil, soil close to the surface, stretches of water, vegetation and fauna. Common standards and methods of data capture, mapping and assessing landscape-ecological parameters are described. In addition, information about the sources of data methods of their processing with the help of geographical information systems (GIS) and a landscape-ecological data catalogue are introduced. This book is an important working instrument for scientists from various disciplines investigating landscape-ecological problems from their own specific viewpoint.

The sources for landscape-ecological data described in 3.4.2 only represent some of the important sources. There exists plenty of other data in various authorities, scientific and other institutions, planning agencies and others. More information can be found in Chapter 8 of this book. The above-mentioned data and data sources show their great number, diversity and variability. In Germany in the field of topographic data, great efforts have been made with the development of the Official Topographic/Cartographic-Information System (ATKIS). This system is a very important step towards the creation of a unique, updated topographic database covering the whole country and suitable for various uses. The standardization of these data enables the investigation of a certain geographical space located in a number of federal states and which crosses federal boundaries.

Thematic data are available in different geographical dimensions, on different scales and with different accuracies, in analogue and/or digital form for selected small geographical spaces, certain administrative units, individual federal states and the whole country. In contrast to the topographic information, in most cases one main problem is the separate creation of thematic information without any coordination concerning content, accuracy, scale or the geographical dimension between different institutions in different federal states. This results in thematic maps or databases which do not match up at borders between federal states.

The conscientious combination and analysis of data of the same or comparable geographical dimension is essential for scientifically sound landscape-ecological investigations, including landscape assessments and landscape planning. In this field, irrespective of the discipline involved, each scientist must use the available data in a careful, responsible manner.

## 3.5 Remote sensing data

Human activities affect the land surface, water and the atmosphere. Mankind and the environment, mankind and the natural balance coexist in a complex relationship and need to be observed at different levels of research and by different means. From the ecological viewpoint, human interventions need to be planned and sometimes even channelled, and the accompanying conditions, changes, and dynamics need to be recorded. Causes and effects can only be determined insufficiently by ground data as environmental changes partly occur in large spaces and sometimes even in inaccessible areas. Remote sensing data allow processes of large dimensions to be captured and their changes to be monitored. Information concerning the physical condition of the environment can be gathered by taking ground measurements. However, to investigate and observe large or inaccessible areas, remote sensing data play an important role and deliver the spatial data needed, which can then be added to the ground measurements. Apart from its comparatively large spatial cover, remote sensing has also undergone technical and methodological improvements during the past few decades.

### 3.5.1 Aerial photography and satellite imagery

Remote sensing is the recording of the earth's surface from a certain distance. It comprises photographically documented aerial pictures and satellite imagery mostly gained in digital form. Recently aerial scanner data have completed this series by combining aerial flight altitude with satellite scanner techniques. Remote sensing data can be classified either by the carrying platform (e.g. hand, car, aircraft, satellite) or by the type of sensor (e.g. camera, scanner). In contrast to aerial photos, which have been developed and analysed since the early 20th century for civil research purposes, scanned or photographic satellite images have only been established for the last thirty years. Both carrier systems, aircraft and satellites, are the indirect observation instruments used to record and display the earth's surface from a certain distance and with a certain spatial extraction depending on parameters such as the recording instrument, its field of view and the flight altitude. It is only by means of systematic analysis that the data collected can be processed and interpreted for research purposes and used for concrete investigations such as layers of information documenting land cover in landscape ecology. Knowledge about recording techniques, parameters of photos and scanned images, and the repetition rate is fundamental to choosing the right system for the application required, appropriate auxiliary data, and also selecting the suitable methodology to analyse the data.

### 3.5.2 Development of the use of instruments

It was Carl Troll who summarized the applications of aerial photography for geographical field research in 1939 (Troll 1939). During World War I, aerial observation for certain scientific aims started with the sudden emergence of the aircraft sector. As it was at that time that terrestrial field research began to be carried out by aerial photos, series of such photos were produced and put together to form aerial plans and for aerial maps without leaving any land cover unsurveyed. Troll stated that aerial photography unified different earth sciences and called for a precise research methodology for the focused use of such spatial information to support geographical field observation. He realized that based on the complete analysis and evaluation of the aerial data, not only topographical information could be gained but also maps for geology, soil sciences, vegetation, and landscape ecology, and that aerial photos could thus serve spatial planning.

Since then the great demands made on remote sensing and the nature of the tasks have remained more or less the same. What has changed is that nowadays remote sensing techniques belong to many different devices. With the progress made in techniques, greater distances from the earth's surface have become possible using satellites as sensor carriers in addition to aeroplanes. Opto-electronical scanners and radar systems complement the recording devices that started with camera systems. Observation devices have thus changed and become more diverse (Lillesand & Kiefer 1999).

Over the past 20 years, satellite data have become increasingly sophisticated, enabling high-resolution images, and their usage has spread to many fields. Operational systems with scanners and radar devices can register an area in wavelengths beyond the visual range of human beings.

Starting with the US aeronautics and space programme, developments over the past 35 years in the field of remote sensing have accelerated. The first useful colour photos of the earth were taken by an automatic camera aboard the American Mercury capsule in May 1961, followed by photos from the Gemini mission in June 1965 and later from Apollo 9 in March 1969. The first weather satellites were launched in the 1960s by the United States. Meteorological satellites followed in the 1970s with the NOAA, GEOS and METEOSAT missions.

The launch of the US satellite ERTS-1 (Earth Resources Technology Satellite, later renamed Landsat 1) on 23 July 1972 marked the breakthrough for earth observation from space. The whole Landsat series, 1-5 and now 7, has delivered image data of almost all the earth's territories at regular, fairly short intervals. Landsat 5 Thematic Mapper was constructed to take pictures for about 6 or 7 years, but having been in the orbit since 1984 it still delivers data regularly. Thus data availability of more or less the same quality has been enabled for almost two decades. The launch of Landsat 7 ETM (Enhanced Thematic Mapper) continues with a better system and radiometric correction. The new sensors have been sent into orbit with almost the same traditional sensors of Landsat 4 and 5 TM but enhanced with a higher resolution thermal band and additionally equipped with a panchromatic sensor. The thermal band of Landsat TM and ETM is outstanding and not available from other producers of multispectral satellites. However, as other satellite series have started to take images in multispectral and panchromatic modes at the same time, it is the ETM sensor of Landsat 7 that assures competitiveness (see Richards 1999).

From the 1980s onwards, a multitude of national activities for remote sensing applications with improved satellite systems emerged. Important European Space Agency projects include the European radar satellites ERS-1 and ERS-2.

In 1986 France launched its first Spot satellite (Système pour l'Observation de la Terre), which was then followed by another three missions, Spot 2-4. Its HRV Sensor (Haute Résolution Visible/High-resolution Visible Sensor) operates in two modes, namely the multispectral mode and the panchromatic mode. Thanks to its side-viewing features, the HRV instruments are capable of plotting contours by taking stereoscopic images, as well as imaging a particular location more frequently, without waiting for an overhead pass. SPOT-4 carries an additional "Vegetation" instrument in order to provide the capability of small-scale monitoring of the earth's vegetation (pixel size = 1.15 km at nadir).

The first Indian Remote Sensing Satellite was launched in 1988. Named IRS-1A, it was equipped with a sensor called LISS (Linear Imaging Self-Scanning Sensor). LISS-I and LISS-II are both multispectral cameras. There followed IRS-1B, and, with IRS-1C a second generation of remote sensing satellite series was built with enhanced capabilities in terms of spatial resolution and spectral bands. IRS-1C was launched on 28 December 1995 and was followed (Kramer 1996) by IRS-1D three years later. Satellites of the series 1C and 1D carry three sensors, PAN which is a panchromatic camera, LISS-III, and WiFS (Wide Field Sensor). These products compete directly with Landsat TM and ETM as well as with Spot image data (see Tab. 3.2).

One of the latest satellite systems for landscape-ecological investigations on large scales is the Ikonos satellite, launched in September 1999, with a spatial resolution of 4 metres on the multispectral sensor mode and 1 metre on the panchromatic mode. For ecological research, such as the urban nature within building complexes, panchromatic and multispectral images can be combined to be able to interpret near-infrared, for example, with a 1-metre resolution capability. The viewing geometry only offers an 11-kilometre-wide ground track, but this is sufficient for large-scale investigations, especially as the geometric resolution calls for new methodological approaches and a high demand on data storage when working with several composed images.

For earth observation from space, a wide range of instruments is and will continue to be used in combination with other instruments. With their active and passive sensor techniques they cover a large band width of the electromagnetic spectrum. In over 80 missions more than 200 different instruments are to be used and examined during the next 15 years. The sensors can be categorized as follows:

- Instruments for atmospheric chemistry
- Atmosphere explorer
- Cloud profile and rain radar instruments
- Radiometer to record the earth's radiation balance
- High-resolution systems
- Multispectral radiometer
- Active radar
- Lidar (acronym for "light detection and ranging")
- All-round radiometer
- Radiometer for the colour of the sea
- Polarimetric radiometer
- Radar altimeters
- Wind radiometer

| Dimensions of satellite image data | Examples of multispectral sensors (spectral range) | | | Potential applications |
|---|---|---|---|---|
| **Electro-magnetic** | **Landsat-TM / [-ETM]** | **SPOT-MS and Pan** | **IRS-1C /D LISS and Pan** | |
| *spectrometric*: intervals of used spectral fields of a sensor<br><br>*radiometric*: number of discrete values (dependent on the capacity of the available detectors of a system)<br><br>into which the measured signals can be divided<br>(e.g. 128, 256, 1024 grey levels) | Band 1: 0.45-0.52 µm VIS<br>Band 2: 0.52-0.60 µm VIS<br>Band 3: 0.63-0.69 µm VIS<br>Band 4: 0.76-0.90 µm NIR<br>Band 5: 1.55-1.75 µm MIR<br>Band 6: 10.40-12.5 µm (thermal-IR)<br>Band 7: 2.08-2.35mm MIR<br><br>[Pan:<br>Band 8: 0.50-0.90 µm] | Band 1: 0.50-0.59 µm<br>Band 2: 0.61-0.68 µm<br>Band 3: 0.79-0.89 µm<br><br>Pan: 0.51-0.73 µm | Band 1: 0.52-0.59 µm<br>Band 2: 0.62-0.68 µm<br>Band 3: 0.77-0.86 µm<br>Band 4: 1.55-1.75 µm<br><br>Pan: 0.5-0.75 µm | ▪ separating different forms of land cover (e.g. soil and vegetation, deciduous and coniferous forest; coastal mapping)<br>▪ investigation of plant stress (high reflexion for healthy vegetation<br>▪ differentiation of plant species (absorption of chlorophyll)<br>▪ recording of plant vitality; investigation of biomass<br>▪ differentiation: cloud / snow; estimating plant humidity and soil moisture<br>▪ thermal mapping; investigating the influence of heat on plants<br>▪ geological mapping;<br>▪ land use monitoring; spatial planning;<br>▪ Stereoscopic interpretation (digital elevation models - only Spot pan) |
| **Spatial**<br>▪ Size of ground elements (dependent on the geometric characteristics of a sensor)<br><br>▪ Swath width | 30 x 30 m (bands 1-5 / 7)<br>120 x 120 m (band 6 TM)<br>[60 x 60 m (band 6 ETM)]<br>Pan: 15 x 15 m<br><br>185 x 185 km | 20 x 20 m<br><br>Pan: 10 x 10 m<br><br>60 x 80 km | 23.5 m (bands 1-3)<br>70.8 m (band 4)<br><br>Pan: 5.8 m<br>142 x 142 km (LISS);<br>70 km (Pan) | In the multispectral mode only a scale for regional planning is possible. Data fusion (e.g. resolution merge) with panchromatic images allow a large scale level. |
| **In time**<br>Cycle of repetition between two images of the same area (repetition rate) | 16 days | 26 days | 24 days | Resource management for land and water; landscape monitoring |

**Table 3.2:** Dimensions and potential applications of satellite images using selected examples of sensors

The listing shows a wide range of existing sensors. For Central Europe and the demands in landscape-ecological research the most important sensors are the high-resolution systems. These sensors characterize the largest application range of all instrument categories. They serve for the registration of vegetation classes and the erosion of coast lines, but also for geological mapping. For example of opencast mining areas. They will be dealt with later, especially in Chap. 5. Other important systems include hyperspectral scanners, as well as multispectral and passive radar radiometers. Hyperspectral data are still being tested and asre mainly used on aircraft. The multispectral radiometers are important sources for recording processes in the biosphere, because they collect information about influences on the global vegetation which in turn allow conclusions about the ecosystem. The latter 'SAR' (synthetic aperture radiometer) data are especially important for agriculture and forestry, as well as for measurements of snow and ice cover.

The operational sensor data most frequently used for landscape-ecological investigations is presented in Tab. 3.2, their spectral, spatial and radiometric resolution being accompanied by general application possibilities.

### 3.5.3 The importance of resolution

The possibilities for observing details of the earth's surface by means of remote sensing data depend mainly on the resolution capabilities of the different sensors. These dimensions are a source of both possibilities and limitations for image processing, interpretation and analysis ( Bähr & Vögtle 1998).

- Repetition rate
  The resolution in time indicates how often measurements are repeated of the same area. Regarding satellite systems, repetition is determined by its orbit.
- Ground resolution
  The spatial resolution that is the smallest linear (or angular) separation between two objects achieved by a sensor is one of the most important factors for interpreting remote sensing data, because the realization of details is directly dependent on the ground resolution. Great demands are made on spatial resolution, e.g. for updating regional and spatial planning, for agricultural statistics (harvest trends), and for the analysis of forest damage.
- Spectral resolution
  The power of a system to separate single wave lengths is defined as its spectral resolution. It relates to the number and size of the wavelength intervals ('bands') into which the electromagnetic spectrum can be divided and for which the sensor is constructed.
- Radiometric resolution
  The radiometric resolution characterizes the number of levels into which the signal received by the sensor can be divided. For instance, in one spectral band 256 grey levels can be displayed.

Landscape-ecological targets use and apply geographical knowledge especially with respect to regional geographical realization when it comes to solving practical tasks. Applying landscape ecology means investigating actual practical problems of the use, demands, development and protection of a landscape by the human beings. It also serves to assess changes to its natural features and its function. Moreover, there is a close relationship between these ecological tasks and integrated spatial planning.

With regard to the complex subject 'landscape' and the ordinary levels of spatial planning, classifications are undertaken that allow the characterization of systematic and hierarchical structural levels. When taking remote sensing data into account to work on different levels, the resolution plays a crucial role. The ground resolution enables work at certain scales but prohibits work at more detailed scales than those defined. If sensor data possess a ground resolution of 30 metres, these data will only be suitable for planning to a limited extent. Certain other tasks such as protection programmes need a refined land use mosaic and thus the interpretation of a multitemporal data set within the same year and within one phenological phase (e.g. land use monitoring). Hence the repetition rate is important for several high quality imageries within a defined period of time, and it usually takes several years to repeat aerial flights over the same region. The spectral dimension is the basis for classifying differences between different land use types and, the more bands that are available, the more sophisticated a classification scheme and a subdivision of classes can be. Examples include classifying different crops or finding spectral differences within standing water. Conclusions can thus be made for instance for the evaluation of agricultural management systems, or water depth and water quality, pollutant inputs, etc. (Banzhaf & Kasperidus 1998).

### 3.5.4 Landscape balance

Multitemporal analyses help record structural changes as well as general environmental and landscape modifications. Thus the component of process investigation is taken into account. This technique for monitoring a landscape and its land use is one of the advantages within the applications of remote sensing data and their methodologies. One attempt to gain information about the link between processes of the ecosystem and structures is to differentiate structural units (e.g. the forest stands) and the subsequent investigation of the influences of these differentiating elements (in forest stands this would be the tree species) on landscape functions (e.g. groundwater regeneration) (Kratz & Suhling 1997). Such scientific insights are a basic prerequisite for the sustainable prevention of landscape functions. The advantage of this improvement is the multitemporal and spatially extensive possibility of registering parts of landscapes in connection with the use of remote sensing and geographical information systemss. However, it is not possible to provide any evidence about the issue of fluxes and process courses with remotely sensed data (Volk & Steinhardt 1999, pp. 234).

Methods of satellite remote sensing are used to rapidly record large areas. Parameters such as land use, vegetation and soil characteristics, to mention just those of vital importance, can be recorded for landscape-ecological questions. The spectral reflection and degree of emission from the visible to the infrared and microwave range provides information on the leaf area index, temperature and moisture contents of vegetation and its type, damage, biomass and, by applying active radar images, even soil moisture might be estimated. As vegetation is of fundamental importance for the albedo and terrestrial water balance, it mainly influences the atmospheric water vapour content. It makes the hydrological cycle a condition, for example, but is also responsible for land erosion and the composition of the atmosphere. The proportion of precipitation and evaporation defines the character of water balance on the continent. Evaporation depends not only on climatic conditions but also on soil features, tillage and the vegetation cover. This means it is dependent on factors manipulated by the human being and the way land use is carried out. Hence man interferes actively in the water balance of a region and fixes the usable water availability and quantity. For water balance, soil models, and the modelling of other landscape-ecological quantities, it is thus essential to derive the active biomass spatially as well as the land use types in the ecosystems. Analysed image data are essential for spatial investigations, although it should be borne in mind that they are only one information layer for landscape-ecological models (Volk & Steinhardt 1998).

Landscape-ecological models need to be standardized to be able to compare the individual analyses. They can be of great use for regional and global differences and changes because individual point measurements and different devices in different regions neglect the need for comparison. Interpolation methods may help but the more heterogeneous a landscape is, the more doubtful the interpolated point measurements will be. In this context, satellite data are very helpful in minimizing the spatial variability as well as the variability in time to which data registration is often subject because it is collected in a synoptic data set over hundreds of kilometers. It will be shown in Chap. 5 that using the same sensor as one data layer in different investigation approaches will bring stable results, but changing the sensor and consequently inserting different data qualities within the very same analysis will evoke wider variation and thus a worse analysis.

Images from space should aid the forestry commission to survey the stands and damage to the forest. Annual changes in forestry can easily be documented by satellite data and therefore commissions can react more rapidly to negative changes, or those affecting large area. At the same time the ground data need to be continued even if they are much more time-consuming because detailed surveys are necessary to compare them with spatial data and to offer test areas for the calibration and validation of spatial data. Point measurements are generally of great use but they neglect an overview. Therefore, the more detailed landscape-ecological analysis is, the closer the link between point-measured data and spatial data derived from different satellite sensors. And the better this connection works and the more accurate satellite image data is for planning commissions, the less the need for aerial photography. Aerial photography is much more expensive and analysis much more time-consuming than satellite imagary is. A satellite image costs a few thousands of dollars, whereas an aerial flight mission for a small-scale mapping of for example the Erzgebirge mountain range can be performed for about $25,000 (Bundesministerium für Bildung und Forschung 2000).

Another field of interest for the use of satellite images and further new technologies is agriculture. The latest projects have focused on for example the spatial quality assessment of grain during the growing period up to the harvest. Typical regions are taken as test areas for certain fruit/grain to identify and observe its development over the whole vegetation period. Together with additional pieces of information linked in a geographical information system, the yield can be estimated and the quality of the grain can be concluded. Such an economical assessnent can be undertaken while the grain is still growing in the field. Arguments like "The farmer knows the growing conditions on his fields much better than a satellite image can" must still be accepted, at least today regarding agriculture in Central Europe. However, in test regions remote sensing data have started to provide detailed information in fields.

Competitive economics and environmental regulations are two important forces driving precision farming to improve agricultural efficiency by matching inputs - such as water, seed type, fertilizer, and weed, disease, and pest sprays - with soil types and terrain.

The electronics revolution of the past few decades has spawned two technologies that will impact on agriculture in the next decade. These technologies are geographical information systems and the Global Positioning System (GPS). Along with GISs and GPS, a wide range of sensors, monitors and controllers have appeared for agricultural equipment such as shaft monitors, pressure transducers and servo motors. Together they will enable farmers to use electronic guidance aids to direct equipment movements more accurately, provide precise positioning for all equipment actions and chemical applications and, analyse all this data in association with other sources of data (agronomic, climatic, etc). This will add up to a new, powerful toolbox of management tools for the progressive farm manager.

In terms of records and analysis, precision farming may produce an explosion in the amount of records available for farm management. Electronic sensors can collect a lot of data in a short period of time. A lot of new data is generated every year (yields, weeds, etc.). Farmers will want to keep track of the yearly data to study trends in fertility, yields, salinity and numerous other parameters. This means a large database is needed with the capability to archive and retrieve data for future analysis.

Precision farming should not be thought of as only yield-mapping or variable-rate fertilizer application, and evaluated only by one or the other. Its technologies will affect the entire production function (and by extension, the management function) of the farm. Precision farming allows for improved economic analysis by monitoring and fine-tuning production. The variability of crop yield in a field allows risks to be accurately assessed. By knowing the cost of inputs, farmers can also calculate return over cash costs for each acre. Certain parts of the field which always produce below the break -even line can then be isolated for the development of a site-specific management plan.

Precision farming makes farm planning both easier and more complex. There is much more map data to utilize in determining long-term cropping plans, erosion controls, salinity controls and the assessment of tillage systems. But as the amount of data grows, more work is needed to interpret the data - and this increases the risk of misinterpretation. Farmers implementing this technique will probably work closer with several professionals in the agricultural, GPS and computing sciences. What is perhaps more important for the success of precision farming, at least initially, is the increased knowledge that a farmer needs of his natural resources in the field. This includes a better understanding of soil types, hydrology, microclimate, aerial photography and aerial scanner data. The latter is the most important key source of data, and no farmer should start precision farming without it (Goddard 1997).

When in spring 1992 the European Commission resolved to implement a reform of European agricultural policy, a new market policy was launched that was no longer financed by price support but based on spatial compensation payment. Every farmer who produces grain, protein plants or other specific seeds only gets world market prices for his products. They are usually lower than the those the years before with guaranteed prices. By way of compensation for the difference, the farmer receives compensation payments according to the area of his arable land. At the same time the farmer is obliged to set aside 10 percent of his production area. As supervision is hard and individual checks are expensive, remote sensing data offer an ideal instrument for spatial analysis and the control of agricultural production. Image analysis offers opportunities for field scale survey and monitoring provided that three images are available within one growing season.

### 3.5.5 Outlook

We have seen that satellite imagery is subject to improved techniques and that methods of image analysis will need to make progress in order to keep up with this rapid development. Compared to aerial flight missions, satellite images can be acquired much more often, regularly, much more cheaply.

The time intervals between data acquisition and the supply of processed data need to be shortened immensely. Tasks that can be fulfilled with satellite image data were exemplified and illustrated in this chapter. A close link between ground data, individual measurements and remote sensing data is indispensable to calibrate and validate the data and to obtain precise results.

The satellite industry designs devices for all sorts of tasks and challenges, some of which are mentioned above. Mini-satellites are set to become much in demand on the market as they will work more locally and at very low cost, enabling data to be ordered in advance for precisely specified regions. The ground resolution is now approaching 1 metre on various devices, so that decision makers can work on large-scale image maps and with the most up-to-date data.

## 3.6 GIS and models - important tools for landscape analysis and landscape assessment

According to Duttmann (1999, p. 363), landscape-ecological spatial analysis and spatial assessments need both the supply of complex spatial basic data and the availability of substantial methods to analyse and assess these data. Because of their high functionality concerning processing, analysing and representing spatial data geographical information systems have been established more and more as an essential and indispensible tool in the specific field of landscape analysis, landscape assessment and landscape planning during the last ten years. Due to the rapid development of faster and cheaper computers and higher demands on the software the functionality of GIS has been improved and extended especially in the nineties of the last century.

According to Borrough and McDonnell (1998, p. 11) the "tool base definition of a GIS is a powerful set of tools for collecting, storing, retrieving at will, transforming and displaying spatial data from the real world for a particular set of purposes." and "geographical information systems have three important components - computer hardware, sets of application software modules, and a proper organizational context including skilled people - which need to be in balance if the system is to function satisfactorily." (Borrough and Mc Donnell 1998, p. 12)

The main difference between GISs and other kinds of information systems is the spatial reference. This means that in a GIS all the geographical objects and phenomena are registered and processed with the help of two types of data, geometrical data and attribute data. The first ones are information about the geographical location, size, spatial distribution, arrangement and neighbourhood of such objects. The spatial reference can be established by means of different coordinate systems, e.g. geographical coordinates (latitude, longitude) or other spatial reference systems like the Gauss-Krüger coordinate system. The coordinate system need not only be two-dimensional (x-coordinate, y-coordinate); it can be three-dimensional (x-coordinate, y-coordinate, z-coordinate), too. The second ones provide thematic information about an object. This means information can be provided on what kind of object it is. For example, a selected area may represent a certain kind of land use or a specific soil. Linking of both geometrical and attribute data is essential for efficiently and successfully working with a GIS.

Depending on their ability to process certain kinds of data, GISs can be divided into vector-based and raster-based systems. Vector data represent a certain spatial object with the help of their basic elements points, lines and areas. Moreover, relationships with the neighbourhood are described, for example the starting point and the end point of a line and the areas adjacent to it. Raster data do not differentiate between points, lines and areas. Unlike the vector data, they are based on picture elements, or 'pixels'. Pixels are geometrical base elements which form a matrix of rows and columns of uniform elements in the form of squares or rectangles. They are the result of a specific registration process in which for example the earth's surface is scanned with the help of specific scanners from a satellite and the results are satellite images or a thematic map is scanned. Hence a geographical object consists of several pixels. The size of a pixel depends on the quality and resolution of the scanner used. More detailed information about different resolutions of images is contained in part 3.5.3 of this chapter. There are no logical links between the pixels. Each pixel contains values about a specific characteristic in the form of gray values or colour values.

During the last decade, a development towards hybrid systems has taken place. Hybrid GISs can process both vector data and raster data. In view of the large variety of data needed for landscape-ecological investigations, hybrid GISs have been increasingly used in this special field. They allow for example the combination of vector data derived from a topographic or a thematic map and raster data from a satellite image.

A GIS consists of hardware, software modules, data and people who can use the software for various purposes. These four main components form a unit. The hardware required for a GIS comprises digitizers and scanners for registering various information, a computer with a monitor, the keyboard and the mouse, a printer (or plotter) for the analogue output of registered, processed and cartographically or graphically presented data, disk drives, CD-ROM drives, CD-ROM writers and other devices. The software consists of several modules for capturing, creating, storing, administering, processing, manipulating, analysing and representing different spatial data. Spatial data in the form of vector and/or raster data with their geometrical and thematic components are the main part and the core of a GIS. However, to build up and operate a GIS, well-trained personnel are equally import. They must know how to use the different software modules and their interfaces, and be aware of their possibilities and limitations. They have to improve their knowledge permanently. They have to find out about new modules of a certain GIS software and to learn how to use them. New developments in the field of GISs have to be carefully monitored, as only in this way can users remain up-to-date and retain their positions as experts in this very interesting and varied field of work.

There are no set limits to building up a GIS because of the rapid development in hardware. Moreover, various users are calling for the further development of GISs for the organization and manipulation of spatial data, as well as for software improvements.

According to Bill and Fritsch (1991), the software of a GIS consists of four basic modules: data input, data management, data analysis and data representation. Data input serves both the transfer of analogue data - for example, from topographic or thematic maps into the GIS with the help of special devices such as digitizers or scanners - and the integration of available digital data from various sources and institutions with the help of specific interfaces into the GIS. The spatial data have to be stored and managed within the GIS such that interactive manipulations and certain steps of data processing are possible. In this field, the database system with its two components, the database and the database management system, form the basis for organizing and managing spatial data. Depending on the types of data suitable, data structures have to be found which can be reflected in different types of database models like hierarchical models, network models, relational models and object-oriented models. In addition to these models, the latest development tends concern the integration of knowledge-based systems into a GIS.

GISs are distinguished by varied possibilities for analysing spatial data for a certain purpose. According to Zepp and Müller (1999), the main basic analysis functions of a GIS are geometrical, arithmetical and logical data links, statistical analysis, overlays and intersections, data transformations in the form of raster-vector transformation and vector-raster transformation, coordinate transformations from one coordinate system to another, neighbourhood analysis, relief analysis and generalizations. In addition, most GISs have their own macro language or an interface to a higher programming language so that specific applications can be developed for a certain purpose by users themselves.

GISs have numerous possibilities when it comes to visualizing and displaying spatial data or analysis results. They include cartographic representations on the screen and in the form of printed/plotted maps. The basic graphical elements point, line and area can be cartographically modified and displayed with respect to their size, shape, shading, brightness, orientation and colour. Furthermore, maps can and should be augmented by text, for example geographical names, a title, scale, legend, sources and editors. The tools for showing data have so many free options that users can create their own maps with the help of different cartographical display methods corresponding to the topic and purpose, the scale and geographical dimension, the degree of generalization, etc. The quality of data presentation varies greatly. It depends on the knowledge, experience and (unlimited) creativity of people working in this field.

There are many commercial GIS producers throughout the world. In the field of landscape-ecological investigations, both geographical information systems and digital image processing systems are used. Depending on the tasks at hand and the objectives, and in view of the great variety of data, hybrid systems have come to be more used than only vector-based or raster-based systems. Well-known products in the field of GISs include for example ArcInfo and ArcView, which are available for PCs and workstations. In the field of digital image processing systems, Imagine and Easy Pace should be mentioned. Yet besides these high-quality software packages, a host of other systems exist, e.g. SPANS, PolyGIS, AtlasGIS and IDRISI.

The developers of these complex systems strive to increase the functions of their products by creating new modules and interfaces for exchanging data with other systems, and are also keen to make their products more comfortable and easy-to-use. During the last three or four years, efforts have mainly been directed towards improving PC-based GIS systems, with progress being slower for workstations. One reason is the fact that many more institutions run the software on PC as both hardware and software are cheaper. This trend is exemplified by ArcInfo and ArcView by ESRI. While the further development of ArcView (usable for PCs and workstations) has been permanently advanced by the creation of new modules and improving the handling of this software, the further development of ArcInfo has been neglected.

Despite these efforts in further developing GIS -software, two main aspects, the processing of three-dimensional data and the integration of the time factor, have not yet been considered sufficiently.

Three-dimensional representations of data play an important role in landscape-ecological investigations and have long been called for by users. Therefore ESRI has developed a new module within ArcView called the 3D Analyst. According to the company's description, this module enables realistic surface models to be created from multiple input sources. This enables the altitude to be determined at any point, ascertaining what can be seen from an observation point, the calculation of volumetric differences between two surfaces, working with 3D vector features to make realistic models of the 3D world, and visualizing data in 3D. The ArcView 3D Analyst is the first GIS module which allows the processing of three-dimensional data.

The integration of the time dimension remains an unsolved problem. Spatial and temporal changes play an important role for landscape-ecological investigations, especially in investigating the landscape balance and its fluxes of matter and energy. According to Zepp and Müller (1999, p. 368) modelling spatially and temporally variable landscape balance processes need both the new data management concepts (including new data management techniques) and expanded simulation and visualization techniques. The latter are essential for real-time simulations and for spatial representations of such simulated processes with the help of GISs.

In view of the available limited modelling and simulating possibilities within a GIS, the simulation of spatial and temporal processes is carried out with the help of external models linked to the GIS. According to Goodchild (1993), there exist two main ways of linking up a GIS to models. The first one is 'loose coupling' or 'low-level coupling', a simple form of model coupling. Each model needs specific input data in a certain data structure. In the case of loose coupling, the GIS solely has the function of a data server, and serves the preparation and derivation of data needed for the model input. In addition the GIS is used for storing and displaying the results of the modelling process. The simulation itself takes place outside the GIS, in an external model. The GIS and model are only linked up for data exchange. Data processing takes place in separate, mutually independent systems.

The second one is 'tight coupling' or high(er)-level coupling. In this case, the GIS and the model are directly linked. Data transfer takes place automatically between the database and the model. A joint interface between the GIS and the model allows both the modelling process and interactions between GIS and model. The availability of such an interface is essential for linking the GIS to the model. An interface can be programmed with the help of the programming language of the GIS itself or for instance with another macrolanguage. The latest developments are usually based on the second type of model coupling. The great advantage of this type of coupling consists in that fact that the data resulting from the modelling process are written to the GIS database immediately.

Depending on the specific problem and objective, the models are created and used for different purposes in the field of landscape-ecological investigations. They are used to model changes concerning individual landscape-ecological parameters or parts of a landscape over a shorter or longer period. According to Leser (1997), ecosystem models have to be 'limited' spatially and functionally with respect to the structure of the ecosystem model and/or the spatial structure, the investigation period or the observation period, the available technical and/or personal infrastructure. The functional structure of the observed real ecosystem has to be recognized within the model. The two components space and time have to be represented sufficiently in the model. Models need real data for simulation processes and for the validation of the models to identify errors during modelling. The input data must have a specific structure depending on the model used. Chap. 7 contains an overview of specific models in the field of landscape-ecological investigations. Moreover, Chaps. 6 and 7 contain the results of applying models in the field of landscape-ecological research.

The large number of different spatial data needed for landscape-ecological analysis, landscape-ecological assessments and landscape planning require suitable instruments and tools to process them. During the past ten years geographical information systems have become the most important and efficient tools for registering, storing, managing, analysing and showing spatial data in a two-dimensional or three-dimensional form. Most GISs available cannot process the dynamic phenomena which also play an important role in investigating landscapes. Therefore, GISs have recently been linked to models to simulate changes in space and time.

Generally speaking, the development of GISs is leaning more and more towards the creation of user-defined GISs. This means that depending on the specific purpose, the users create their own GISs consisting of specific selected modules of a certain software which are then linked together with interfaces programmed by the users themselves.

## 3.7 References

Akademie der Wissenschaften der Deutschen Demokratischen Republik (Hrsg) (1981) Atlas Deutsche Demokratische Republik. VEB Hermann Haack, Geographisch-Kartographische Anstalt Gotha/Leipzig

Arbeitsgemeinschaft der Vermessungsverwaltungen der Länder der Bundesrepublik Deutschland (AdV) (Hrsg) (1998) Amtliches Topographisch-Kartographisches Informationssystem ATKIS. ATKIS-Dokumentation, Teil D, ATKIS-Objektartenkatalog (ATKIS-OK), Hannover

Bähr H-P, Vögtle T (Hrsg) (1998) Digitale Bildverarbeitung. Anwendung in Photogrammetrie, Kartographie und Fernerkundung. 3rd enlarged edition, Wichmann Verlag, Heidelberg

Banzhaf E, Kasperidus HD (Hrsg) (1998) Erfassung und Auswertung der Landnutzung und ihrer Veränderungen mit Methoden der Fernerkundung und geographischen Informationssystemen im Raum Leipzig-Halle-Bitterfeld. UFZ-Bericht 2/1998

Bastian O, Schreiber K-F (Hrsg) (1999) Analyse und ökologische Bewertung der Landschaft. Spektrum Akademischer Verlag, Heidelberg - Berlin

Bayrischer Klimaforschungsverbund (BayFORKLIM) (1996) Klimaatlas von Bayern. Druck und Verlag Hanns Lindner, München

Bayrisches Staatsministerium für Wirtschaft, Verkehr und Technologie (Hrsg) (1997) Bayrischer Solar- und Windatlas

Benkert D, Fukarek F, Korsch H (Hrsg) (1998) Verbreitungsatlas der Farn- und Blütenpflanzen Ostdeutschlands. G. Fischer Verlag, Jena-Stuttgart-Lübeck-Ulm

Bill R, Fritsch D (1991) Grundlagen der Geo-Informationssysteme. Band 1 Hardware, Software und Daten. Wichmann-Verlag, Karlsruhe

Bundesanstalt für Geowissenschaften und Rohstoffe (Hrsg) (1995) Bodenübersichtskarte der Bundesrepublik Deutschland 1:1,000,000. Karte mit Erläuterungen, Textlegende und Leitprofilen, Hannover

Bundesanstalt für Geowissenschaften und Rohstoffe, Niedersächsisches Landesamt für Bodenforschung, Geowissenschaftliche Gemeinschaftsaufgaben (1999) Der Metadaten-Katalog von BGR, NLfB und GGA (www.bgr.de)

Bundesanstalt für Landeskunde und Zentralausschuss für deutsche Landeskunde (Hrsg) Karte "Naturräumliche Gliederung Deutschlands mit Höhenschichten, 1:1,000,000" von E. Meynen und J. Schmithüsen

Bundesministerium für Bildung und Forschung (Hrsg) (2000) Der Blick für das Ganze. Nutzen der Fernerkundung für den Menschen. BMBF Publik

Burrough PA, Mcdonell RA (1998) Principles of Geographical Information Systems. Oxford University Press, Oxford

Duttmann R (1999) Geoökologische Informationssysteme und raumbezogene Datenverarbeitung. In: Zepp H, Müller MJ (Hrsg) (1999) Landschaftsökologische Erfassungsstandards. Ein Methodenhandbuch. Forschungen zur Deutschen Landeskunde, Band 244, Deutsche Akademie für Landeskunde, Selbstverlag, Flensburg, pp 363 - 437

Frietsch G (1997) Ergebnisse der CIR-Biotoptypen- und Landnutzungskartierung und ihre Anwendungsmöglichkeiten in der Naturschutzpraxis. In: Ergebnisse der CIR-Biotoptypen- und

Landnutzungskartierung und ihre Anwendungsmöglichkeiten in der Naturschutzpraxis. Tagungsband, Sächsische Akademie für Natur und Umwelt, pp 7 - 11

Goddard T (1997) What is Precision Farming? In: Proceedings: Precision Farming Conference, January 20 - 21, 1997, Taber, Alberta, Canada.

Goodchild MF (1993) The state of GIS for environmental problem-solving. In: Goodchild MF, Steyart LT, Parks BO (eds) Environmental modeling: progress and research issues. GIS World Book, Ft. Collins CO

Institut für Angewandte Forschung "Landschaftsentwicklung & Landschaftsinformatik" (IAF) der Fachhochschule Nürtingen (Hrsg) (1996) Digitaler Landschaftsökologischer Atlas Baden-Württemberg. Institut für Angewandte Forschung "Landschaftsentwicklung & Landschaftsinformatik" (IAF) der Fachhochschule Nürtingen

Kratz R, Suhling F (Hrsg) (1997) Geographische Informationssysteme im Naturschutz: Forschung, Planung, Praxis. Verlag Wolf Graf von Westarp

Kramer HJ (1996) Observation of the Earth and Its Environment. Survey of Missions and Sensors. 3rd enlarged edition, Springer Verlag, Berlin-Heidelberg-New York

Landesamt für Landesvermessung und Datenverarbeitung Sachsen-Anhalt (Hrsg) (1999) Amtliches Vermessungswesen im Land Sachsen-Anhalt. Geobasisinformationssystem. Teil: Geotopographie mit dem Verzeichnis der Topographischen Landeskartenwerke.

Landesamt für Umweltschutz Sachsen-Anhalt (Hrsg) (1996) Karte "Naturschutz im Land Sachsen-Anhalt. Karte der Schutzgebiete." 1:200,000

Leser H (1997) Landschaftsökologie.Verlag Eugen Ulmer, Stuttgart

Lillesand TM, Kiefer RW (1999) Remote Sensing and Image Interpretation. J. Wiley & Sons, New York, Chichester, Brisbane, Toronto, Singapore

Meynen E, Schmithüsen J (Hrsg) (1953-1962) Handbuch der naturräumlichen Gliederung Deutschlands. 2 Bände, Bad Godesberg

Mosimann T (1999) Angewandte Landschaftsökologie - Inhalte, Stellung und Perspektiven. In: Schneider-Sliwa R, Schaub D, Gerold G (Hrsg) Angewandte Landschaftsökologie. Grundlagen und Methoden., Springer-Verlag, pp 5 - 23

Neef E (1963) Dimensionen geographischer Betrachtungen. Forschung und Fortschritt 37: 361 - 363.

Ogrissek R (Hrsg) (1983) Brockhaus abc Kartenkunde. Brockhaus, Leipzig

Richards JA, Jia X (1999) Remote Sensing Digital Image Analysis. An Introduction.Springer Verlag, Berlin-Heidelberg-New York

Sächsisches Landesamt für Umwelt und Geologie (Hrsg) (1992) Gewässergütekarte 1991. Freistaat Sachsen. 1:400,000

Sächsisches Landesamt für Umwelt und Geologie (Hrsg) (1996) Landnutzungskarte des Freistaates Sachsen 1:100,000

Sächsisches Staatsministerium für Umwelt und Landesentwicklung (Hrsg) (1992) Karte "Naturschutz. Schutzgebiete in Sachsen." 1:200,000

Steinhardt U (1999) Die Theorie der geographischen Dimensionen in der Angewandten Landschaftsökologie. In: Schneider-Sliwa R, Schaub D, Gerold G (Hrsg) Angewandte Landschaftsökologie. Grundlagen und Methoden., Springer-Verlag, Berlin-Heidelberg-New York, pp 47 - 64

Troll C (1939) Luftbildplan und ökologische Bodenforschung. Zeitschrift der Gesellschaft für Erdkunde zu Berlin: 241-298

Volk M, Steinhardt U (1998) Integration unterschiedlich erhobener Datenebenen in GIS für landschaftsökologische Bewertungen im mitteldeutschen Raum. Photogrammetrie - Fernerkundung - Geoinformation 6: 349-362

Volk M, Steinhardt U (1999) Fazit: Ableitung dimensionsspezifischer Indikatoren für die Landschaftsbewertung. In: Steinhardt U, Volk M (Hrsg) Regionalisierung in der Landschaftsökologie. Teubner, Stuttgart-Leipzig, pp 233-235

Zepp H, Müller MJ (Hrsg) (1999) Landschaftsökologische Erfassungsstandards. Ein Methodenhandbuch. Forschungen zur Deutschen Landeskunde, Band 244, Deutsche Akademie für Landeskunde, Selbstverlag, Flensburg

# 4 Researching state and dynamics in landscape using remote sensing

Ellen Banzhaf

This chapter looks at the usage of remote sensing data for practical research. Remote sensing data can be used for individual applications and research studies as well as routine work, and basic data sets can be provided as vital pieces information for landscape ecological investigations. Experimental work such as tests related to sensor development, signature research and non-operational digital analysis methods are not dealt with here as this chapter is more concerned with applied remote sensing. Nevertheless, the importance of basic research is self-evident and is much respected for efficient, rational applications.

The main fields of investigation that use remote sensing are disciplines that ascertain the Earth's surface, especially its land cover, sealed or unsealed, with or without vegetation, and its use and possible conservation, as well as its sometimes vast changes and high dynamics. Attention is focused on local, regional and general environmental planning, mainly in connection with opencast mining, agriculture, (sub)urban development, and information needed for surveys and planning authorities. Rather than listing all the activities carried out during the past few years, instead useful examples will be given of practical applications that have been undertaken. The remote sensing data collected for all applied landscape ecological investigations are based on electro-optical scanner images. Generally used sources of spatial data are listed in Table 4.1.

## 4.1 Image processing

### 4.1.1 Image correction, enhancement, and manipulation

Geometric and radiometric corrections, image enhancements and manipulations are applied to data sets that have already been geometrically and radiometrically corrected. These processing steps mainly depend on the purpose of image analysis and can be divided into basic radiometric and geometric corrections or enhancements, and digital image processing steps with changes to data set content.

All these operations transform input data by means of digital image processing and mathematical functions to produce new, changed output data.

**Table 4.1:** Example sources of spatial data

| Point | Line | Area |
|---|---|---|
| Monospectral and multispectral data (geometric, spectral and radiometric information for each single raster cell) | Road maps | Landscape units |
| Topography | Contour lines | Land use maps |
| Climatic measurements (rainfall, temperature, evapotranspiration, etc.) | Administrative units | Land cover maps |
| Soil samples | Watercourses | Soil type maps |
| Individual spectrometric measurements in test areas | Railway network | Biotope maps |
| | | Geological maps |
| | | Town plans |
| | | Watersheds |
| | | Hydrogeological maps |

With respect to **geometric corrections**, the form and location of the pixels (= picture elements) are changed and transformed into a defined spatial reference system without altering the grey levels. Such transformations are necessary for:

- Mathematical links between several image matrices that are to be made congruent, e.g. multispectral, multitemporal or multisensoral image data. This is the main field for landscape-ecological questions dealing with land cover dynamics, changes and change detection.
- Overlaying the input data with other thematic information derived by digitized maps, e.g. landscape units, political boundaries, planning networks or contour lines. For visual interpretation, for example, it is essential that satellite image data can be overlaid by further information often gained by digitized vector data, as a satellite image itself can be the subject of a regional map. Another example is if image information is to be transferred into a GIS or any other related landscape ecological model, when the same geometrical correction of all input data sets is compulsory for proper analysis.

In **radiometric corrections and enhancements**, grey values of pixels are transformed for system corrections or image restorations necessitated by technical shortcomings and a lack of information of the sensor system or data transmission. In contrast to such corrections of raw data, specific radiometric enhancements are carried out on system-corrected data to allow better data analysis and to optimize the data quality for individual demands. They include:

- Pixel-based grey value manipulation: This manipulation is carried out to produce linear contrast-stretching or contrast-smoothing, to reduce atmospheric influence, and to diminish bright differences caused by relief. These options change the input data and are usually included in the preprocessing work when a certain operation is to follow. For multispectral image analysis, the data sets of each individual band necessary to produce colour composites are worked on separately and linked afterwards with all the calculations based on the grey value histograms of each individual data set. The same procedure can be carried out before a digital image is classified to increase the input data quality.

**Atmospheric correction:**

Frequently, detailed correction for the scattering and absorbing effects of the atmosphere is not required and often the necessary ancillary information such as visibility and relative humidity is not readily available. In these cases, if the effect of the atmosphere is judged to be an imagery problem, approximate correction can be carried out to remove haziness so that the dynamic range of image intensity is improved. Consequently the procedure of atmospheric correction is frequently referred to as haze removal (Richards & Jia 1999).

- Threshold procedure: This procedure, also known as density slicing, either comprises equivalent densities in which grey level slices are aggregated to form one grey value so that the different slices can be shown distinctively, or alternatively it comprises non-equivalent grey level classes that represent certain object classes in a monospectral data set. The latter is a method added to a hierarchical image classification as the classification result is not accurate enough and certain classes need to be characterized, for example by an additional image. In a monospectral data set, the threshold procedure defines two or more grey value levels: if the resulting image only contains two values, a binary image is produced. Hence important image segments are masked and inserted into further classification steps. This procedure fulfils two aims: the calculation process is reduced, and disturbing or competing objects are excluded (Hildebrandt, 1996).
- Digital filters: Digital filters change not only individual grey values in an image but more importantly the neighbourhood of grey values in defined window sizes such as the most commonly used windows 3x3, 5x5 or 7x7 (number of pixels in both directions: columns and rows) with each pixel being at the centre of the kernel and thus the centre of the filter operation. Apart from offering radiometric correction by substituting missing pixels or bad lines, these filters are used to emphasize or to smooth grey value differences due to specific object characteristics. High-pass filters like the edge enhance filter make local contrasts (streets in contrast to buildings, contours in open landscapes, etc.) more evident, whereas low-pass filters suppress such contrasts.
- Another form of radiometric manipulation is carried out when data fusion is calculated. For better image interpretations, it is useful to merge a multispectrally high resolution image with a geometrically high resolution image. Since IRS-1C and Landsat-7-ETM went into operation, both sensors - multispectral

and panchromatic - have been installed on the same satellite system, ensuring that the acquisition date and time are the same. Otherwise such merges were and still are carried out with different satellite systems, e.g. Landsat-TM or Spot multispectral images being transformed with panchromatic images such as those from a Spot panchromatic sensor. The disadvantage is that two acquisition dates are involved and thus phenological variations, atmospheric differences and many other deviations impair the merged product.

### 4.1.2 Visual image interpretation

Satellite image maps are produced every five years for the area of Leipzig, Halle and Bitterfeld. They are all drawn to a scale of 1:100,000, show the regional landscape structure with its dynamics from 1989 onwards, and cover an area of about 5,000 km². A large, complex area is presented with a spatial resolution of pixels measuring 15 metres by 10 metres pixel sizes with a valuable information content for regional planning. The colouring has been chosen such that the contents are similar to those on physical maps in atlases (Banzhaf 1998). As towns are shown in red, woodlands and floodplains in green, and rivers and lakes are printed in dark blue, the satellite image map can easily be read by non-remote sensing experts, too. This satellite image is the result of Landsat bands 1, 5, and 7 being assigned to red, green, and blue, and also being enhanced by means of image processing techniques such as contrast-stretching and histogram manipulation. The merged panchromatic band, taken either by Spot (for the year 1994, shown here) or Landsat-7-ETM (for the year 1999) was spatially enhanced by an edge enhance filter before resolution merge was carried out. The merger of the two different data sets only succeeded because both images underwent the very same geometric correction and therefore are a perfect spatial match. Ancillary data such as names of towns and federal states, motorways, landscape elements (e.g. opencast mining areas and nature reserves) help make the map readable, while a legend with interpreted characteristic landscape types boost its comprehensibility. As it is printed as an analogous product on a DIN A0 Format, it has become a useful remote sensing product and thus an instrument for planners, who do not need to know much about remote sensing techniques (such as the name and equipment of the satellites used) or the image processing methodology (image enhancement, resolution merge method; see chapter above) on which the map is based (Seger 1998).

As this image shows a landscape at a regional level without the information limitations often stemming from the involvement of administrative units, it gives an insight into small details and overriding correlations and is able to dispense with generalization.

In contrast to mathematically oriented operations including digital classification, such a visual analysis works without statistical results, diagrams or thematic evaluation. It still provides information on land use patterns and distributions and is a tool for decision-makers. What has to be taken into consideration is that phenology, atmospheric conditions and the date of acquisition influence the visual impression and the information details of each individual map. This makes a major difference to updating any thematic presentation or topographic maps where a chosen colour stands for the same type of legend in each copy of a time series.

In the case of visually interpreting satellite image maps the possess the same pattern recognition as on aerial photos just covering a larger area. Such patterns are areas, or lines or point information that form the basis of an image (or photo) analysis and they are equivalent to a certain land use area and, beyond that, they can represent normative space categories of estate parcels. Patterns differ by color and texture from their surroundings, and these characteristics as well as their outline and size gives an idea on the structure of land use. Its analysis and assessment lead to political decisions that are usually made by restricted local information. The use of parcels mirrors socio-economic and ecological aspects, land use classes represent real eco-systems, and spatial patterns dominate the use of a certain landscape, its aesthetical equipment, and its temporal dynamics.

As the maps have shown a time series since the roots for the reunification in Germany were laid in 1989 they show the extremely high dynamics of an important conurbation area in East Germany. During the first years the urban centers Leipzig and Halle spread into its surroundings and an overdevelopment took place outside the cities first of all connected with huge building-complexes of commercial sites and later on also with new quarters of buildings for housing. Infrastructure has changed with the construction of new motorways, the extention of the airport Leipzig-Halle, a new freight traffic center and a new central market. The land use functions cannot be interpreted in the map, this is reserved for people with a profound regional knowledge, as local decision makers usually are. What can be derived is that land use has changed rapidly and that the sealed area outside the large cities has increased. Another phenomenon is the land use change connected with the traditional open pit mining: as most of the mining activities were stopped in the early 1990's and flooding started the whole landscape changes from a devastated type to a landscape type with lots of lakes and possible recreational sites. Such vast changes are very well documented in such satellite image maps as the regional scale offers the possibility to get a detailed overlook over an "initial" land cover distribution and its variation and dynamics usually connected with a socio-economic and ecological change of the regional land use and function.

### 4.1.3 Quantitative analysis of classified images

Computer interpretation of remote sensing data is known as quantitative analysis because of its abilities to identify pixels based upon their numerical properties and to count pixels for area estimations. It is also generally called classification, which is a method by which labels may be attached to pixels depending on their spectral character. This labelling is performed by a computer trained to recognize pixels with spectral similarities.

The image data for quantitative analysis must be available in digital form. This is an advantage with image data types, such as that from Landsat, Spot, IRS, etc. compared to more traditional aerial photographs, as the latter require digitization before quantitative analysis can be performed. Detailed procedures and algorithms for quantitative analysis are the subject of specific remote sensing handbooks and will not be discussed here. Instead an outline of the essential concepts in classification will be followed by a number of study cases.

Recognition that image data exists in sets of spectral classes, and identification of these classes as corresponding to specific ground cover types, is carried out using the techniques of mathematical pattern recognition or pattern classification and their more recent neural network counterparts. The patterns are the pixels themselves, or strictly speaking the mathematical pixel vectors that contain the sets of brightness values for the pixels. Classification involves labelling the pixels as belonging to particular spectral (and thus informational) classes using the spectral data available. There are two broad classes of classification procedure, and both are applied in the analysis of remote sensing data. One is referred to as 'supervised classification' and the other as 'unsupervised classification'. These can be used as alternative approaches but are often combined into hybrid methodologies.

### *Unsupervised classification*

Unsupervised classification is a means by which pixels in an image are assigned to spectral classes without the user having foreknowledge of the existence or names of these classes. It is usually performed using clustering methods. These procedures can be used to determine the number and location of the spectral classes into which the data fall and to determine the spectral class of each pixel. The analyst then identifies these classes a posteriori, by associating a sample of pixels in each class with available reference data, which could include maps and information from ground visits. Unsupervised classification is therefore useful for determining the spectral class composition of the data prior to detailed analysis by the method of supervised classification.

### *Supervised classification*

Supervised classification procedures are the essential analytical tools used for the extraction of quantitative information from remotely sensed image data. An important assumption is that each spectral class can be described by a probability distribution in multispectral space. The distribution found to be most useful is the normal or Gaussian distribution. It is robust in the sense that classification accuracy is not overly sensitive. The multidimensional normal distribution is completely specified by its mean vector and its covariance matrix. Consequently, if the mean vectors and covariance matrices are known for each spectral class, the set of probabilities that describe the relative likelihoods of a pattern at a particular location belonging to each of these classes can be computed. It can then be considered as belonging to the class which indicates the highest probability. Before classification can be performed, a representative set of pixels, commonly called a training set, is defined for each of the classes and referred to as supervised learning. The view of supervised classification adopted has been based on the assumption that the classes can be modelled by probability distributions and are consequently described by the parameters of these distributions. Other supervised techniques also exist, in which neither distribution models nor parameters are relevant. More recently, new classification methods and combinations such as neural network non-parametric classification or pre-classificational segmentation, followed by an object-based classification including a fuzzy logic algorithm plus expert knowledge, have been shown to be promising for remote sensing applications. However, as they are still in the test phase, no study case will be presented here.

### 4.1.4 Examples of quantitative approaches to image classifications

#### *Image enhancements and methodological improvements*

One target of landscape ecology is to obtain spatial information on land use distribution, either for planning purposes or as a basic input data set for various model applications. This example refers to the urban and industrial planning of Bitterfeld and Wolfen, two small towns in eastern Germany with vast industrial sites, including their changes during the past decade since German reunification and their possible future development. It discusses some aspects of an undergraduate dissertation (Ihl, 1999, pp.1-58).

In order to document land use and its rapid changes, several time intervals of satellite imagery are used. These images form the instrument to derive the spatial information needed for this highly dynamic region and to compare land use at different points in time in order to characterize the total consumption of sealed areas and the changes of opened-up and sometimes resealed areas.

Methodological procedures are applied on the basis of three different satellite sensor systems, i.e. Landsat-5-TM, Spot and IRS-1C. As they are listed in Table 4.1, it only needs to be mentioned that they possess different spectral and spatial resolutions and that comparison of the data sets for their application to urban planning is of major interest. Therefore, first of all certain preprocessing steps are carried out before classification takes place. In a further image interpretation step, land use differences are calculated for the points in time photographed, synthetic images are produced showing the spatial changes, and statistical analyses of this extremely dynamic industrial region are presented to urban decision-makers. They form the basis for partners in local administrations and regional authorities to work out action programmes together with industrial enterprises.

Remote sensing tasks dealing with the research of urban spaces need to have an especially high precision for details such as building complexes, land use parcels, edges and other high frequency information. As this precision depends on the geometric resolution, image processing is geared towards enhancement to derive spatial details.

#### *Data merging of multispectral and panchromatic images*

Combining panchromatic and multispectral image data boosts the geometric resolution of the latter. Various methods have been developed for data fusion: most of them have the disadvantage that the radiometric resolution is changed, especially outside the visible spectra. This phenomenon can easily be neglected when visually interpreting the image, but for classification processes and resulting comparisons it proves to be a serious problem.

#### *Color transformation methodology*

In the IHS method, three bands are extracted from the multispectral data set in the red, green and blue (RGB) colour space and transformed into an image of intensity, hue, and saturation (IHS). As intensity is equivalent to the sum of the colour components, it is statistically adapted by histogram matching. This intensity band can now be substituted by the stretched panchromatic image. This manipulated multispectral data set is retransformed into the original red, green and blue, and possesses the same geometric resolution as the panchromatic image.

### *Principal component analysis*

Multispectral data sets have a multidimensional data space with different bands. As the spectral attributes of surface objects contain a certain grey level in one band, it can be related to the grey value interval in a second band. The band correlation is high, meaning that the gain in information is little and thus mainly redundant. The aim of principal component analysis is to transform the grey values of the different bands so that the covariances, which express the semantic context of bands, diminish. The mathematical procedure first calculates the eigenvalues and standardized eigenvectors and then manipulates the variance-covariance matrix to obtain covariances with the value of zero. The new bands possess uncorrelated information and redundancies are avoided. The main advantage of this data transformation is data compression, where the first principal component is equivalent to the total intensity of the image which can be used for data fusion such as the 'resolution merge'. The disadvantage is that certain structures and textures are only found in higher and usually reduced principal components.

### *Filter operations*

Digital filters are algorithms that change the original frequency within a chosen window box and cause radiometric transformation. The high-pass filter method is used to try and separate the spectral from the spatial information. The addition of the spatially and spectrally stressed data sets results in a new, improved image. The panchromatic image is transformed into a high-pass filter image by subtracting the low-pass filter image from the original data. Thus a high-frequency image with a spatial component, mostly edges, emerges. Low-pass components usually carry spectral information that show high variability with less correlation between the individual bands.

### *Comparison between the different methods*

Visual comparison reveals that the IHS method and principal component transformation lead to a distinct improvement in the optical resolution. Even industrial building complexes can be visually made out . The opposite of this enhancement is the high-pass filter method, where the results are much less distinctive. Due to radiometric alteration, the IHS method results in the severe fluctuation of the optical colours in the near and middle infrared, making it unsuitable for classification.

In contrast to other studies, a relatively small correlation coefficient is calculated for the Spot and the Landsat-5-TM images. This is due to phenological changes between the acquisition data of 16/05/94 (SPOT) and 21/07/94 (TM) resulting in different spectral reflections. In spring the fields still have no green vegetation, whereas in midsummer arable crops predominate. To obtain better quality information, difference images are calculated by subtracting the products taken from the data fusion from the original image. The comparisons are performed for the bands 3, 4, 5 which show distinct changes in the grey values for the principal component procedure. Especially in the near infrared band, the distortion is evident, whereas in the visible red (band 3) and in the infrared (band 5) the deviation is still visible but weaker. Thus woodlands and green spaces could not be separated in a classification approach. It is only the high-pass filter method that does not show such changes. The distortions are reduced to the edge enhancements which implies no phenological deviations as seen in the other images. Thus the high-pass filter method will be used for classifying the images.

### *Hierarchical supervised classification*

When carrying out hierarchical classification, the image is segmented step by step according to different criteria. In this case the first step is to generate a binary image by defining a certain threshold. This segmentation is restricted to a single image or to one spectral band. Such preparation is useful for the following masking of the multispectral data set. After segmentation, the actual classification can start. For classification, the panchromatic and the multispectral image are merged (high-pass filter) and masks for main classes typical for the region are generated by the above-mentioned binary image. Then the maximum-likelihood classification is calculated for all three masked images. After this mathematical operation, the three sub-classifications are reassembled and presented in one map for statistical analysis. The following Table 4.2 shows the classification accuracy for each point in time investigated and for all defined classes.

**Table 4.2:** Classification accuracy for the land use of Bitterfeld and Wolfen (Ihl 1999, p.30)

| | **1989** (TM & Spot sensors) | | **1994** (TM & Spot sensors) | | **1997** (IRS-1C LISS & pan) | |
|---|---|---|---|---|---|---|
| **Class name** | Producer's accuracy | User's accuracy | Producer's accuracy | User's accuracy | Producer's accuracy | User's accuracy |
| **Built-up** | 90.2 | 95.8 | 84.9 | 88,2 | 87,9 | 89,4 |
| **Open area** | - | - | 80.6 | 86,2 | 86,6 | 83,8 |
| **Water** | 93.4 | 97.7 | 100.0 | 100,0 | 90,6 | 85,2 |
| **Open pits** | 94.7 | 90.0 | 96.8 | 96,8 | 70,3 | 63,3 |
| **Wood-land** | 83.6 | 83.6 | 94.3 | 98,0 | 74,5 | 88,3 |
| **Field / meadow** | 95.5 | 90.1 | 93.2 | 90,1 | 76,4 | 69,3 |
| **Field** | 83.7 | 85.7 | 88.2 | 78,9 | 73,5 | 83,3 |
| **Total accuracy** | | **90.3 %** | | **91.3 %** | | **80.0 %** |

***Derivation of vegetation indices for agricultural differentiation***

This study case deals with land use classification for the area south of Leipzig. This region is dominated by two main economic sectors, for as well as historically being an area of intensive agriculture, it has also been exploited for lignite mining, as a result of which vast open pits have become characteristic for the whole district. From the 1950s until 1990, farming was economically and politically organized in the form of collective farms in which equipment and production were shared. To manage this form of cultivation, paths, hedges, and baulks were eliminated and large fields were created. In particular, recultivated areas reclaimed from opencast mining industries were endlessly vast treeless plains with poor-quality soil. After 1990 opencast mining was almost completed ceased. The re-naturation of the mining landscape is closely linked to agriculture and forestry.

The target of land use classification is to divide land use classes within agricultural use. Rather than produce a detailed classification of the opencast mining areas, the aim is to derive a differentiated data set as a basic level to be used for simulation models of farms and vegetation matters (e.g. nitrogen, phosphorus).

For multispectral classification, a Landsat-5-TM image is taken, acquired on 21/07/94. The field maps of the agricultural holdings on the acquisition date are available as ground truth information. In the first step feature space images are generated that can be used for the selection of training sites. These features are derived from the principal component analysis of the seven TM bands for which the first three principal component values describe 98% of the variance. The spectral separability of crop species is shown in Table 4.3.

**Table 4.3:** Spectral separability of cultivated plants

| land use | leaves / shrub | grassland | stubble | hay | spring barley | winter crop | sun-flowers | lucerne | rape | onions |
|---|---|---|---|---|---|---|---|---|---|---|
| leaves / shrub | x | x | x | x | x | x | -- | x | x | x |
| grass-land | | x | x | x | x | x | -- | -- | x | x |
| stubble | | | x | x | x | x | x | x | x | x |
| hay | | | | x | x | x | x | x | x | x |
| spring barley | | | | | x | -- | x | x | -- | -- |
| winter crop | | | | | | x | x | x | -- | x |
| sun-flowers | | | | | | | x | -- | x | x |
| lucerne | | | | | | | | x | x | x |
| rape | | | | | | | | | x | -- |
| onions | | | | | | | | | | x |

Two classification algorithms are tested and their results compared on qualitative and quantitative levels. The minimum distance classification is qualitatively weaker with respect to sunflowers and alfalfa, although it offers a fairly high quantitative precision of about 82%. Maximum-likelihood classification with a 1st Pass Parallelepiped preclassifier attains a total accuracy of about 90%. The latter is taken to derive further information on textural analysis and vegetation indices.

From the qualitative point of view, mistakes occur in the assignment of the classes 'built-up area' and 'leaves/shrubs'. Furthermore, not all of the standing water spaces can be clearly distinguished from woodlands. For these two groups of classes, an attempt is made to establish unambiguous separability by means of a textural filter. In addition, an attempt is made to distinguish between certain crop types that cannot be recognized in the multispectral classification by testing vegetation.

### *Texture analysis*

Texture analysis is taken as a possible step when optimizing a multispectral classification in this case study. By masking the two groups of classes ,'built-up area' against 'leaves/shrubs' and 'water' against 'woodland', two masks are generated and transformed into a binary image for each of them. Table 4.4 shows the separation of the spectrally similar but texturally different class types inserted into the results of the spectral analysis.

The procedure of this textural analysis is based on co-occurrence matrices supporting textural characteristics. These are a specific type of histogram which is not based upom a standardized distribution of grey values as is assumed for the maximum-likelihood classification; instead it is based on the relative distribution within a certain defined space, the co-occurrences of the grey values. The multinomial distribution is taken to characterize features as a probability distribution. In this case, which parameters lead to a positive result is empirically tested.
The following parameters are defined:

| | |
|---|---|
| *Number and spectral frequency of used bands:* | *band 4,5,3* |
| *Choice of the searching directions:* | *horizontal, vertical, left diagonal and right diagonal* |
| *Choice of window size:* | *(1) 5x5; (2) 9x9* |
| *Choice of grey value differences:* | *(1) 5 grey levels; (2) 10 grey levels* |
| *Number of feature spaces:* | *four* |

The test phase shows that the separability of the above-mentioned classes are characterized best when taking choice number (1) with a window size of 5x5 pixels, six grey levels and all four searching directions. The most important factor for achieving a good result is that the textural differences between the distinguished objects are obvious even though they are spectrally very similar. As some of the agricultural crops like alfalfa, rape, sunflower and grassland are spectrally very similar and cannot be separated by either standardized maximum-likelihood classification or textural analysis, these classes must be merged into superordinate classes implying a loss of information. This prompts the testing of vegetation indices for better assignment.

**Table 4.4:** Spectral analysis (x principal component values) and application of a textural filter (TF)

| Land use | leaves/ shrub | grass-land | stubble | winter crop | spring barley | Hay | wood-land | water | built-up area | open-cast mining | embank-ment along opencast mining |
|---|---|---|---|---|---|---|---|---|---|---|---|
| Leaves / shrub | x | x 1/3 | x 1/2 | x 1/3 | x | X | x 1/3 | x 1/3 | TF | x | x |
| Grass-land | | x | x | x 1/3 | x | X | x | x | x 1/3 | x | x |
| Stubble | | | x | x | x | X | x | x | x | x | x |
| Winter crop | | | | x | x 1/3 | X | x | x | x 1/3 | x | x 1/2 |
| Spring barley | | | | | x | X | x | x | x | x | x 2/3 |
| Hay | | | | | | X | x | x | x | x | x |
| Wood-land | | | | | | | x | TF | x 1/2 | x | x |
| Water | | | | | | | | 120 | x | x | x |
| Built-up area | | | | | | | | 120 | x | x | x 1/2 |
| Open-cast mining | | | | | | | | | | x | x |
| Embank-ment along opencast mining | | | | | | | | | | | x |

### *Vegetation indices*

It is assumed that the application of different vegetation indices allows thresholds to be defined that lead to an unambiguous division between certain crop types. The exemplified type 'alfalfa' can be classified as either grassland or a winter crop. This assumption is tested and proved in the classified image. Hence altogether three different indices are chosen to distinctively assign the crop type 'alfalfa'. One of the indices is a ratio index known as NDVI (normalized difference vegetation index), another is the perpendicular index (PVI), and the third is a multidimensional index (tasseled cap).

For each land use class, areas of interest are defined and masked so that the spectral information only exists for the areas of interest. Thus single images are produced for each of the class types and then the vegetation indices are calculated. This calculation enables the user to check whether thresholds are exceeded for other classes and serves to estimate a complete index classification. The procedure is then continued on the object classes 'winter crop' and 'grassland' as 'alfalfa' is assigned to 'winter crop' in the multispectral analysis.

After having calculated the NDVI, a clear distinction between 'grassland' and 'alfalfa' can be made out thanks to the threshold. The classes 'winter crop' and 'alfalfa' are generally separable with a small overlap area between the maximum value of 'winter crop' and the minimum value of 'alfalfa'. This is the best result gained and is therefore used to improve the multispectral classification (see Fig. 4.1). The tasseled cap index allows 'alfalfa' to be distinguished from either 'winter crop' (brightness) or 'grassland' (greenness and wetness), but not from both simultaneously. The perpendicular vegetation index (PVI) does not meet the required demands; although 'grassland' and 'alfalfa' can be separated, the area overlapping with 'winter crop' is too high. The following tables show the result of the different indices tested including the thresholds for the exemplified classes (see Table 4.5, Table 4.6 and Table 4.7).

Some steps in classification are tested that could improve the standardized multispectral classification method. Generally speaking, methodological improvements are needed to achieve more unambiguous classification results and to be able to work with more subordinate classes to meet the demands in agricultural management and control and to be able to supply distinguished data sets for landscape-ecological models.

**Table 4.5:** Threshold value for the normalized difference vegetation index (NDVI)

| | winter crop | grassland | lucerne |
|---|---|---|---|
| min | 0.067 | 0.488 | 0.250 |
| max | 0.297 | 0.669 | 0.434 |
| mean | 0.155 | 0.604 | 0.341 |

**Table 4.6:** Threshold value for tasseled cap

| | winter crop | grassland | alfalfa |
|---|---|---|---|
| min | 123.00 | 141.00 | 144.00 |
| max | 146.00 | 171.00 | 162.00 |
| mean | 133.14 | 156.40 | 153.98 |

**Table 4.7:** Threshold value for perpendicular vegetation index (PVI)

| | winter crop | grassland | alfalfa |
|---|---|---|---|
| min | 0.167 | 0.243 | 0.197 |
| max | 0.229 | 0.323 | 0.225 |
| mean | 0.197 | 0.295 | 0.211 |

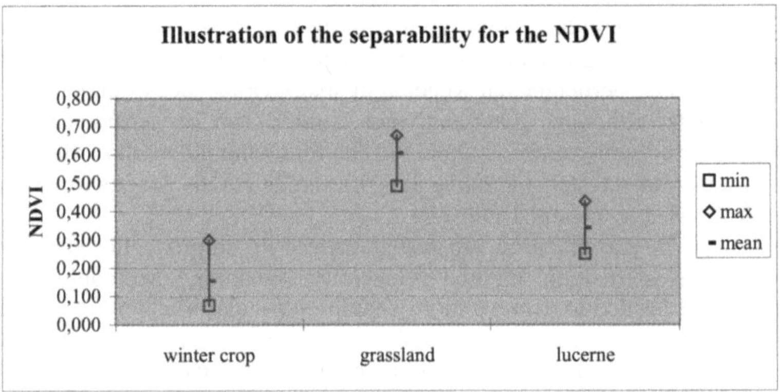


**Fig. 4.1:** Separability for the NDVI

### ***Unsupervised classification to typify objects for a floodplain area***

The various environmental authorities need more and more up-to-date and spatial data to carry out their statutory tasks. Conventional aerial photo interpretation is beset by the dual problems of remaining up-to-date and high costs when large areas have to be covered. As the implementation of satellite imagery necessitates intensive image processing and analysis, the direct use of primary data is impossible for public authorities. What the authorities need is a way of conveniently using preprocessed satellite images. If this could be achieved, satellite data sets could provide a genuine alternative to aerial photography in the day-to-day work of planning authorities. A project for the federal state of Saxony-Anhalt started in 1999 aims to develop catalogues (satellite products, procedures, features characteristic of environmental information) to meet the demands of different end-users. The results are transferred into a user-friendly software so that the decision-makers can perform their tasks independently using an intuitive GIS environment.

The project covers landscape types such as opencast mining, agriculture, forestry, open spaces, urban landscapes and floodplains. The part of the project dealing with floodplains is presented where the condition and the change of this land cover type is investigated on the basis of high resolution satellite imagery. Attention is focused on classifying different data phenologically (during one year) and chronologically (over the years) and, if data is available, also for flooded rivers. Different classification procedures are tested and interpretation is subsequently carried out in a GIS to derive parameters relevant for nature preservation and for landscape evaluation. To detect changes the satellite image data are compared with the biotope mapping and land use typification on the basis of colour infrared photography produced in 1992/93 for the whole state.

One methodological target is to develop a concept to record landscape changes. Standardized classification is difficult for this type of landscape because with a high water level the phenology changes so rapidly during the year and moreover cannot be easily made out from one year to the next either as precipitation and groundwater levels keep on changing irregularly. Tree-trunks and leaves are employed for the first land use type to be characterized. The test site is the middle part of the river Elbe between the towns of Wittenberg and Dessau; the data basis is provided by three Landsat-5-TM images taken in June, August and September 1997. Apart from initial visual interpretation, unsupervised classification is carried out to find out more about the spectral variations of this specific land use type. Reference areas taken from the biotope mapping are only partly accurate as extensification has taken place in the use of grassland.

To register the spectral characteristics of floodplain grassland, an ISODATA cluster analysis is calculated for all three TM data sets. As it is an unsupervised classification, 25 spectral classes are defined in a preprocessing step before classification takes place. A very high number of clusters was found in the separated grassland biotope types. The signature mean values of the TM bands show the heterogeneous spectral character of grassland. The unambiguous assignment of the calculated spectral classes to one of the defined biotope types is not possible (Banzhaf & Fistric 2001).

When aggregating the spectral classes to a single target class named grassland, the problem of mixed classes occurs. Such incorrect assignment mainly concerns fields and deciduous forest, although the spectral overlaps vary depending on the points in time captured. Fig. 4.2 shows the spectral variation for the band combination 3,4,5 (visible red spectra, near infrared, and shortwave infrared). In order to obtain a better result and higher separation accuracy, a much higher number of spectral classes must be defined before a cluster analysis is calculated. Furthermore, a multitemporal approach must be investigated for grassland including additional synthetic bands to increase separability.

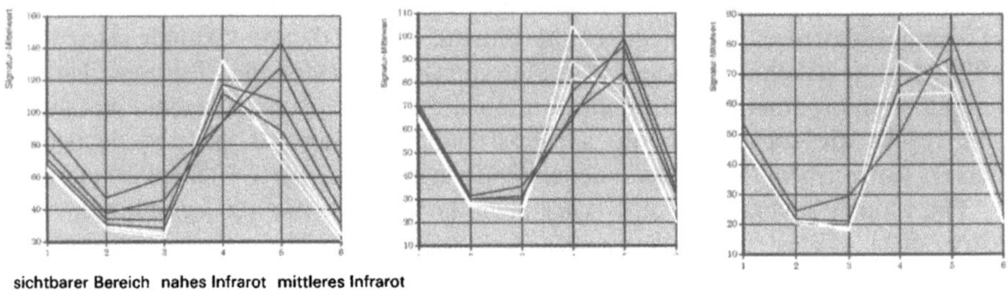


**Fig. 4.2:** Signature mean values of certain spectral classes for grassland (visible red spectra, near and short wave infrared) in June (left), August (centre), and September (right)

The signature mean values presented are calculated using ISODATA cluster analysis of Landsat-TM images from 1997 (June, August, September) to derive the class grassland. The thermal band was not included in this unsupervised classification, and so the displayed spectra are equivalent to TM bands 1-5 and 7.

For a supervised classification procedure, the standardized maximum-likelihood algorithm will hardly be applied alone as the spectra allows too many differentiations to obtain a single class as a result. Thus a new direction for classifying this land use type needs to use a multi-step operation possibly starting with segmentation, followed by object-based classification with fuzzy logic and/or maximum-likelihood algorithms and including a knowledge-based approach.

## 4.2 Possibilities and limitations for land use monitoring with remote sensing data and GIS methods

### 4.2.1 The role of remote sensing data in urban and regional monitoring

Since reunification, the redevelopment process in east German urban regions has resulted in structurally modified and often diffuse settlement structures in an extraordinarily short period of time. This highly dynamic impact has led to a clear reevaluation of the surrounding countryside and to an absolute deconcentration of population and places of employment. Simultaneously, population density and employment figures indicate extensive social, economic and ecological consequences for the urban and peri-urban spaces.

The degree of anthropogenic influence has reached its peak in urban landscapes. It increases in the suburban region compared to almost natural and agricultural landscapes with the growth of overdevelopment. With respect to the rapidly expanding suburbanization and settlement dispersion in German (and other European) urban-suburban regions, importance is shifting from town centres to the suburbs. This process is closely connected with an enhancement of the suburban status and with an increasing deconcentration of population and working-places.

Suburbanization processes need to be observed by scientific investigations, although this has sadly not been the case so far. Such a monitoring could serve to analyse and evaluate the complex and widespread development processes in the suburban area, and could form a basis for deriving models and recommendations for action plans. This recording and description of spatial and dynamic processes calls for the application of new quantitative methods and evaluation approaches.

If remote sensing data are only used for landscape monitoring and land-use classification, the experiences of the methodological tools of satellite data will only be partially exploited. Due to the improved geometrical resolution of new sensors, the analysis of settlement areas is faced by new challenges. Depending on higher monitoring scales, analysis could even involve the identification of individual objects if only whole settlement areas of towns and villages need to be derived. Because of such extended possibilities, several current research efforts are trying to develop new strategies for very heterogeneous and rapidly changing urban and suburban areas from different points of view. Increasing precision means that predicting for example abiotic and biotic components in landscapes and the sealing degree in urban agglomerations is possible.

In addition to the latest satellite-based remote sensing data (e.g. Landsat-TM-7, IRS-1C&D, IKONOS), aerial photographs (e.g. CIR photographs taken to produce country-wide biotope mapping) are available for many urban regions, which can be used as further sources of information. The use of fuzzy logic classifications and the application of textural parameters allow special classification methods that enable the better exploitation of different data. Furthermore, the combination of remote sensing data and spatial models allow predictions that can be essential for urban and regional planning.

### *Landscape structure analysis as a new tool for natural area potential analysis*

According to Plachter (1991) and Fiedler et al. (1996), the dimensions of human influence on spatial structures are so fundamental that land use is entitled to an indicator function for the detection and valuation of a social influence (Schönfelder 1984). The appearance of a landscape is characterized by its natural features including its complex effects on the one hand and the social demand expressed by intensive land use and multipurpose land demands on the other. Therefore landscape monitoring is understood as a system of observations showing modifications in the state of landscape under human impact, and referring to landscape components such as vegetation and soil cover, land use and landscape structure.

Landscape monitoring includes

- The observation and evaluation of factors which have an influence on the landscape, its state and dynamics
- The estimation and evaluation of such influencing factors
- Forecasting and estimation showing the development of the state of a landscape (Bastian & Schreiber 1999, Zierdt 1997).

Integrative, spatially oriented landscape monitoring requires the entry of a landscape's historically grown variety, landscape structure, substantial landscape functions, and also their consequences such as land use modification, fragmentation, diffuse settlement, the modification of the spatial structure of a landscape, as well as the loss of habitats. Such structural features of the terrestrial land cover are directly or indirectly linked with a multiplicity of functions. Landscape metrics can be taken as indicators to analyse, describe, and quantify patterns, compositions and configurations of a landscape type and its compartments.

In particular in the peri-urban cultural landscape, nature is described by indicators such as structure (linear or planar expansion, fragmentation, island areas, etc.), dynamics (entry of the modification processes) and texture (neighbourhood relations to other land use forms). This is based on the identification and computation of static and dynamic indicators that help provide a synthetic assessment of peri-urban landscapes. The indicators also allow the comparison of the environment's condition in different conurbations. The static indicator includes the proportion of urban land uses at different points in time, of land use interrupted by the road network, but also the fragmentation of recreational sites within metropolitan areas, and of built-up areas within green spaces in peri-urban areas. Dynamic urban area indicators refer to the typology of changes and the transition from one land use class to another.

The landscape between the agglomerations of Leipzig and Halle (east Germany) is gripped by a process of rapid transformation by anthropogenic, often uncontrolled interventions. Thus new landscape structures are established which enable new development and process cycles in the region. So far structural investigations have addressed flora and fauna. Using the available research concept, criteria are to be compiled for the need for local recreation and for nature-related conditions concerning the quality of life.

Initially a detailed inventory of the different areas examined should be compiled depending on data availability and the entry of the natural-space configuration which is as up-to-date as possible (e.g. nature protection areas, watercourses, woodlands, forest/field proportion, cycle track configuration, etc.). On the basis of these empirical investigations indicators can be derived to evaluate landscape features, attractions and deficits regarding the following characteristics:

- natural area potential
- suitability for recreation
- aspects of nature protection
- aspects of cultural protection

The rapid change of current modifications and processes simultaneously require and enable the execution of landscape monitoring. Using the latest methods and geo-information data, an important contribution can be made in particular to monitor and evaluate developments being carried out. Landscape monitoring dedicates itself to check and forecast state and dynamics from natural to technical ecological systems. It refers to landscape components such as vegetation and soil cover, land use and spatial landscape structure (Bastian & Schreiber 1999).

### *Change detection classification based on Landsat TM-5 and TM-7 images*

The area between the two conurbations Leipzig and Halle is classified for the years 1989 and 1999 in order to identify the development of human interventions immediately following the Peaceful Revolution and German reunification. As IRS data have only been available for Germany since 1996 and as the two classifications need to be compared as precisely as possible, both dates are classified on the basis of Landsat-TM data. Several preprocessing steps are necessary on the images before classification can take place. The older image was taken by Landsat-5-TM (7 July 1989) and contains bad lines that need to be eliminated and veils of clouds that are diminished by applying the histogram minimum method. The latter image was taken by Landsat-7-TM on 4 September 1999 and does not contain any disturbances.

Atmospheric conditions of the two images are corrected by means of the histogram minimum method. In this preprocessing step histograms of all TM bands are computed for the full image, which generally contains some areas of low reflectance (e.g. clear water, deep shadows or exposures of dark basalt). These pixels will have values very close to zero in the short-wave infrared band (TM band 4). If the histograms of TM bands 1 to 3 are plotted they will generally be seen to be offset progressively towards the higher grey levels. The lowest pixel values in the histograms of these three bands is a first approximation to the atmospheric path radiance, and these minimum values are subtracted from the respective images (Mather 1987). In both images band 6 is eliminated and a synthetic NDVI band calculated and attached.

Both classifications are calculated using the maximum likelihood classifier with the non-parametric rule of the parallelepiped optimization put first. A hierarchical classifications needs to be generated as different settlement densities and opencast mining are spectrally very similar, as well as fields without crops and unsealed ground (e.g. airport) are difficult to distinguish.

The change detection for this region is shown in Table 4.8. It is obvious that the settlement density and sealed areas have increased at agricultural land's expense. As rather natural wetlands have remained under conservation, their share has not diminished during the decade. Based on these classifications, buffer zones along the motorway A14 are calculated for the first three kilometres adjacent to the motorway linking Leipzig and Halle. This limitation stems from an investigation in the federal state of Thüringa where most changes along the motorway have taken place within this small range of kilometers. The results shown in Table 4.9 express the immense increase in sealed areas within the first few hundred metres and the reduction of the agriculturally used land.

Hence buffer zones or other GIS methods are an important tool for analysing rapid, vast development structures. Quantified analysis is a first step towards investigating land use changes and, by supplementing it by GIS methods, modifications at this scale can be structured.

**Table 4.8:** Classified Land Use for the Region between Halle and Leipzig

| Image Acquisition Date and Sensor Land use classes | 07.07.1989 TM-5 [ % ] | TM-7 [ % ] | Change detection [ % ] |
|---|---|---|---|
| Disperse settlement | 7.6 | 14.1 | + 6.5 |
| Dense settlement | 3.5 | 4.2 | +0.7 |
| Sealed area (e.g. roads) | 3.5 | 7.1 | +3.6 |
| Area without green vegetation | 9.0 | 5.5 | -4.5 |
| Fields with crop | 35.3 | 31.8 | -3.5 |
| Green top and bush vegetation | 10.9 | 15.3 | +4.4 |
| Pasture and meadow land | 21.1 | 9.7 | -11.4 |
| Forest | 7.2 | 11.0 | +3.8 |
| Water | 1.9 | 1.1 | -0.8 |

**Table 4.9:** Buffer Zones along the Motorway A 14

| Classes [%] | Year | Buffer Zones [m] | | | | | | | |
|---|---|---|---|---|---|---|---|---|---|
| | | < 100 | < 300 | < 500 | < 800 | < 1100 | < 1600 | < 2100 | < 3000 |
| Sealed area | 1989 | 27.0 | 8.5 | 3.4 | 5.6 | 5.5 | 4.4 | 5.7 | 9.6 |
| | 1999 | 80.0 | 23.8 | 11.9 | 12.2 | 9.0 | 6.0 | 6.8 | 14.4 |
| Unsealed area | 1989 | 0.0 | 0.9 | 0.5 | 0.3 | 0.5 | 0.4 | 1.5 | 1.8 |
| | 1999 | 3.8 | 7.6 | 2.7 | 3.0 | 3.4 | 3.0 | 2.5 | 1.2 |
| Agriculturally used area | 1989 | 71.0 | 87.8 | 91.4 | 89.0 | 87.6 | 87.6 | 83.1 | 81.3 |
| | 1999 | 13.5 | 66.5 | 81.1 | 79.5 | 81.2 | 82.7 | 80.3 | 78.0 |
| Woodland | 1989 | 1.2 | 2.0 | 1.5 | 1.4 | 2.1 | 2.9 | 3.4 | 1.2 |
| | 1999 | 0.9 | 0.9 | 0.8 | 1.1 | 1.9 | 2.7 | 3.7 | 1.1 |
| Water | 1989 | 0.0 | 0.0 | 0.0 | 0.0 | 0.2 | 0.3 | 0.5 | 0.4 |
| | 1999 | 0.8 | 0.4 | 0.3 | 0.3 | 0.5 | 0.3 | 0.4 | 1.3 |

* mean values are taken from the summarized buffer zones and rounded to one decimal place

### *Binary classification concentrating on green spaces by means of IRS-1C LISS Data*

In this first classification phase, a conventional, multispectral classification is applied to the IRS data. The intermediate result produced provides a set of spectrally rather homogeneous land cover classes, and thus it can be reliable used to identify land cover classes, such as water or woodlands. A multi-step, hierarchical procedure is then carried out of the type developed in earlier projects to classify both satellite-based and airborne, multispectral scanner data (Netzband 1998, Netzband et al. 1999). In a first step, an unsupervised classification (i.e. without signature analysis by the analyst) is executed which supplies 15 classes. These classes have to be assigned to land-use types by interactive, visual checks and postprocessing or, if necessary, aggregated. Furthermore, it is important to separate individual classes that are spectrally unique. The class separation is performed by a multispectral, supervised classification in which each identified class is 'extracted' by masking it in the intermediate result, in order to exclude it from the following classification steps. For the classification, a parallelepiped classifier is used. In this procedure pixels are not classified which do not belong to clusters of the spectral signatures, and pixels in the overlap area of two clusters are classified according to the maximum likelihood method. The resulting classes can be overlaid as masks on the image finally resulting and can be stored as independent layers.

For the following calculation process especially two classes could be separated (also compare Fig. 4.3 and Fig. 4.4):

- Forest, stand of woods (larger trees),
- Allotments as well as grassland and meadow surfaces in the inner and peri-urban areas.

### *Calculating the green spaces according to the Ring-Sector-Model*

To evaluate the green area distribution by classified satellite image data in the peri-urban area, the 'ring-sector-model' is suggested. This space reference model was developed by Simon (1990) to analyse intra-regional occupation commuter relations in Switzerland. It is based on the dimensional grid (same distances) of conurbations, by superposing any number of concentric sets and sectors over a region.

The model guarantees that a uniform external limitation of different test areas is given. Additionally, it serves to describe intra-regional characterizations of features. Gradients between the town centre and outskirts can be analysed and quantified with the ring-sector model in a differentiated manner (see Fig. 4.3 and Fig. 4.4).

In this case the greenery distribution is investigated in the region of Leipzig using four radii (5km, 10km, 15km, 20km) with eight selected sectors. . The following figure shows the calculated green proportions of concerning woods (forest, larger trees) and of allotments and shrub vegetation. In each case it is calculated for all radii and sectors for the classified IRS-1C LISS image data.

Distinct differences are recognizable in the distribution:

- Forest and large tree vegetation is mainly concentrated in the north-west and south/south-west of Leipzig (area of wetlands)
- The internal ring-road (5 km) is best furnished with large vegetation
- Large deficits exist in the northern peri-urban area
- The eastern to southern environment has a higher stand of large vegetation
- The provision of the suburban landscape with small trees and bushes is generally very poor
- Theses small trees and bushes are distributed relatively evenly throughout the city
- Small trees and bushes tend to be concentrated in the eastern surroundings
- The highest values appear towards the city centre

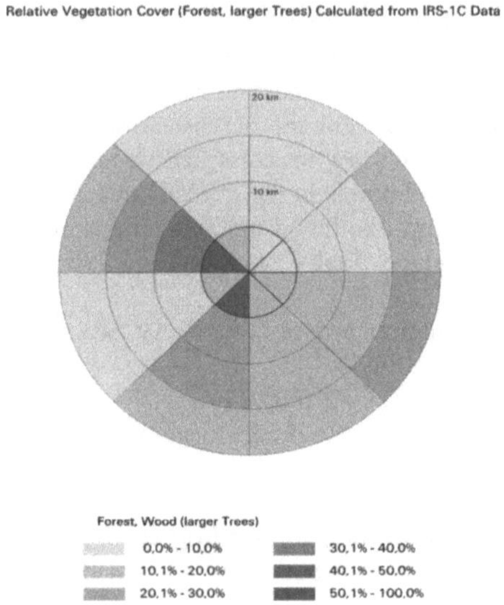


**Fig. 4.3:** Relative Vegetation Cover (Forest and larger trees) calculated from IRS-1C LISS

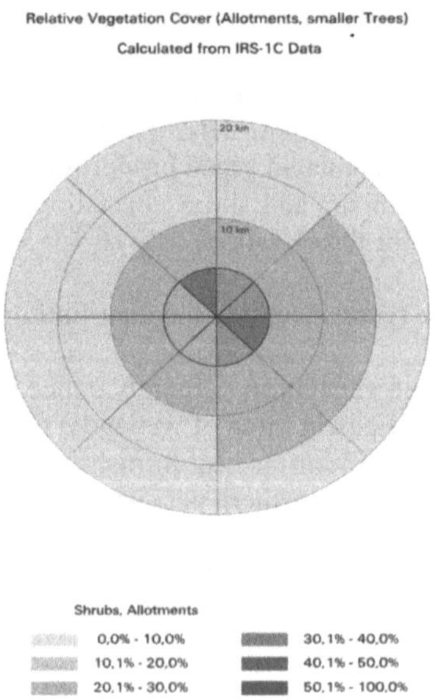


**Fig. 4.4:** Relative Vegetation Cover (Allotments, smaller trees) calculated from IRS-1C Data

The main characteristics of this landscape type are:

- The suburban landscape as a whole is poorly furnished with small trees and bushes
- These small trees and bushes are distributed relatively evenly throughout the city
- Small trees and bushes tend to be concentrated in the eastern surroundings
- The highest values appear towards the city centre

The patch density (PD) (Turner 1990) calculated for the ring-sector model delivers the following information (see Fig. 4.5, Fig. 4.6 and Fig. 4.7):

- PD equals the number of patches of the corresponding patch type (NP) divided by the total landscape area, multiplied by 10,000 and 100 (to convert to 100 hectares).
- It facilitates the comparison of landscapes with different sizes.
- The peak of forest patches is in the 5-10 km zone and an increase in shrubs and smaller trees can be identified towards the city center.

A historical land use comparison for the conurbation of Leipzig was carried out between 1930 and 1998 to monitor the changes in landscape over several decades. With the exemplified structural metrics 'patch density' and 'edge density' it can be seen that linear structures of green spaces such as alleys, rows of trees and hedges have diminished enormously. The reasons are the vast, monostructured agricultural landscape and the devastation of large sections of the landscape by opencast mining activities.

With respect to the structural analysis of landscape metrics, it can be stated that the ability to quantify landscape structure is essential for the study of landscape function and change. For this reason, much emphasis has been placed on developing methods to quantify the landscape's structure.

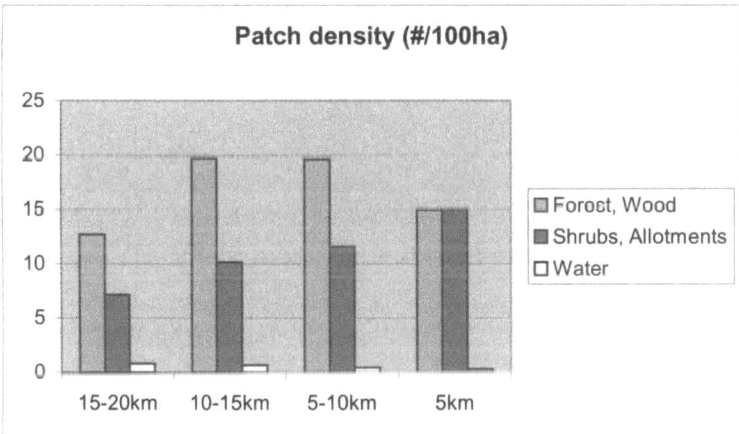


**Fig. 4.5:** Land use monitoring and structural analysis for the patch density

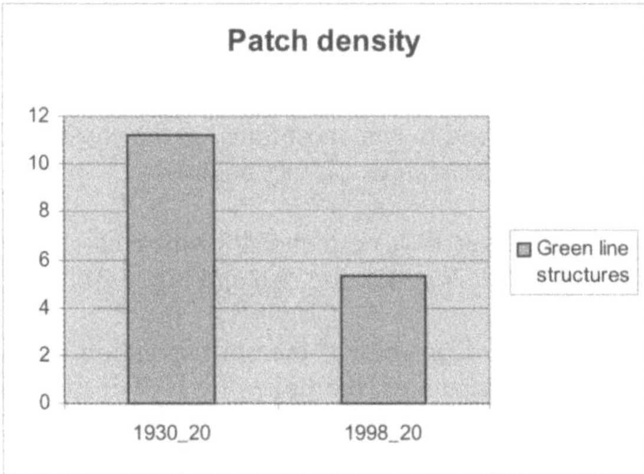


**Fig. 4.6:** Patch density for the land use classifications of the conurbation of Leipzig

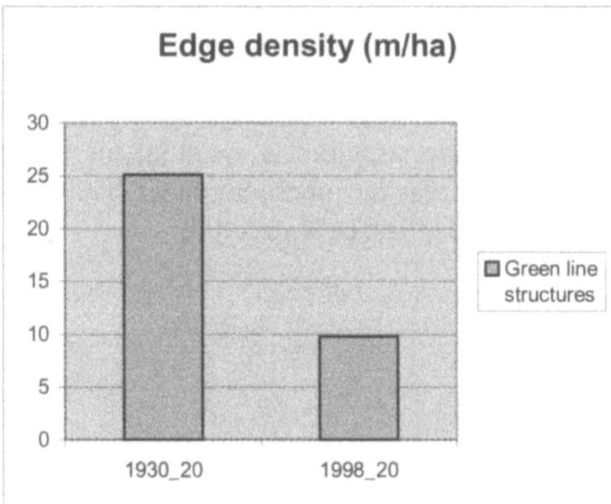


**Fig. 4.7:** Edge density for the land use classifications of the conurbation of Leipzig

## 4.3 Conclusion and outlook

Methods for the efficient spatial and temporal classification of the urban natural environment using geo-information are discussed. Multicriteria evaluations of urban nature are compiled for the suburban area regarding the quality of life, biotope protection, biodiversity, natural aesthetics and recovery. A balance and a modelling of the dynamics in nature and a conflict analysis will follow.

Nature, in particular, in the suburban cultural landscape is to be described and analysed with respect to indicators such as structure (linear or areal propagation, fragmentation, etc.), dynamics (entry of the modification processes) and neighbourhood relations with other land use forms, such as with housing estates or business parks.

Initial results for drawing up the inventory of the supply of natural areas clarify the possibilities offered by modern monitoring methods using geo-information to evaluate the natural area potential in the peri-urban landscape. The development of applicable indicators for planning processes is of particular importance.

## 4.4 References

Banzhaf E (1998) Die Satellitenbildkarte "Raum Leipzig-Halle-Bitterfeld" für das Jahr 1994. In: Erfassung und Auswertung der Landnutzung und ihrer Veränderungen mit Methoden der Fernerkundung und geographischen Informationssystemen im Raum Leipzig-Halle-Bitterfeld, UFZ-Bericht Nr. 2/1998, pp 71-73

Banzhaf E, Fistric S (2001) Operationalisierung von Fernerkundungsdaten für die Umweltverwaltung im Land Sachsen-Anhalt - Teilvorhaben Auenlandschaften. In: Dech S, Bettac (Hrsg) DFD-Nutzerseminar in Neustrelitz; DLR-Mitteilungen, in print.

Bastian O, Schreiber K-F (1999) Analyse und ökologische Bewertung der Landschaft. Gustav Fischer Verlag, Jena

Fiedler HJ et al. (1996). Umweltschutz. G. Fischer Verlag, Jena-Stuttgart

Hildebrandt G (1996) Fernerkundung und Luftbildmessung für Forstwirtschaft, Vegetationskartierung und Landschaftsökologie. Wichmann Verlag, Heidelberg

Ihl T (1999) Veränderung der Flächennutzung im Stadt- und Industrieraum Bitterfeld/Wolfen mittels Klassifikation von multisensoralen und multitemporalen Satellitenbildern. In: Banzhaf, E. (Hrsg): Bitterfeld / Wolfen als Beispiel für den Wandel einer Industrieregion in den neuen Bundesländern. Untersuchung von Flächennutzungsänderungen der beiden Städte und Umstrukturierung ausgewählter Altindustriestandorte. Stadtökologische Forschungen Nr. 20. UFZ-Bericht Nr. 6/1999, pp 1-58

Mather PM (1987) Computer processing of remotely-sensed images - an introduction, J Wiley and Sons

Netzband M (1998) Möglichkeiten und Grenzen der Versiegelungskartierung in Siedlungsgebieten. Dissertation. IÖR-Schriften 28, Universität Dresden

Netzband M et al. (1999) Classification of settlement structures using morphological and spectral features in fused high resolution satellite images (IRS-1C). Int. Archives of Photogrammetry and Remote Sensing, Vol. 32, Part 7-4-3 W6

Plachter H (1991) Naturschutz. Gustav Fischer Verlag, Stuttgart

Richards JA, Jia X (1999) Remote Sensing Digital Image Analysis. An Introduction. Springer Verlag, Berlin-Heidelberg-New York

Schönfelder G (1984) Grundlagen für die Vorhersage (Prognose) von Landschaftsveränderungen als geographischer Beitrag für die Landschaftsplanung. Dissertation, Universität Halle

Seger M (1998) Die Satellitenbildkarte "Raum Leipzig-Halle" für das Jahr 1989. In: Erfassung und Auswertung der Landnutzung und ihrer Veränderungen mit Methoden der Fernerkundung und geographischen Informationssystemen im Raum Leipzig-Halle-Bitterfeld. UFZ-Bericht Nr. 2/1998, pp 63-69

Simon M (1990) Das Ring-Sektoren-Modell: Ein Erfassungsinstrument für demographische und sozio-ökonomische Merkmale und Pendlerbewegungen in gleichartig definierten Stadt-Umland-Gebieten. Grundlagen, Methodik, Empirie; Bern.

Turner MG (1990) Spatial and temporal analysis of landscape patterns. Landscape Ecology 4: 21-30

Zierdt M (1997) Umweltmonitoring mit natürlichen Indikatoren: Pflanzen - Boden - Wasser - Luft. Springer-Verlag, Berlin-Heidelberg

# 5 The analysis of spatio-temporal dynamics of landscape structures

Angela Lausch[1], Hans-Hermann Thulke[2]

## 5.1 Concept of landscape ecology

The development of landscape ecology is directly related to the discussion over the concept of landscape (Finke 1994). This goes back a long way and was chiefly shaped by the works by Bobek & Schmithüsen (1949), Neef (1963, 1964), Leser (1997), and Formen & Godron (1986). Within the sphere of landscape physiology, the concept emerged that the landscape is a synthesis of a large number of individual elements. This concept later became established within the division into units of natural areas and acquired key importance for landscape ecology (Finke 1994). Although attention initially focused on surveying and describing the spatial distribution patterns of ecosystems, the limitations of this highly isolated approach later became apparent, precipitating concentration on questions of how ecosystems mutually influence each other, and how ecological neighbourly relations take place over space and time. Finke (1978) regarded the central task of landscape ecology to be surveying the spatial distribution patterns and the spatio-functional combination of ecosystems.

In contrast to the *classical* approaches of German-language landscape ecology such as by Bobeck & Schmithüsen (1949) and Neef (1963, 1964), the approach of *quantitative landscape ecology* emerged in the 1980s and 1990s, which was in particular developed by the North American landscape ecologists Forman & Godron (1986), Urban et al. (1987), Turner (1989), Turner & Gardner (1991a,b), Hansen & Castri (1992), O'Neill (1995), Hansson et al. (1995), and Wickham et al. (1997). These approaches were based on the definition of landscape ecology put forward by Forman & Godron (1986), which states:

*"Landscape ecology explores how a heterogeneous combination of ecosystems - such as woods, meadows, marshes, corridors and villages - is structured, functions and changes. From wilderness to urban landscapes, our focus is on:*

*a) the distribution patterns of landscape elements or ecosystems*

[1]UFZ - Centre for Environmental Research Leipzig/Halle; Department of Applied Landscape Ecology, Leipzig, Germany, lausch@alok.ufz.de

[2]UFZ - Centre for Environmental Research Leipzig/Halle, Department of Ecological Modelling, Leipzig, Germany, hanst@oesa.ufz.de

b) *the flows of animals, plants, energy, mineral nutrients, and water among these elements and*
c) *the ecological changes in the landscape mosaic over time."*

Landscape ecology hence investigates three essential features of the landscape (Forman & Godron 1986):

1. **Structure** refers to the spatially related properties of elements of the ecosystem and their spatial interrelationship within the landscape. They are used to describe the distribution of energy, material and type with respect to the size, shape, number, type and arrangement (configuration) of ecosystems in a landscape.
2. **Function** describes the existing interaction between the spatial elements of the ecosystem, which is expressed in exchange processes of energy, material and substances.
3. **Dynamics** is exhibited by the change to structures, to functions of the landscape structure, and to the landscape mosaic over time.

## 5.2 The purpose of the spatio-temporal analysis of landscapes

The conceptual and theoretical basis of landscape ecology put forward by Forman & Godron (1986) provided an important foundation for understanding the links between landscape structure, landscape function, and their dynamics. The view that "environmental problems are nothing more than the result of disturbed interrelations between different processes and structures" (Tischendorf 1995, p. 3) shows that for the sake of future environmental protection, a landscape-based approach is necessary which enables landscape structures, functions and their dynamics to be quantitatively determined.

Studying the landscape, its current state (structure) and its future changes (dynamics) enables understanding of the ecological mechanisms and processes that drives changes in the landscapes.Thus, the spatio-temporal analysis of landscape is a necessary basis for a mechanistic linkage between particular species or human being and the changing characteristics of the landscape.

It should be emphasized that for each management indicator species the connection with characteristics of the landscapes will be different, especially when human society is at the focus of investigations.

*Linking landscape to management indicator:*

With respect to flora and fauna, the main questions concern the isolation, fragmentation and networking of habitats, isolation, space requirements, biotope quantity, ecotone theory, corridors, barriers and the invasion of plant species. Recently, it has been increasingly recognized that effective species and population protection can be improved by integrating an approach which takes into account a certain area with its landscape features. Publications by Tischendorf (1995), Kuhn (1997) and Samietz (1998) confirm that modelling populations within their habitats necessarily relies on the consideration of both the landscape structure and its dynamics.

Linking landscape to demands of human society:

The appearance of the landscape is on the one hand moulded by its "natural features and their complex effects as well as on the other hand by the demands made on it by society ... in the type and manner of land usage" (Bastian & Schreiber 1994, p. 170). The demands placed on ecosystems by humans comprise the creation of landscape structures from a mosaic of natural and anthropogenically formed landscape elements, whose size, shape and arrangement vary depending on type as well as the intensity and duration of an anthropogenic influence (Burgess & Sharpe 1981, Forman & Godron 1986, Urban et al. 1987, O'Neill 1988, Turner & Ruscher 1988, Bastian & Schreiber 1994). The resulting changes are "often subordinated to assessments which are adequate solely with respect to society. This is expressed for example in the linearity of landscape-changing measures (arable strips, watercourse regulation, roads, corridors, etc.), which must be interpreted as a mirror of untiring efforts towards more efficiency" (Tischendorf 1995, p. 4). Hence Krummel et al. (1987), O'Neill (1988) and Turner (1990) showed by analysing shape complexity (*SHAPE, fractalness*) that anthropogenically influenced landscapes exhibit simpler patterns compared to natural landscapes. Krummel et al. (1987) as well as Wickham & Norton (1994) found a clear relationship between the landscape metrics (*LSM*) *fractal dimension (FD)* and the proportion of anthropogenic land usage. According to Plachter (1991) and Fiedler et al. (1996), the degrees of human impact on spatial structures and patterns are so drastic that land use acquires an indicator function for characterizing and evaluating human impact (Schönfelder 1984). The combinations of features documented in land use are an expression of a recent or past effect structure, and provide an opportunity to ascertain the processes and functional states taking place without having to extensively investigate them (Symader 1980). Hence studying the landscape structure is an important parameter for assessing socially relevant landscape functions.

## 5.3 Concept of landscape metrics

In order to investigate the structures of the landscape, they must be named and categorized. The term 'structure' is a frequently used concept, whose employment however often leads to misunderstandings. 'Structure' refers to the spatially related features of elements of the ecosystem and their spatial interrelationship within the landscape. They are used to describe the distribution of energy, material and type with respect to the size, shape, number, type and arrangement of ecosystems in a landscape (Forman & Godron 1986). The spatially related characteristics and the resulting spatial relations are expressed in the *composition* and *arrangement (configuration)* of the landscape elements of an ecosystem. Hence landscapes can be differentiated in terms of their landscape composition and/or their landscape arrangement.

Landscape composition refers to the number, proportional frequency and diversity of landscape elements within the landscape, with the concrete spatial relations being neglected. The composition of the landscape can for example be described by parameters such as the ratio of the area of elements to the total area of the landscape (*largest patch index, LPI*) or the number of elements of different types occurring within a standardized patch (*patch richness density, PRD*).

The arrangement of landscape elements, landscape arrangement or *configuration* covers all the spatially related characteristics and primarily describes the spatial position and the spatial distribution of the elements within a landscape. In this work for example, concrete neighbourhood relations, the distances between elements (*nearest neighbour distance, NEAR*) and the alignment of borders to each other *(interspersion and juxtaposition index, IJI*) are important for the description of the landscape configuration.

According to Forman & Godron (1986), the composition and arrangement of landscape elements (landscape element structure), biotopes (biotope structure) and the entire landscape (landscape structure, landscape pattern, landscape fabric) represent key aspects of landscape ecology.

The ecosystem or the landscape itself has a very complex nature. This results in the necessity of dividing up a landscape ecosystem into manageable sub-systems which can be dealt with methodically (Leser 1997). According to Paffen (1953, p. 16), the division of the landscape should "proceed from small to large by starting with building blocks and - as in construction work - putting them together to produce larger and larger units." The acceptance and application of the *hierarchy of landscape ecosystems* hence integrated the question concerning the *hierarchy of the spatial units to be investigated.* Just what are the smallest building blocks or the smallest spatial units? The decisive factor in distinguishing the smallest spatial units proved to be the criterion of homogeneity. According to Neef (1964, p. 1), a geographical area can be described as homogeneous "if it has the same structure and the same activity structure and therefore a uniform material balance - and hence exhibits the same ecological behaviours."

The large number of terms discussed so far such as *landscape cell* (Paffen 1953), *tiles* (Schmithüsen 1948), *physiotope* (Richter 1965, Herz 1974) and *ecotope* (Neef 1963) for the smallest homogenous spatial unit shows that the practical implementation and usage of the concept of homogeneity varies greatly and that therefore the terms used need to be defined.

Forman & Godron 1986, Wiens 1989, McGarigal & Marks 1994 regard the smallest individual, largely homogeneous element of a landscape as a *landscape element* or *patch. Patches* represent discrete areas (functional areas) with independent characteristics such as size, perimeter and shape, which are qualitatively described depending on the scale of surveying and observation.

For analysis of the landscape using remote sensing methods, a *patch* is formed by the pixels which belong together of *one land-cover class* of the classified satellite picture (Fig. 5.1) as well as of a biotope type and a land usage class of the biotope type mapping (Fig. 5.1). The total number of *patches* of the same type form a class (*patches*, landscape elements class landscape, Fig. 5.1). According to Urban et al. (1987), landscapes comprise a pattern of *patches*. Hence the characteristics or the spatial position relations of *patches* sharing the same characteristics (distance, position vis-à-vis each other) and the qualitative neighbour relations decide the composition and configuration of the landscape.

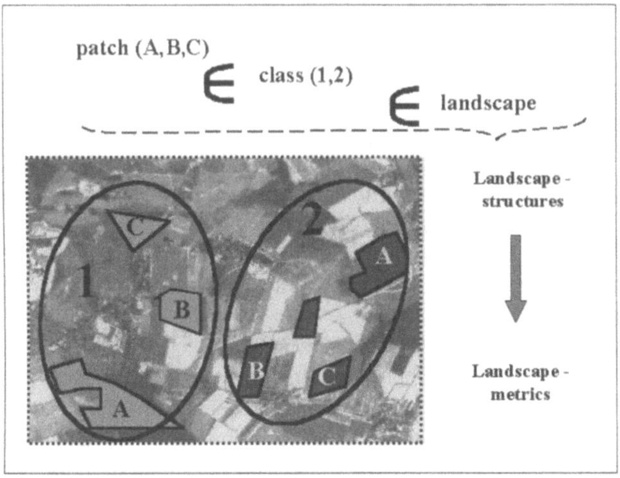


**Fig. 5.1:** The spatial units patch - class - landscape

## 5.4 Types of landscape metrics and processing of satellite data

Types of landscape metrics

Table 5.1 describes the indexes for quantifying the biotope and landscape structure of the areas investigated.

*Processing of satellite data*

Image data from the Spot-XS sensor for the study area in the time segments 1990, 1994 and 1996 underwent hierarchical classification using the Maximum Likelihood Algorithm. Eleven types of land cover were distinguished. By integrating the "Landscape Development Concept" (Regionaler Planungsverband Westsachsen 1996a,b) for the 2020, conclusions concerning the area's future development can be drawn beyond the year 1996. Owing to the limited geometrical resolution (20 metres per pixel) of the multi-spectral Spot-XS data, the existing transport network could only be partly visually documented, and could not be satisfactorily classified. Because of the importance of linear landscape elements (e.g. the transport network) for calculating landscape metrics (Lausch & Menz 1999, Lausch 2000), the classification results were converted to a raster cell size of 10 metres per pixel, and the transport network (also with a cell size of 10 metres) as captured by the vector biotope type mapping of Saxony was juxtaposed on the land-use data. The landscape metrics from the land-use data of the southern region of Leipzig were calculated on a raster basis using the structural analysis program FRAGSTATS (Vers. 2.0, McGarigal, Marks 1994).

Process monitoring for the southern region of Leipzig (Fig. 5.2) is based in methodological terms on the components shown in Fig. 5.2.

**Area metrics**

The area metrics quantifies the composition of the landscape. However, this provides no information concerning the arrangement of the landscape elements. Area metrics frequently provides a basic unit for calculating a large number of indexes at the level of landscape and class. Hence the minimal area size and the landscape perimeter provide important parameters for quantifying the structural indicators concluded.

Area metrics is of increasing importance for assessing of species diversity, area occupation and species distribution patterns, as well as in particular for investigations into landscape change.

*Patch density, patch size and variation of patch sizes:*

These LSM are used to quantify the number or density of *patches*, the average size of *patches* and the change to the *patch* size. This allows important information to be derived concerning aspects of landscape arrangement. The number of areas and the area density are important when investigating questions of the spatial heterogeneity of the landscape mosaic, the fragmentation of landscape, and the influence of ecological processes. The number of *patches* and the average *patch* size provide direct information about the landscape pattern. According to Gardner et al. (1993), they are an important criterion for relating the loss of biodiversity in a district to the decrease in a habitat's area.

| | | |
|---|---|---|
| Class / Landscape | *TA* | *Total landscape area (ha)* |
| Class | *%LAND* | *Percent of landscape (%)* |
| Class / Landscape | *LPI* | *Largest Patch index (%)* |
| Class / Landscape | *NP* | *Patch density (#/100ha)* |
| Class / Landscape | *MPS* | *Mean Patch size (ha)* |
| Class / Landscape | *PSCV* | *Patch size coefficient of variation (%)* |

**Edge metrics**

Edges (*edges, borders*) form the boundaries of *patches*. They are very informative when it comes to describing the configuration (landscape arrangement) of a landscape structure. Hence according to Miller et al. (1997), an increase in anthropogenic influence and the fragmentation of land cover types increases the amount of edges. The absolute length of perimeters and edge density are crucial for a variety of ecological phenomena. Many ecological processes take place at or even across boundaries and so are directly dependent on them. The fragmentation and heterogeneity of a landscape (e.g. an agricultural landscape) are estimated by taking into account the edge metrics (Forman & Godron 1996; Reese & Ratti 1988).

| | | |
|---|---|---|
| Landscape elements | *VEDCON* | *Vertical edge contrast index (%)* |
| Class / Landscape | *MVECI* | *Mean vertical edge contrast index (%)* |
| Landscape elements | *PERIM* | *Perimeter (m)* |
| Class / Landscape | *ED* | *Edge density (m/ha)* |
| Class / Landscape | *ESI* | *Edge structure index (m)* |

**Metrics for shape and forms (formdescriptors)**

Analysing the shape and form of landscapes has become an important subject of investigation for research into processes of landscape ecology (Forman & Godron

| | | |
|---|---|---|
| 1986, McGarigal & Marks 1994). For example, shape indexes are especially important for questions of population ecology concerning the occurrence, abundance and migration processes of species (Hamazaki 1996; Hawrot et al. 1996, Samietz, 1998). The form metrics calculated in FRAGSTATS (Vers. 2.0, McGarigal & Marks 19994) are based on an analysis of the perimeter-to-area of the landscape elements. The limitations of these indexes are that in raster depictions, the length of the perimeter may be higher than with vector data owing to the "stepping" of line segments and depends on the resolution of the data. The perimeter-to-area method is relatively insensitive to differences in the *patch* morphology. After all, *patches* may have different shapes but nevertheless identical areas and perimeters, and hence the same index value. As a result, this metrics is not suitable for assessing element morphology as such, but is more useful for estimating the complexity of the shape of a landscape element. Consequently, these LSM quantify the landscape arrangement with respect to the complexity of the *patch* shape. | | |
| Landscape elements | *SHAPE* | *Shape index* |
| Class / Landscape | *LSI* | *Landscape shape index* |
| **Nearest neighbour metrics** | | |
| This LSM shows the distance between a *patch* and its nearest *patch* in the same class, with the distance being measured from edge to edge. Hence this metrics quantifies the landscape arrangement (landscape configuration). This knowledge is of particular interest for investigating migration processes and population dynamics, especially in order to survey the isolation and fragmentation of landscape elements. The determination of landscape configuration is also of great importance concerning the monitoring of landscapes. | | |
| Class / Landscape | *MNN* | *Mean nearest-neighbour distance (m)* |
| Class / Landscape | *NNCV* | *Nearest-neighbour coefficient of variation (%)* |
| Landscape elements | *PROXIM* | *Proximity index* |
| Class / Landscape | *MPI* | *Mean proximity index* |
| **Metrics of diversity and distribution** | | |
| The diversity indexes quantify the composition of the landscape investigated. They are influenced by the two components *richness* and *evenness*. *Richness* refers to the number of classes occurring in the landscape, whereas *evenness* describes the distribution of the various classes. The richness of existing classes depends on the scale chosen. Thus large areas are frequently richer as they have a higher degree of heterogeneity of inventory than comparable small areas. In order to investigate diversity, a number of indicators can be used whose applicability, interpretability and informativeness are the subject of controversial discussion. | | |
| Landscape | *PRD* | *Patch richness density (#/100ha)* |
| Class / Landscape | *IJI* | *Interspersion and Juxtaposition index (%)* |
| Landscape | *SHDI* | *Shannon's diversity index* |
| Landscape | *SHEI* | *Shannon's evenness index* |

**Table 5.1:** Description of LSM calculated using FRAGSTATS (Vers. 2.0, McGarigal & Marks 1994)

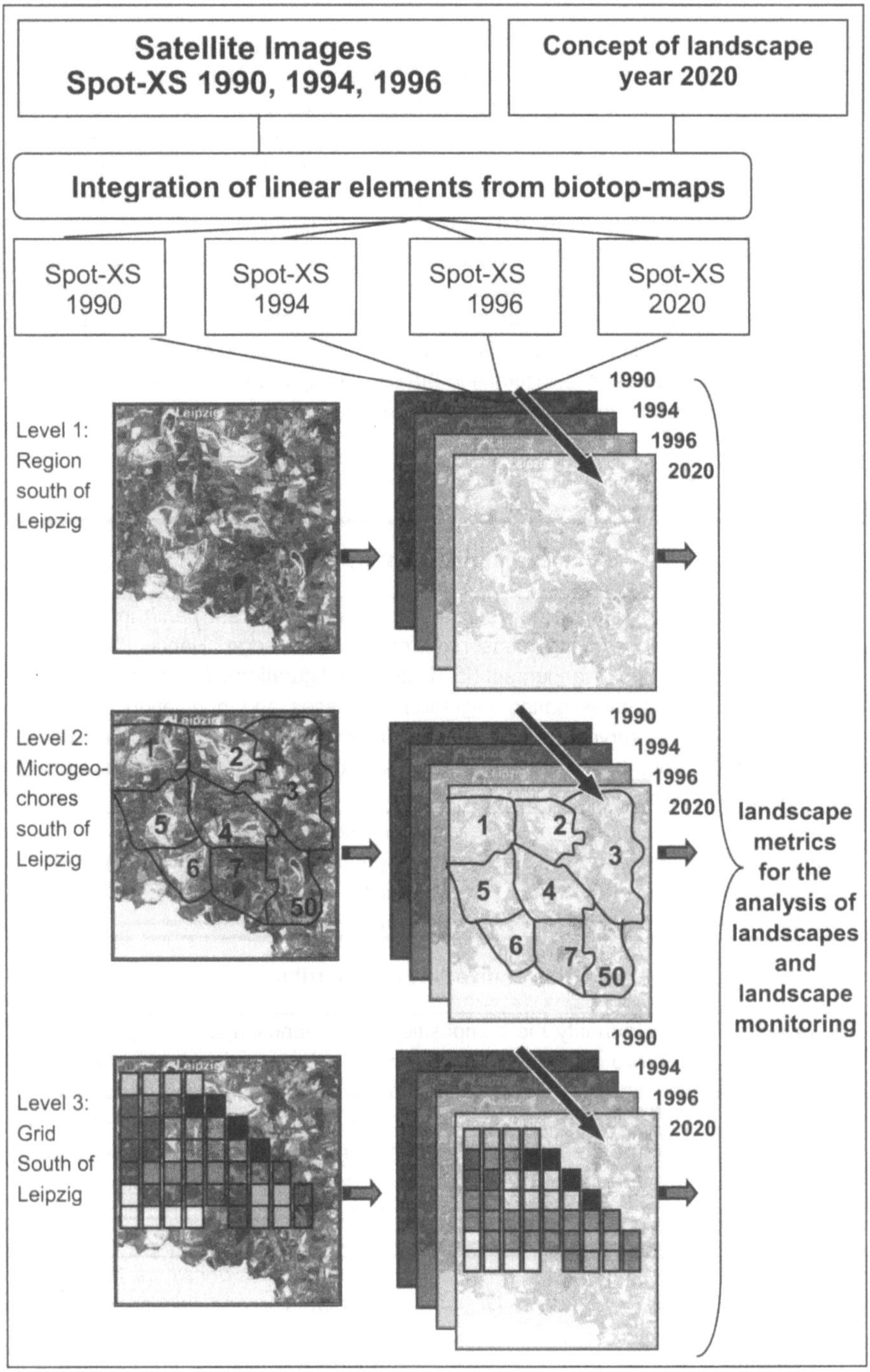


**Fig. 5.2:** Method of hierarchical spatio-temporal landscape monitoring in Leipzig's southern region 1990-2020

## 5.5 Demonstration of the hierarchical approach for a particular mining region in Germany

The analysis of landscape structures is to be demonstrated using the region south of Leipzig by way of example.

Opencast mines, one of mankind's most radical interventions in landscapes, are now serving as the basis of new biotopes and land use structures. In the region south of Leipzig (western Saxony), lignite surface mining has scarred several hundred square kilometres. Mining began in the mid-19th century and reached its peak in the 1980s. Its sharpest impact on the landscape was the almost complete destruction of river wetlands. Nowadays, many of the landscape functions destroyed in Leipzig's southern region are undergoing regeneration. This still 'unstable' landscape offers a unique opportunity to test the suitability of landscape structural indicators for landscape monitoring.

The project had the following aims:

1. To develop and standardize a methodology to quantify and assess changes of land use, landscape structure and biodiversity in the region south of Leipzig by means of remote sensing and a GIS;
2. To assess changes to landscape structure and land use on several spatial and temporal scales in the study region;
3. To analyse current patterns and their change of land use in the region south of Leipzig;
4. To apply models that integrate data on both land-use change as well as current and past land use and landscape patterns in order to understand anthropogenic effects on landscapes.

Numerous examinations for landscape monitoring consider the special value of such analysis and comparisons only for different time series of the same region. With this examination we want to show that, in addition, it is necessary not only to use digital data of different points in time but also to investigate different spatial scales. Therefore the research into landscape change in the region south of Leipzig is based on the analysis of spatio-temporal changes of landscape structure. Different methods for landscape changes are linked to each other (area-use analysis, density examinations, quantification of landscape structure measures such as the number, size, shape and arrangement of patches). By including different methods that analyse the changes of landscape structure this approach guarantees that a wide spectrum of features obtained for a landscape will be taken into account allowing a relevantly and concrete spatio-temporal assessment of various changes in landscape (Lausch & Herzog 1999). The methodological tools applied include remote sensing techniques, geographical information systems (GIS) and mathematical methods for the quantitative assessment of landscape patterns by means of landscape metrics.

The survey and description of the current state of landscape and also its spatio-dynamic processes entails using new quantitative methods and evaluation approaches to capture, quantify and show spatial patterns (Turner 1989). Surveying the spatial structure provides information on the current state and changes of

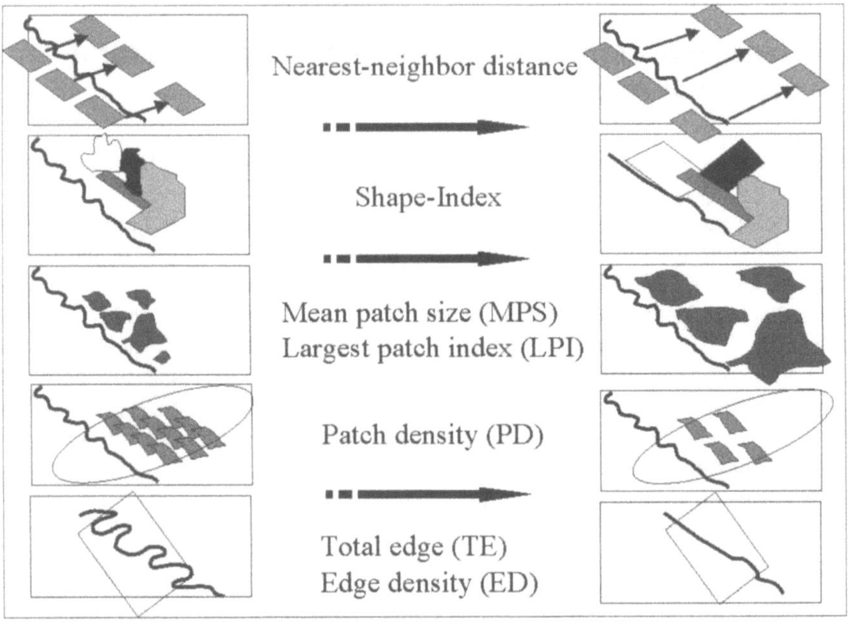


**Fig. 5.3:** Spatio-temporal dynamic of landscape structures - quantification and analysis with landscape metrics

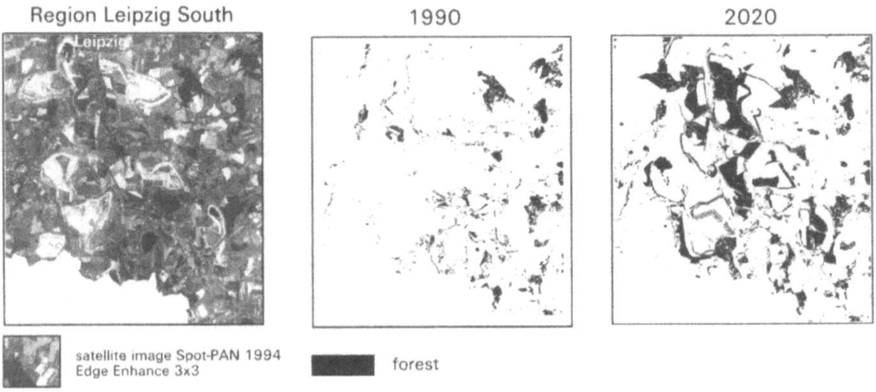


**Fig. 5.4:** Study of land use in Leipzig's southern region - changes in the land cover-class , Deciduous Forest for the period 1990-2020

landscape elements, biotopes and the entire landscape. This in turn necessitates suitable indicators which can be used to capture changes to the spatial structure. In order to characterize the landscape structure, the approaches taken by the landscape ecologists Forman & Godron (1986), Turner (1989), Turner & Gardner (1991a,b) for the calculation of LSM prove suitable. LSM are indicators which enable the pattern, composition and consideration of patches to be analysed, described and quantified. Important conclusions concerning the change to the landscape can be drawn from the spatial-temporal change to the LSM (Fig. 5.3).

Remote sensing data and also the vectorial mapping of biotope and land use structures provide sets of important basic data for investigating the landscape structure (Fig.5.2). For example for Leipzig's southern region, the LSM for regionally typical land use classes (open land without vegetation, water, copses, forest, farming land, built-up areas, overall landscape) were determined for the time segments 1990, 1994 and 1996, as well as for the situation as planned in 2020 (Fig. 5.2). However, landscape changes and developments take place in different ways. Hence the diversity of the spatio-structural changes of a landscape require the inclusion of LSM with a varying degree of informativeness such as area metrics, edge metrics, shape metrics, nearest neighbour metrics, and diversity and distribution metrics.

However, the LSM calculated for Leipzig's southern region exhibit significant variations for the large but nevertheless heterogeneous landscape south of Leipzig. Thus the survey of the landscape structure by LSM is directly related to the spatial resolution and the scale used for investigation. The LSM have specific qualities of informativeness for the region investigated south of Leipzig for the interpretation of the structural provision of biotope and land-usage structures. Owing to the high variability of structural features in the area, the LSM surveyed at the level of classes and landscape can so far only be incorporated into the appraisal as mean values. Area-specific quantitative and qualitative information can solely be concluded by looking at smaller areas.

In order to obtain concrete spatio-temporal information on changes to the landscape structure quantified by the landscape metrics, a hierarchical spatio-temporal approach is proposed for analysing landscape structures with varying spatial units of reference:

- *Level 1:* *Calculation of the landscape structure of the entire region*
- *Level 2:* *Calculation of the landscape structure at the level of small, demarcated landscapes (units of natural areas - microgeochorological areas)*
- *Level 3:* *Investigations of the landscape using a small raster format*

*Level 1: Calculation of the landscape structure of the entire region*

Concrete information about the area statistics in Leipzig's southern region can be derived by evaluating satellite pictures (i.e. classifying the Spot-XS data). The proportions and changes of important biotope and land usage elements of the southern region are shown in Table 5.2. Whereas an increase in size was ascertained for the land use classes Woods, Deciduous Forest (Fig. 5.4) and Water in the period 1990-2020, a sharp reduction in the class of Open Land without Vegetation (opencast mining areas) was predicted.

**Table 5.2:** Analysis of land use change of selected classes 1990-2020

| Land-cover classes | 1990 [ha] | 1994 [ha] | 1996 [ha] | 2020 [ha] |
|---|---|---|---|---|
| Pioneer Vegetation | 1.677 | 4.174 | 3.087 | * |
| Woods | 387 | 693 | 1.123 | 1.191 |
| Deciduous Forest | 3.884 | 4.681 | 4.880 | 11.583 |
| Water | 1.145 | 1.459 | 1.826 | 5.022 |
| Open Land without Vegetation (opencast mining area) | 8.231 | 5.941 | 4.800 | 2.264 |

* No data available

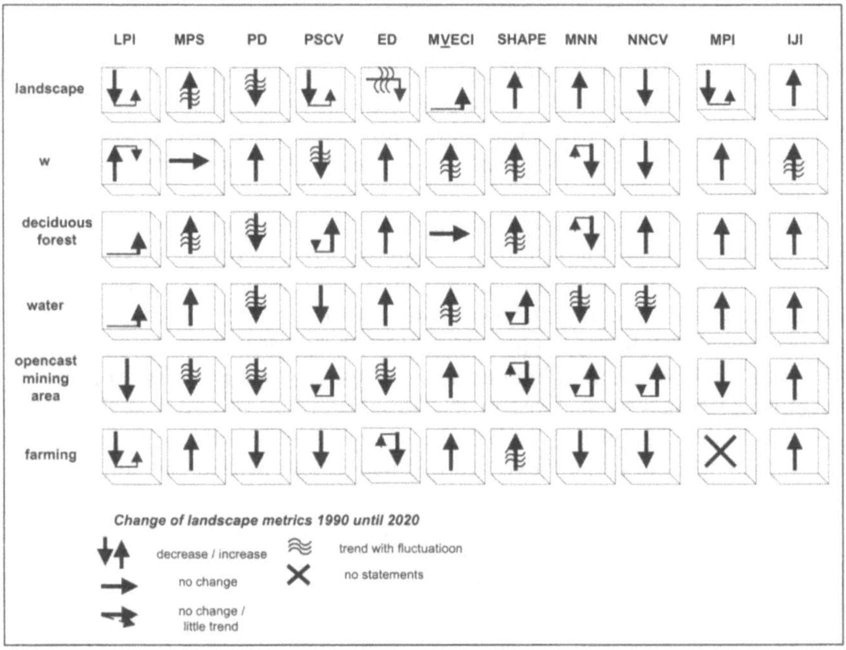


**Fig. 5.5:** Survey of the development of landscape metrics for the whole landscape and the land-cover classes characterising individual patches in Leipzig's southern region 1990-2020

The many different types of spatio-structural changes to the landscape required the inclusion of area metrics (LPI, MPS, PD, PSCV), edge metrics (ED, MVECI), shape metrics (SHAPE), nearest-neighbour metrics (MNN, NNCV, MPI), and metrics of diversity and distribution (IJI, Fig. 5.5) Our studies show that a set of 12 indicators suffices to quantitatively capture the changes to landscape elements, the land-cover classes classifying each patch (Woods, Deciduous Forest, Water, Open Land without Vegetation [opencast mining areas], Grassland and Arable Land), as well as the landscape of Leipzig's southern region (Herzog et al. 2000). In order to derive findings concerning changes related to each individual indicator they were not summarized using factor analysis. This approach enables information to be derived on the spatio-structural development of the land-cover classes studied.

The investigations of the landscape structure in the region between 1990 and 2020 enable the spatio-structural changes to be assessed. Hence the following information can be concluded about both the region and the individual land usage classes studied.

*Landscape elements of Leipzig's southern region (whole landscape)*

- Decrease in dominant individual patches (*LPI*_land)
- Increase in the average patch size (*MPS*_land)
- Decrease in patch density (*PD*_land)
- Decrease in the variability of area size from average area size (*PSCV*_land)
- Decrease in edge density in 2020 (*ED*_land)
- Increase in regional height contrast in 2020 (*MVECI*_land)
- Increase in shape complexity (excluding qualitative change) (*SHAPE*_land)
- Increase in the average distance between landscape elements (*MNN*_land)
- Decrease in the variability of distances from the average distance (*NNCV*_land)
- Increase in the isolation of landscape elements in space (*MPI*_land)
- Increase in the proportional distribution of landscape elements (*IJI*_land)

*Deciduous Forest*

- Increase in dominant individual patches in 2020 (*LPI*_deciduous forest)
- Increase in average area size (*MPS*_deciduous forest)
- Decrease in the density of deciduous forest patches (*PD*_deciduous forest)
- Increase in the variability of patch size from average patch size (*PSCV*_deciduous forest)
- Increase in the edge density of deciduous forest patches (*ED*_deciduous forest)
- No change in the height contrast in the surroundings of deciduous forest patches (*MVECI*_deciduous forest)
- Increase in shape complexity (*SHAPE*_deciduous forest)
- Decrease in the average distance between deciduous forest patches (*MNN*_deciduous forest)
- Increase in the variability of distances from the average distance (*NNCV*_deciduous forest)
- Decrease in the isolation of deciduous forest patches (*MPI*_deciduous forest)
- Increase in the proportional distribution of deciduous forest areas (*IJI*_deciduous forest)

*Open Land without Vegetation (opencast mining areas)*

- Decrease in dominant individual areas (*LPI*_opencast mining areas)
- Decrease in the average area size (MPS_opencast mining areas)
- Decrease in density of opencast mining areas (*PD*_opencast mining areas)
- Increase in the variability of area size from average area size (PSCV_opencast mining areas)

- Decrease in the edge density of opencast mining areas (*ED*_opencast mining areas)
- Increase in the height contrast in the surroundings of opencast mining areas (*MVECI*_opencast mining areas)
- Decrease in the shape complexity of opencast mining areas (*SHAPE*_opencast mining areas)
- Increase in the average distance between opencast mining areas (*MNN*_opencast mining areas)
- Increase in the variability of distances from the average distance (*NNCV*_opencast mining areas
- Increase in the isolation of opencast mining areas (*MPI*_opencast mining areas)
- Increase in the proportional distribution of opencast mining areas (*IJI*_opencast mining areas)

*Level 2: Calculation of the landscape structure at the level of small, demarcated landscapes (units of natural areas - microgeochorological areas)*

The investigations of the landscape metrics show that the categorization and aggregation of spatial units within mean value formation causes information to be lost. As a result, the information provided is generalized. In order to obtain concrete spatio-temporal analysis of the behaviour of landscape metrics and to assess the regional trend development, analysis of the landscape metrics are required at the spatial reference basis for each individual microgeochorological area (Level 2). The reference unit microgeochorological area (MG) permits information about the features of small areas and their change over time to be expressed. Compared to simple rectangular demarcation of the landscape, the units of natural areas used proved to be more suitable spatial units for the determination of potential characteristics and for determining the influence of factors disrupting landscape ecology. Thus the investigation, quantification and modelling of connections in order to determine degrees of strain, the sustainability of usage changes and usage intensification is only possible on the basis of such spatial units (Haase 1996).

In order to obtain information on the development and change of the landscape structure of all the microgeochorological areas examined (50 MG), trend or time-series analysis proved necessary. Trend or time-series analysis enables the nature of the phenomenon revealed by the series of observations to be identified, and also allows the prognosis of the variables of interest to be investigated. This entails identifying and describing the pattern of the time-series data observed. Although there exists a whole range of techniques for carrying out time-series and trend analysis, we could not use traditional methods of time-series analysis and trend determination as described by Wei (1989), Kendall & Ord (1990), Montgomery et al. (1990) and Walker (1991) in this investigation as we only had four time segments. Instead, a method was developed (Lausch 2000) to qualitatively capture trend dynamics via a hierarchical decision tree.

*Trend*

*exists (yes)*

*Positive trend (increase)* *Intensity*

*Negative trend (decrease)* *Intensity*

*Trend*

*does not exist (no)*

*No change (stagnation)*

*Variations (fluctuation)*

*Intensity*

Information is concluded using three metrics (M1-M3) which cumulate in different ways the increases in LSM between the individual time segments.
The following abbreviations are used:

A = LSM 1990
B = LSM 1994
C = LSM 1996
D = LSM 2020

In the following, for each microgeochore the percentage increases (PZ) in the respective LSM between two time segments is calculated as follows:

PZ1 = (B-A)/A
PZ2 = (C-B)/B
PZ3 = (D-C)/C

These values are used to construct the following three metrics for trend determination.

a) "Intensity metrics" (M1)

This metrics is designed to measure the intensity of trend development over the four time segments. For this purpose, the absolute amounts of the individual percentage increases are cumulated and then standardized by averaging over the three time steps

**M1 := (Abs(PZ1)+Abs(PZ2)+Abs(PZ3)) x 100 / 3**

b) "Balance metrics" (M2)

The following metrics allow the alignment of the potential trend development (e.g. positive/negative alignment) to be determined. Hence the signed sum of the actual percentage increases are cumulated and standardized. Hence the measure depicts the overall trend which is not balanced over time.

**M2 := (PZ1+PZ2+PZ3) x 100 / 3**

c) "Trend indicator" (M3)

A strict trend or non balanced structural changes can be identified by comparing the two previous metrics. For this purpose, the total trend (M2) must correspond to the absolute intensity (M1).

**M3 := Abs (M1 - Abs (M2))**

The three metrics enable the hierarchical decisions sought to be reached as follows. If any structural change occurred, and hence we can inquire about possible trends, M1 necessarily must be greater then 0.2.

*Trend yes/no?*

*M3* $\leq$ *0.2* -> *M1* $\approx$ *Abs(M2)*
*i.e. all important increases in the same direction*
*"Trend exists"*
*Trend direction?*
*M2* > *0 i.e. main increases (PZ) are positive*
*"Trend positive and intensity matches value of M2"*
*M2* < *0 i.e. main increases (PZ) are negative*
*"Trend negative and intensity corresponds to value of M2"*

*M3* $\geq$ *0.2* -> *M1* $\neq$ *Abs(M2) i.e. the main increases are balanced*
*"Trend does not exist"*
*Fluctuation dynamics?*
*M1* < *5 i.e. increases are on average very small*
*"Stagnation i.e. 'no' change"*
*M1* $\geq$ *5 i.e. increases are on average considerable*
*"Fluctuation and intensity correspond to the value of M1"*

The decision threshold used (0.2% and 5%) were established with the help of the quantiles of the frequency distribution across the values of the individual landscape metrics.

In order to take into account statistical uncertainty, an LSM is for example identified as 'stagnating' at a level being quantified (landscape, class, patch) if the average change per year is less than 5%.

The findings of the trend analysis for the landscape metrics are shown Fig.5.6 for the microgeochorological areas of Leipzig's southern region. On the basis of these assessment matrices, the landscape dynamics can now be determined in their spatial relationship. In particular, the questions can now be answered concerning the areas (units of natural areas - microgeochorological areas) in which the landscape structures exhibit high dynamics, a trend in development (positive/negative), or which only undergo a very limited change with respect to the LSM investigated. Moreover, there now exists the possibility to compare all the spatial units investigated in terms of their landscape structure and development. The studies of trend development of landscape structure in small investigation areas allows spatially more concrete information compared to the investigations of landscape structure at level 1.

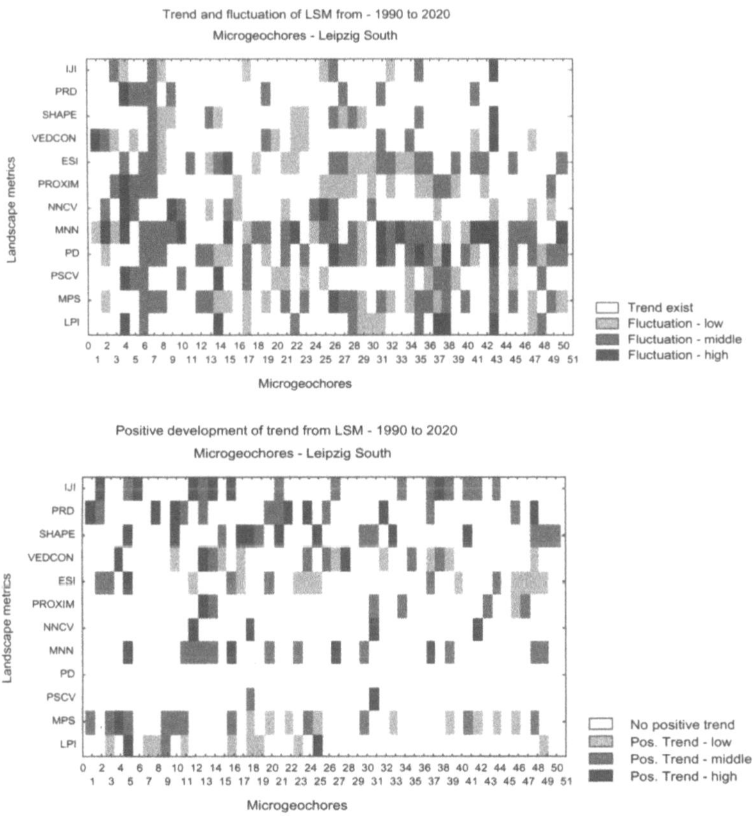


**Fig. 5.6:** Trend detection by LSM for microgeochorological areas in Leipzig's southern region 1990-2020

*Level 3: Landscape investigations in small-scale raster format*

In addition to land-use analysis, evaluating the land-use density and its changes provides another way of assessing the alteration of landscape elements and landscape structures in the study area. For Leipzig's southern region, the densities were ascertained for the following land-cover classes characterizing the area: Open Land without Vegetation (opencast mining areas), Woods, Forestland and Built-up Areas. The proportion of object classes was determined for a reference patch measuring 1 ha and then transformed into Density Classes. In order to conclude information on the spatio-temporal changes of the density of each class, subtraction was used to calculate and graphically present the changes in density of the various Density Classes between 1990 and 1996 (Fig. 5.7). Despite its low proportion of about just 1%, the object class Woods is especially important for the study area as it enables the recultivation measures to be observed and assessed especially clearly. Owing to the small size of these patches and the resulting problems of mixed pixels, its quantitative evaluation suffers from a high error rate.

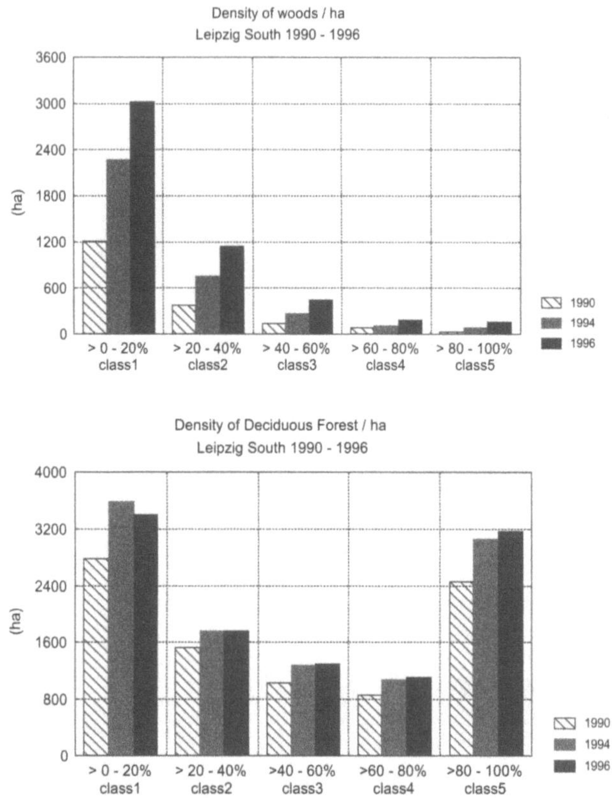


**Fig. 5.7:** Analysis of the density/ha of the classes Woods and Deciduous Forest from 1990 until 2020 for Leipzig's southern region - density classes

The fact that Density Class 1 predominates is immediately apparent from merely visual comparison. In 1990, for instance, 1,203 ha of woods land was assigned to Density Class 1, while around 381 ha of land was classified as Density Class 2. Study of the temporal change of woods density reveals a sharp increase between 1990 and 1996 from about 1,203 to 3,017 ha for Density Class 1, which is largely accounted for by spontaneous succession and afforestation. Examination of the spatio-temporal changes clearly reveals that the changes to density increase of woods are especially pronounced in the mining landscape areas of Leipzig's southern region.

To analyse the density of forest, the land-cover classes of Deciduous and Coniferous Forest were regarded as a single class, as Leipzig's southern region only contains a small proportion of coniferous forest. The existing forest density studies enable specific patch information to be concluded concerning the success of the afforestation of opencast mines currently underway. The existing investigations show that in particular Density Class 1 (patches with low forest density) and Den-

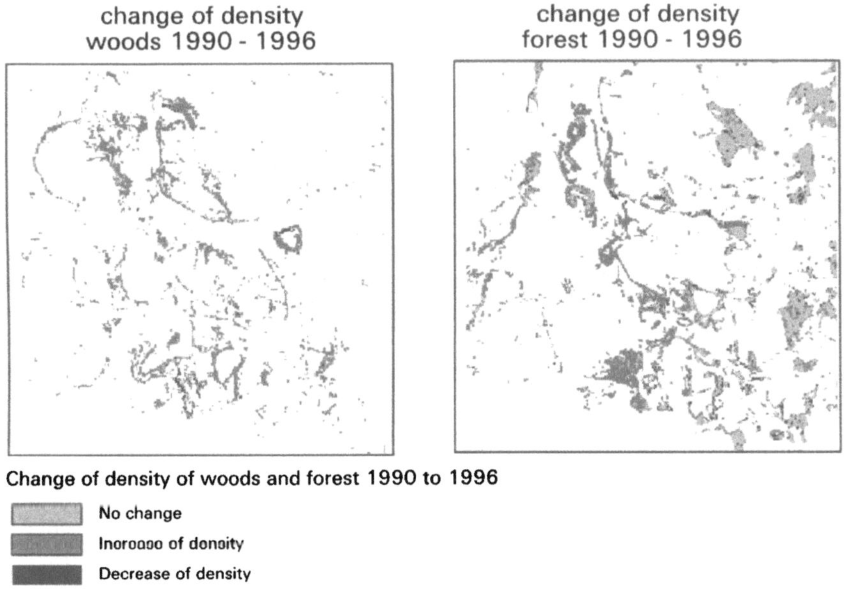


**Fig. 5.8:** Changes in density/ha of the classes Woods and Deciduous Forest between 1990 and 1996 in Leipzig's southern region

sity Class 5 (patches with very dense forest) account for the lion's share of forestland, far exceeding Density Classes 2,3 and 4 in Leipzig's southern region. Studies into the temporal development of forest density also indicate that all density classes expanded between 1990 and 1996, with the areas containing a very high density of forest undergoing the greatest increase between 1990 and 1994. Fig. 5.8 shows the spatial distribution of different forest density classes and their change from 1990 to 1996. For example, the analysis reveal that deciduous and coniferous forestland in the mining landscape is subject to high dynamics. Density chiefly rose in the central areas of the mining landscape, whereas the established areas of forest only underwent slight (if any) increase.

## 5.6 Possibilities and limitations when quantifying landscape metrics

There is already an enormous amount of publications on spatio-structure metrics. They mainly use aerial and satellite photographs to analyse spatial structures. However, despite the urgent need to define standard conditions for quantifying landscape-ecological indices, there are currently no methods of doing so, and consequently comparing the results of spatio-structure metrics is only possible to a limited extent. Analysing LSM is accompanied by factors which may qualitatively and quantitatively distort the capture of landscape elements and the derivation of LSM, causing erroneous assessments. These include in particular:

*Problems of remote sensing:*

- The extent to which landscape elements can be digitally captured depends on the sensors' spectral sensitivity and geometrical resolution (Moody & Woodcock 1995, Qi & Wu 1996, Baruth 1998).
- Depending on the raster cell size used (Haberäcker 1987), the problem of mixed pixels may cause difficulties during the digital capture of small and linear landscape elements, causing the landscape's pattern and fragmentation to be inaccurately reflected by remote sensing (Lausch & Menz 1999, Lausch 2000).
- The quantity of only a few LSM is independent of the scale of capture and the raster cell size used (Wickham & Riiters 1995, Rami 1997).
- Different phenological stages result in different classes and class aggregations.
- The selection of classes should be primarily based on the land-cover classes which can be spectrally captured, not on the topological of investigation. This makes population-ecological studies more difficult with respect to the relations to LSM.
- The information captured records land cover, not land use.
- Faulty classifications cause quantitative changes to the LSM.
- Different processes (elimination, filter methods) are used to eliminate tiny patches and individual pixels.
- Patches on the edges of a study area are sometimes fragmented.

*Problems of raster data:*

- The spatial unit 'patch' is frequently not the optimum unit of reference for landscape-ecological research (discrepancy from spatial units such as habitat and biotope).
- The demarcation of the smallest areas of reference ('patches') takes place differently.
- Raster may cause various systematic false assessments of border length and overestimates of line length (McGarigal & Marks 1994)
- If different raster cell sizes are used to examine the landscape index, quantitative differences arise in the landscape metrics (Wickham & Riiters 1995, Qi & Wu 1996, Rami 1997, Lausch 1999a,b, Herzog et al. 1999, Herzog & Lausch 1999).

## 5.7 Conclusion

Structural features of terrestrial land cover are directly or indirectly connected in a complex manner with a large number of functions. Consequently, obtaining information about the landscape structure provides indications of both the status and the dynamics of the landscape. The main findings are summarized below.

Landscape monitoring of the landscape (example region -opencast mining landscape "southern region of Leipzig") using remote sensing methods was extensively investigated. It was shown that good results can only be achieved in landscape monitoring when just one sensor type is used (SPOT-XS), and the land is broken down into classes which can be spectrally well captured and which accurately characterize the area. Hence only the classes Woods, Deciduous Forest, Open Land without Vegetation (mining areas), Water, and sometimes also Built-up Areas, Grassland and Arable Land, were included in all the subsequent analysis. The inclusion of multi-temporal data sets (SPOT-XS 1990, 1994, 1996) and additional data (landscape development concept for the year 2020) enables the processes of landscape development to be monitored. This in turn permits the temporal and spatial reconstruction of the dynamics of the landscape, including the biotope and landscape structures it contains. In order to carry out landscape monitoring, traditional methods of land balancing and density analysis were combined with new methodological approaches to quantify landscape structures by means of landscape metrics. The investigations demonstrated that in addition to significant temporal change, the landscape also exhibits high regional differentiation. Using a hierarchical approach to landscape monitoring proved decisive. For example, information on the entire development of the southern region of Leipzig over time was derived, and patch-specific and regional development trends and tendencies were identified - the latter being especially important for deciding what action needs to be taken locally.

Landscape metrics may be used as indicators to describe, characterize and quantify the pattern, composition and configuration of the biotope and landscape structure of the southern region of Leipzig on various spatial and temporal scales. The miscellany of these spatio-structural changes necessitates the inclusion of area, edge and shape metrics, nearest-neighbourhood metrics, and diversity and distribution in order to characterize changes to biotope and landscape structures.

The study shows that not just one landscape metric but also a set of metrics is suitable for quantitatively capturing and describing the dynamics of biotope and landscape structures of the landscape. Here, too, the hierarchical approach (levels 1-3) of quantifying biotope and landscape structures from different aggregation levels proved generally suitable for obtaining concrete information about certain areas.

The techniques presented here for monitoring landscape elements, classes and the total landscape provide - depending on the scale, the necessary spatio-temporal resolution, and class-specific resolution - a good basis for investigating and answering the following questions:

a) What changes do the landscape, the class and the landscape elements exhibit for a certain period investigated?
b) In what investigation areas do landscape structures show high dynamics, a trend in development (positive/negative) or only very slight change with respect to the parameters of landscape structure investigated?
c) What landscape structures change in a fashion which can be measured in space and time?
d) What landscape structures may be used as indicators for landscape changes?

Numerous works emphasize the importance of LSM for questions of landscape ecology, and use methods of remote sensing in order to capture them (Krummel et al. 1987; Luque et al. 1994; Menz 1997; Zheng et al. 1997; Koch & Werder 1998; Baruth 1998, Lausch et al. 1999, Lausch & Herzog 1999, Lausch 2000, Herzog et al. 2000). There already exists an enormous amount of work on LSM, which in particular use data achieved from aerial and satellite photographs for the analysis of spatial structure. Owing to the lack of standardization for the capture, analysis and evaluation of LSM (apart from the existing formulae), their comparability with data found in the literature is limited. Hence in addition to the new and further development of landscape metrics, there also exists a demand for the standardization of important model parameters. Only by their creation can indicators from different works be compared with each other, thus making the information they provide easier to analyse. Furthermore, there is also a need to study the structural parameters of landscapes with different natural features and anthropogenic influences using unified model parameters, so that the informativeness of indicators from our own investigations can be classified and assessed.

## 5.8 References

Baruth B (1998) Satellitendaten für den Natur- und Artenschutz. Geographische Rundschau 50 (2): 84-88

Bastian O, Schreiber, K-F (1994) Analyse und ökologische Bewertung der Landschaft. - Gustav Fischer Verlag, Jena

Bobeck H, Schmithüsen J (1949) Die Landschaft im logischen System der Geographie. Erdkunde 3: 112-120

Burgess RL, Sharpe DM (1981) Forest island dynamics in man-dominanted landscapes. Springer Verlag, New York

Fiedler HJ, Große H, Lehmann G, Mittag M (1996) Umweltschutz. G. Fischer Verlag, Jena-Stuttgart

Finke L (1978) Landschaftsökologie - was sie ist, was sie will, was sie kann. Umschau 78: 563-571

Finke L (1994) Landschaftsökologie. Westermann Verlag, Braunschweig

Forman RTT, Godron M (1986) Landscape Ecology. Wiley and Sons, New York

Haase G (1996) Geotopologie und Geochorologie - Die Leipzig-Dresdner Schule der Landschaftsökologie. In: Haase G, Eichler E (Hrsg): Wege und Fortschritte der Wissenschaft. - Sächsische Akademie der Wissenschaften zu Leipzig, Akademie Verlag, pp 201-229

Haberäcker P (1987) Digitale Bildverarbeitung: Grundlagen und Anwendungen. Carl Hanser Verlag, München-Wien

Hamazaki T (1996) Effects of *Patch* shape on the number of organisms. Landscape Ecology 11(5): pp. 299 – 306

Hansen JA, di Castri F (eds) (1992) Landscape boundaries. Consequences for biotic diversity and ecological flows. Springer Verlag, New York

Hansson L, Fahrig L, Merriam G (eds) (1995) Mosaic landscapes and ecological processes. Chapman & Hall, London

Hawrot R, Niemi GJ (1996) Effects of edge type and Patch shape on avian communities in a mixed conifer-Hardwood forest. The Auk 113(3): 586-598

Herz K (1974) Strukturprinzipien in der Landschaftssphäre. Ein Beitrag zur Methodologie der physischen Geographie. Geogr. Ber. 71: 100-108

Herzog F, Lausch A, Müller E, Thulke H-H, Steinhardt U, Lehmann S (2001) Landscape metrics for the assessment of landscape destruction and rehabilitation. Environmental Management 27/1: 91-107

Herzog F, Lausch A (1999) Prospects and limitations of the application of landscape metrics for landscape monitoring. In: Maudsley M., Marshall J. (eds) Heterogeneity in Landscape Ecology: Pattern and Scale. Aberdeen, IALE(UK), pp. 41 - 50

Kendall M, Ord, J K (1990) Time series. Griffin, London

Koch B, Werder U (1998) Detecting and describing landscape structures in the palatine nature park using remote sensing. International Archives of Photogrammetry and Remote Sensing, Vol XXXII, Part 7: 370-375

Krummel JR, Gardner RH, Sugihara G, O'Neill RV, Coleman PR (1987) Landscape patterns in a disturbed environment. Oikos 48: 321-324

Kuhn W (1997) Flächendeckende Analyse ausgewählter ökologischer Paramter. Bewertung von Habitateignung und -isolation mit Hilfe eines Geographischen Informationssystems. Dissertation, Institut für Landschaftsplanung und Ökologie der Universität Stuttgart

Lausch A (1999a) Möglichkeiten und Grenzen der Einbeziehung von Fernerkundungsdaten zur Analyse von Indikatoren der Landschaftsstruktur - Beispielregion Südraum Leipzig. In: Steinhardt U, Volk M (Hrsg) Regionalisierung in der Landschaftsökologie, B. G. Teubner-Verlagsgesellschaft, Stuttgart-Leipzig, pp 162-180

Lausch A (1999b) Methodik zur Erkundung der Biotop- und Landschaftsdiversität in der Braunkohletagebaufolgelandschaft mit Fernerkundungsmethoden. Abschlußbericht DARA GmbH, FKZ 50 EE 95 12

Lausch A, Herzog F (1999) Applicability of landscape metrics for the monitoring of landscape change: issues of scale, resolution and interpretability. Proceedings of the 2. International Conference INDEX 99 "Indices and Indicators of sustainable development", 11th - 16th July 1999 in St. Petersburg, Russia, Ecological Modelling.

Lausch A, Menz G (1999) Bedeutung der Integration linearer Elemente in Fernerkundungsdaten zur Berechnung von Landschaftsstrukturmaßen. PFG 3: 185-194

Lausch A (2000) Raum-zeitliches Monitoring von Landschaftsstrukturen in der Tagebauregion Südraum Leipzig mit Methoden der Fernerkundung und Geoinformation. Dissertation, Rheinischen Friedrich-Wilhelms-Universität Bonn, UFZ-Bericht 12/2000, Leipzig

Lausch A, Krönert R, Herzog F (1999): Landscape diversity in the brown coal mining region South of Leipzig by means of remote sensing and GIS application. Proceedings of the Conference "DGPF" 28. April 1999 in Halle

Luque SS, Lathrop RG, Bognar JA (1994) Temporal and spatial changes in an area of the New Jersey Pine Barrens landscape. Landscape Ecology 9(4): 287-300

Leser H (1997) Landschaftsökologie. Ulmer, Stuttgart

McGarigal K, Marks B (1994) Fragstats - Spatial pattern analysis program for quantifying landscape structure. Forest Science Department, Oregon State University, Corvallis, OR 97331, 67 p.

Menz G (1997) Landschaftsmaße und Fernerkundung - neue Instrumente für die Umweltforschung. Geographische Rundschau 49: 1-7

Miller JN, Brook RP, Croonquist MJ (1997): Effects of landscape patterns on biotic communities. Landscape Ecology 12: 137-153

Montgomery DC, Johnson LA, Gardiner JS (1990) Forecasting and time series analysis. McGraw-Hill, New York

Moody A, Woodcock CE (1995) The influence of scale and the spatial characteristics of landscapes on land-cover mapping using remote sensing. Landscape Ecology 10(6): 363-379

Neef E (1963) Topologische und chorologische Arbeitsweisen in der Landschaftsforschung. PGM 107(4): 249-259

Neef E (1964) Zur großmaßstäbigen landschaftsökologischen Forschung. PGM 108: 1 78

O'Neill RV (1995) Recent development in ecological theory: Hierarchy and scale. Annual conference on surveying and mapping. American Society of Photogrammetric Engineering and Remote-Sensing, Charlotte, North Carolina, 27 February-2 March 1995

O'Neill RV (1988) Hierarchy theory and global change. In: Rosswall T, Woodmansee RG, Risser PG (eds) Scales and global change. Spatial and temporal variability in biospheric and geosperic processes, John Wiley & Sons, New York, pp. 29-45

Paffen KH (1953) Die natürliche Landschaft und ihre räumliche Gliederung. Eine methodische Untersuchung am Beispiel der Mittel- und Niederrheinlande. Forschungen zur deutschen Landeskunde Band 68, Remagen

Plachter H (1991) Naturschutz. Gustav Fischer Verlag, Stuttgart

Qi Y, Wu J (1996) Effects of changing spatial resolution on the results of landscape pattern analysis using spatial autocorrelation indices. Landscape Ecology 11(1): 39-49

Rami M (1997) Landschaftsstrukturmaße und Satellitenbildfernerkundung. Entwicklung des Programms METRICS und seine Anwendung auf LANDSAT- und NOAA-Szenen aus dem Bereich Schwarzwald/Oberrhein. Unveröffentlichte Diplomarbeit, Universität Bonn

Reese KP, Ratti JT (1988) Edge effect: A concept under scrutiny. North America Wildlife and Natural Resources - Conference Transactions, Bd. 53, pp. 127-136

Regionaler Planungsverband Westsachsen (Hrsg) (1996a) Regionalplanung in Westsachsen. - Entwurf vom 09.08.1996, Grimma

Regionaler Planungsverband Westsachsen (Hrsg) (1996b) Regionalplanung in Westsachsen. - Entwurf vom 09.08.1996, Grimma, -Karte Entwicklungskonzept Landschaft, Stand Januar 1995, Maßstab 1 : 75 000

Richter G (1965) Bodenerosion. Schäden und gefährdete Gebiete in der Bundesrepublick Deutschland. FDL 152, Trier

Samietz J (1998) Populationsgefährdungsanalyse an einer Heuschreckenart - Methoden, empirische Grundlagen und Modellbildung bei Stenobothrus lineatus (Panzer). Cuvillier Verlag, Göttingen

Schmithüsen J (1948) Fliesengefüge der Landschaft und Ökotop. Vorschläge zur begrifflichen Ordnung und Nomenklatur in der Landschaftsforschung. BDL, Bd. 5, pp 74-83

Schönfelder G (1984) Grundlagen für die Vorhersage (Prognose) von Landschaftsveränderungen als geographischer Beitrag für die Landschaftsplanung. Dissertation B, Universität Halle

Symader W (1980) Zur Problematik landschaftsökologischer Raumgliederungen. Landschaft und Stadt 12/1, Stuttgart, pp 81-89

Tischendorf L (1995) Modellierung von Populationsdynamiken in strukturierten Landschaften. Dissertation, Philipps-Universität Marburg

Turner MG (1989) Landscape ecology: The effect of pattern on process. - Annual Review of Ecology and Systematics 20, pp. 171-197

Turner MG (1990) Spatial and temporal analysis of landscape patterns. Landscape Ecology 4(1): 21-30.

Turner MG, Gardner RH (1991a) Quantitative Methods in Landscape Ecology. Springer-Verlag, New York

Turner MG, Gardner RH (1991b) Quantitative Methods in Landscape Ecology: An introduction. Ecological Studies 82, Springer, New York, pp. 3-17

Turner MG, Ruscher CL (1988) Changes in the spatial patterns of land use in Georgia. Landscape Ecology 1: 241-251

Urban DL, O'Neill VO, Shugart HH (1987) Landscape ecology: A hierarchical perspektive can help scientist understand spatial pattern. BioScience 37: 119-127

Walker J S (1991) Fast fourier transforms. Boca Raton, FL: CRC Press

Wei WW (1989) Time series analysis: Univariate and multivariate methods. Addison-Wesley, New York

Wickham JD, Norton DJ (1994) Mapping and analyzing landscape patterns. Landscape Ecology 9(1): 7 - 23

Wickham JD, O'Neill RV, Ritters, KH, Wade TG, Jones KB (1997) Sensitivity of selectesd landscape pattern metrics to land-cover misclassification and differences in land-cover composition. PE & RS 63 (4): 397-402

Wickham JD, Riiters KH (1995) Sensitivity of landscape metrics to pixel size. International Journal of Remote Sensing 16(18): 3585-3594

Wiens J (1989) The ecology of bird communities. Vol. 2 - Processes and variations. - Cambridge

Zheng D, Wallin DO, Hao Z (1997) Rates and patterns of landscape change between 1972 and 1988 in the Changbai Mountain area of China and North Korea. Landscape Ecology 12: 241 - 254

# 6 Scales and spatio-temporal dimensions in landscape research

Uta Steinhardt, Martin Volk

## 6.1 Introduction

The human factor 'land use' affects the interactions between water, soil, geomorphology, vegetation, etc. on several spatial and temporal scales in different manners and intensities. The implementation of strategies for sustainable land use assumes specific research concepts from the local to the global scale (micro-, meso- and macroscale). Therefore, landscape ecology science has to provide investigation methods for all these different scales. A number of papers from different scientific disciplines deal with the hierarchical organization of nature (Burns et al. 1991, O'Neill et al. 1986). The hierarchical concept was introduced into German landscape ecology by Neef (1963, 1967) and continued by several other landscape ecologists (Leser 1997). An overview of hierarchical concepts in landscape ecology is given by Klijn (1995). These concepts are mainly focused on the hypothesis, that each of the scale levels (micro-, meso- and macroscale) is characterized by specific temporal and spatial ranges. As a consequence, each scale level needs specific investigation methods as well as data layers with suitable spatio-temporal resolution on the one hand, and which provide specific knowledge on the other (Steinhardt & Volk 2000).

Due to the increased application of GIS over the past few years, this is often reduced to the spatial resolution of the data layers. This paper stresses the necessity of considering scale-specific investigation methods in landscape ecological research. In connection with this, the difficult question of regionalization will be treated. Several examples will be given of proposals for considering scales and spatio temporal dimensions in landscape research, as well as of scale-specific problems within process-oriented or structurally oriented investigations. One of the main topics is the definition of a linkage between the different scales. The authors will present a hierarchical approach, their main hypothesis beeing that the basic components for most landscape-ecological processes are similar at all scale levels. It is only the importance of the factors (and the factors themselfs) which changes for each scale and have to be defined (Helming & Frielinghaus 1999, Steinhardt & Volk 2000, Volk 1999). This hypothesis will be discussed in detail in Chap. 7.

## 6.2 Theory of geographic dimensions

The fundamental idea of a hierarchical organization of nature follows the holistic axiom, that the whole is more than the sum of all of its parts. It was first mentioned by Smuts (1926) and introduced into ecology by Egler (1942).

There are many theoretical approaches to this problem in the literature, and they will be mentioned and discussed first. Afterwards, we will direct the spotlight on a more applied approach - taking into account the hierarchy of both nature and spatial planning.

### 6.2.1 The terms 'scale' and 'dimension'

There are probably few linguistic obstacles to understanding the German approach to problems of scale and hierarchy. Many new words have been coined (e.g. Nanochore, Microchore, Macroregion), that are some times difficult to understand even for geographers and even more so for scientists from neighboring disciplines, and which are a nightmare to translate into other languages. On the other hand, landscape ecology claims to be interdisciplinary. Using terms like micro-, meso- and macroscale instead of topological, chorological, regional, and geospherical dimension in (German) landscape ecology (Fig. 6.2) would promote better acceptance and appreciation from other (bio)ecological and geosciencies. Moreover approaches in German landscape ecology would recieve attention abroad, too (Steinhard 1999). Let us tackle this issue with a specific problem of German landscape ecology: the use and definition of the terms 'dimension' and 'scale'. 'Dimension'was introduced into (German) landscape-ecological research by Ernst Neef 1963, who defined dimensions as "... scale levels bearing identical informations in relation to the contents." If a change in scale leads to a new level of geographic reality a change in geographic dimension occurs, thus enabling different information to be gained. By contrast, the term 'scale' is - especially in the English language literature - used in several contexts. Its meaning varies widely between disciplines and communities (Goodchild & Quattrochi 1997). To landscape ecologists, scale might connote 'grain', a measure of patch sizes in a landscape fragmented into discrete habitats. To a cartographer, scale is defined simply as the ratio between a distance on the map and the distance on the ground, this usage often beeing qualified as 'metric scale'. This issue is further complicated by the use of 'scale' as a basic dimension of generalization. Often generalization adds information rather than reducing it, because some kinds of geographic phenomena can only become apparent from large scale observations. But to a scientist, the representation of topography at 1:10,000 is clearly more accurate than one at 1:100,000. One effect of generalization is growing uncertainty in the representation of real phenomenon that could only be mapped perfectly at a much smaller scale[1]. However to most scientists the term 'scale' is

[1] It has to be mentioned, that there is a completely opposed understandig of "small scale" and "large scale" in German and English or American literature: German landscape ecologist and geographers use the term "scale" in terms of cartographers: So 1:100,000 is a smaller scale than 1:10,000. So *small scale* connotes to a *large area* and vice versa. English and American ecologists use the scale terms contrarily: A small scale is coupled to a small area; a large scale

likely to imply some aspect of the small linear dimension discussed above. The term should be used here as an order in the sense of spatial and temporal spheres. Depending on the specific scale level specific investigation methods need to be applied. Table 6.1 compares levels of geographic dimension and scale.

### 6.2.2 Hierarchy theory

A hierarchy can broadly be defined as 'a partial ordering' of entities (Simon 1973). Complexity frequently takes the form of hierarchy, whereby a complex system consits of interrelated subsystems that are in turn composed of their own subsystems, and so on, until the level of elementary components is reached. The choice of the lowest level in a given system depends not only on the nature of the system, but also on the research question. This corresponds to Herz's (1973) hierarchy of landscape units (Fig. 6.1). The problem of heterogeneity will be discussed in detail in Chap. 6.2.3.

In the literature of hierarchy theory, the subsystems that comprise a level are usually called 'holons' (from the Greek word *holos* = 'whole' and the suffix *on* = 'part' as in proton or neutron; coined by Koestler, 1967). The word holon has been widely adopted mainly because it conveys the idea that subsystems at each level within a hierarchy are 'Janus-faced': they act as 'wholes' when facing downwards and as 'parts' when facing upwards (Wu 1999). With respect to planning practice the scientific term 'holon' should be substituted by the more common term '(landscape) unit'. It is known, that (landscape) units - considered as subsystems at specific scale levels - can be distinguished and mapped by specific criteria.

A hierarchical system has both a vertical structure composed of levels and a horizontal structure consisting of (landscape) units. Hierarchical levels are always separated, by characteristically dominant structures and different process rates. The boundaries between levels and (landscape) units can be considered as surfaces (comparable to layers with barriers). Surfaces filter the flows of matter, energy and information crossing them, and can thus also be perceived as filters. The relationship between subsystems (units) can be distinguished by the degree of interactions among components. Thus, components interact more strongly or more frequently within than between subsystems or surfaces. These characteristics of hierarchical structure can be explained by virtue of 'loose vertical coupling', permitting distinction between levels, and 'loose horizontal coupling', allowing separation between subsystems (units) at each level (Simon, 1973). The existence of vertical and horizontal loose couplings is precisely the basis enabling complex systems to be broken down (e.g. the feasibility of a system to be disassembled into levels and units without a significant loss of information). While the word 'loose' suggests 'can be broken down', the word 'coupling' implies resistance to breakdown. In a- landscape ecological sense

to a large area. For a consistent understanding we will adopt to the English and American scientific community.

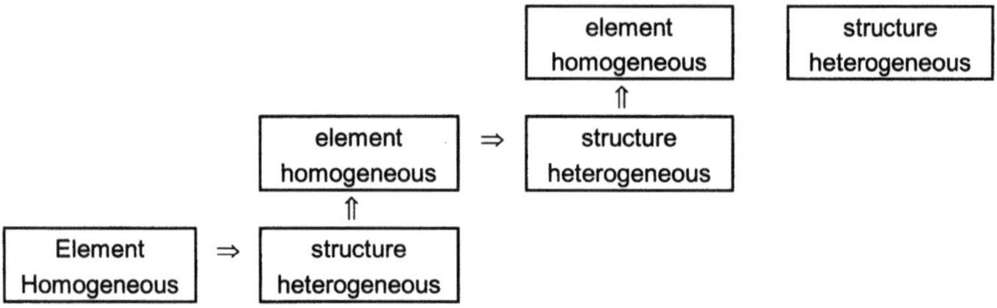


association (⇒) and transformation (⇑ abstracting, concreting)

**Fig. 6.1:** Investigation scheme concerning the hierarchy of landscape units (after Herz 1973)

breakdown means with respect to either considering specific subsystems (soil, water, climate, etc.) or to specific subsets of the earth's surface (landscape units).

According to the principle of breakdown, for a given study focusing on a particular level, constraints from higher levels are expressed as constants or boundary conditions. By contrast, the rapid dynamics at lower levels are filtered (smoothed out) and only manifested as averages or equilibrium values. For a specific problem it is not only possible but also useful to 'scale off' (Simon, 1973) relevant levels from those above and below, thus achieving greater simplification and better understanding. It appears that the tenor reflected in the statement "everything is connected to everything else" often encountered in ecological literature is ultimately unhelpful and perhaps even misleading for understanding complex systems or developing scaling theories. Evidently, for any given phenomenon in this world, some things are more closely connected than others, and most things are only negligibly interrelated with each other (Simon, 1973). Hierarchy theory suggests that when a phenomenon is studied at a particular hierarchical level (the focal level, often denoted as Level 0), the mechanistic understanding comes from the next lower level (Level -1), whereas the significance of that phenomenon can only be revealed at the next higher level (Level +1). It should be pointed out that higher level (Level +1) processes proceed slower and can be considered quasi-constant, while lower level (Level -1) system behaviour operates faster and will be integrated as a mean value. Interestingly, Baldocchi (1993) called the three adjacent scales the reductionist (Level -1), operational (Level 0), and macro (Level +1) scales, respectively. This three-level structure is sometimes referred to as the "triadic structure" of hierarchy (O'Neill, 1989). Thus, three adjacent levels or scales usually are necessary and adequate for understanding most of the behaviour of ecological systems (O'Neill 1988, 1989; Salthe, 1991). The definition and delimitation of a specific hierarchical level is an important step in the problem solution process. The scale level selected determines the main attention to be focused on a specific organizational level of the system being investigated.

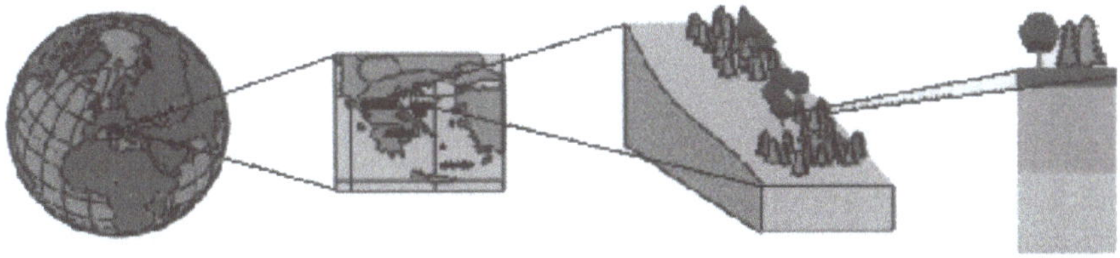


**Fig. 6.2:** Geographical Dimensions (after http://www.geog.uni-hannover.de/phygeo/trianet/Grafik/Dimensionen.html, 2000)

Starting from this point of view, we can formulate the following premises for further discussions:

1. Spatial and temporal scales are fundamentally interlinked.
2. Complex systems can be broken down in time and space simultaneously.

This is be supported empirically by the fact that many physical and ecological phenomena are arranged along the 45° line in a space-time scale diagram (Fig. 6.3).

Beside hierarchies of space and time there are hierarchies in directions of fluxes of matter and energy, and patterns of nested systems, too. All these hierarchies can be classified in two categories: hierarchies of structures and hierarchies of processes.

Patterns and processes have components that are reciprocally related, and both patterns and processes, as well as their relationship, change with scale. Different patterns and processes usually differ in the characteristic scales at which they operate. Again, this relates to the near-breakdown ability of ecological systems, and explains why they _can_ be studied at a variety of scales, and why they _have_ been studied at a variety of scales. To link patterns with processes at the same scale, or to translate them across scales, domains of scale (usually corresponding to hierarchical levels) need to be identified correctly.

The traditional approach was concentrated to a 'vertical' perspective in which a system is viewed as spatially homogeneous and, hence, the internal processes and function is highlighted. In contrast, the landscape approach is directed towards a more 'horizontal' view since it focuses on the spatial distribution of and interactions among ecological entities (Rowe 1961). The vertical perspective promotes a process- or function-based approach (e.g. ecophysiology, population and ecosystem dynamics, etc.), whereas the horizontal perspective tends to encourage a structural, pattern-oriented or geographic approach.

**Table 6.1:** Scale Levels in Geosciences

| **Scale Levels of Landscape Ecological Research** (after Neef 1967, Barsch 1975) | | **Spatial and Temporal Determined Dimensions for Describing Hydrological Processes** (after Dyck 1983) | | | | **Scale Levels in Hydrology** (after Becker 1992) | | |
|---|---|---|---|---|---|---|---|---|
| *Dimension* | | | | | | characteristic Distances | Areas | *Scale Level* |
| *Geospheric* | Geosphere Zone | | | | Climatological level | ≥ 100 km<br>30 - 100 km | ≥ $10^4$ km²<br>$10^3$ - $10^4$ km² | *Macroscale* |
| *Regionic* | Macroregion | climatic zones | | | Climatological mean values, ocean currents, large continental areas | | | |
| | Microregion | large river catchments regions | | | | | | |
| | Macrochore | | | | glacier cover, atmosphere | 10 - 30 km | $10^2$ - $10^3$ km² | |
| | Mesochore | | | hydrological level | | 1 - 10 km | 1 - $10^2$ km² | *Mesoscale* |
| *Chorological* | Microchore | catchments | | hydrology of catchments | | 0,1 - 1 km | 0,1 - 1 km² | |
| | Nanochore | small river catchments | | | | | | |
| | Geoecotoep | | hydrodynamic level | | | | | |
| *Topological* | | river section | | | | 30 - 100 m | 0,001- 0,1 km² | *Microscale* |
| | Physiotope | ground water table hydrotope | hydraulics of rivers and geohydraulics | | | ≥ 30 m | ≥ 0,001 km² | |
| | | | $10^{-6}$ $10^{-4}$ | $10^{-2}$ $10^{0}$ | $10^{2}$ $10^{4}$ | | | |
| | | | | years | | | | |

**Fig. 6.3:** Spatio-temporal hierarchies of landscape (after Wilmking 1998)

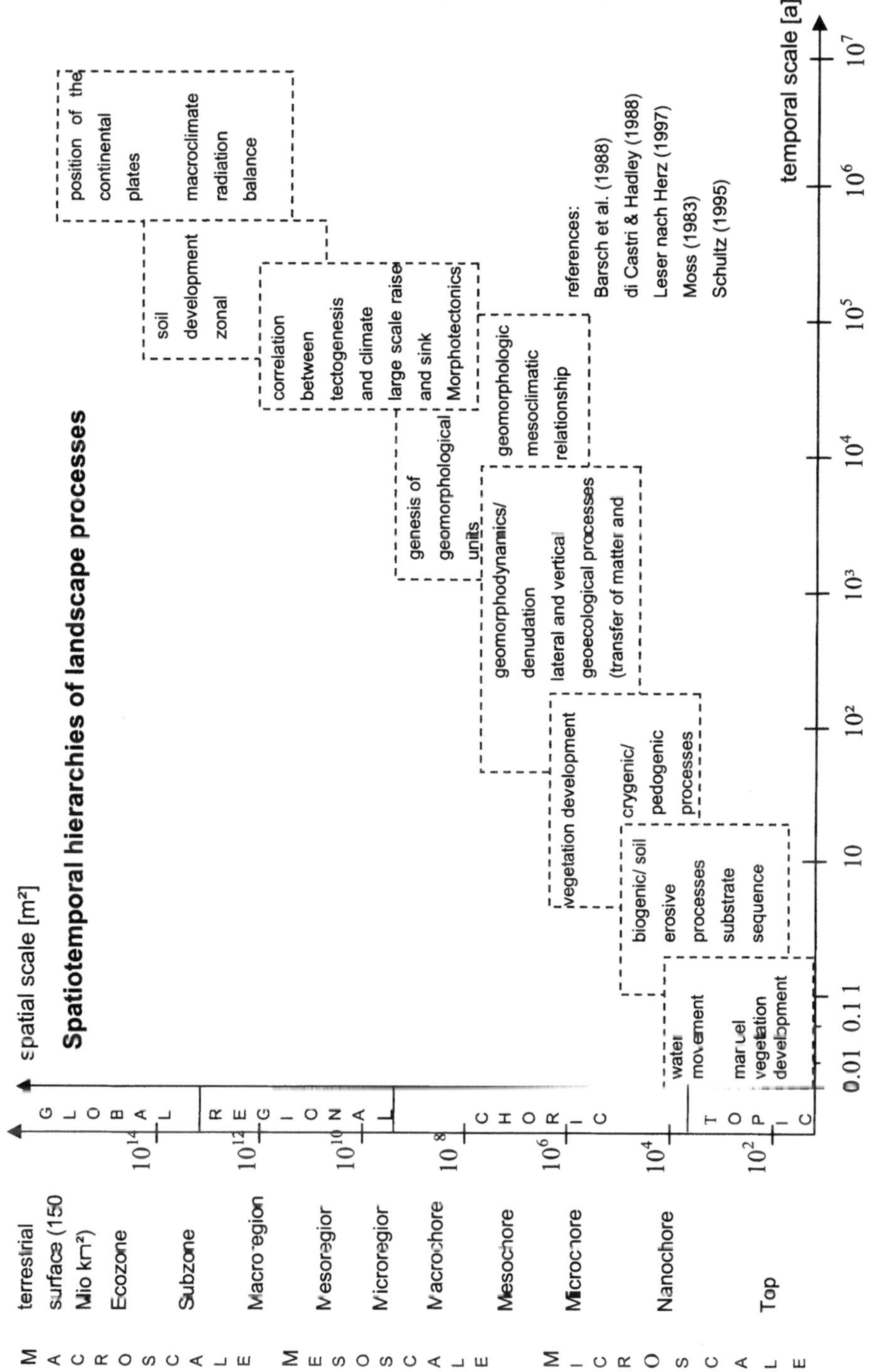

### 6.2.3 The problem of homogeneity and heterogeneity

Neef (1963) reserved the property *homogeneous* for the topological dimension. He described a top as "the basic and indivisible (landscape) unit characterized by a homogeneous combination of all features." By contrast, all other dimensions are determined by a heterogeneity. However, this approach to homogeneity and heterogeneity needs to be reconsidered. Kolasa and Picket (1992) gave a more conceptual definition of heterogeneity: "A system is heterogeneous in time and/or space if a specific temporal interval and/or different locations is characterized by different values." This definition implies that heterogeneity can be achieved at each temporal and spatial level consideration. Herz (1973) also contributed to overcoming the rigid separation between homogeneity and heterogeneity. According to his theoretical studies, homogeneity and heterogeneity can be used to define the 'border' between individual levels of hierarchy: Taking into account the synthetic aspects of transformation and association homogeneity can be attained at each dimension level.

In addition to the main features of geosystems (connected to the earth's surface, high degree of complexity, and spatiotemporal differentiation (Neef 1967), Herz (1984) also developed the principles of landscape ecology (areal-structural principles) taken up by other authors (Bailey 1996):

1. The *principle of correlation* describes the existence of correlative coherences between the partial complexes of the landscape and results in the vertical structure of the landscape (Fig. 6.4).
2. The *principle of areality* assumes that all of these feature combinations are spatially bounded. Taking into account that landscape boundaries are boundaries in a continuum with the character of hemlines (ecotones) and do not isolate parts of the earth's surface, the horizontal or lateral structure of the landscape can be characterized (Fig. 6.5).
3. Following the *principle of polarity* we can observe a constitutional neighborhood between units of the same hierarchical level as well as their dynamic coupling (similarity or contradistinction) - hence the resulting source-sink relations (Fig. 6.6a,b).

Last but not least, the principle of hierarchy allows the delimited units to be classified or subdivided.

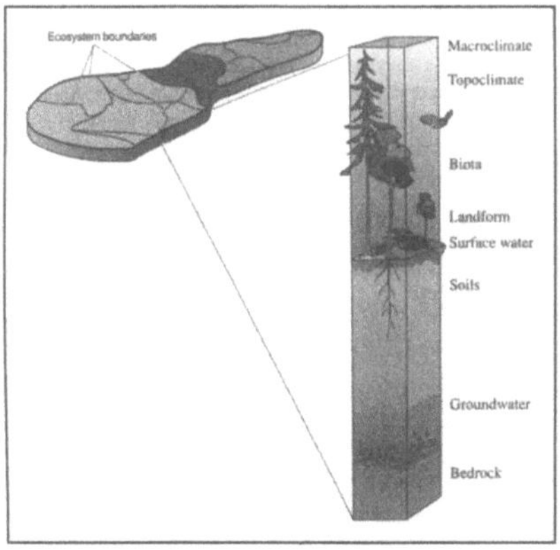


**Fig. 6.4:** Vertical landscape structure

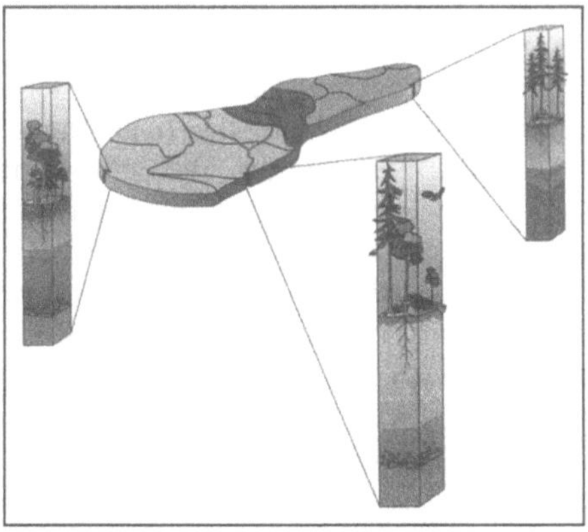

**Fig. 6.5:** Horizontal landscape structure (boundaries / ecotones)

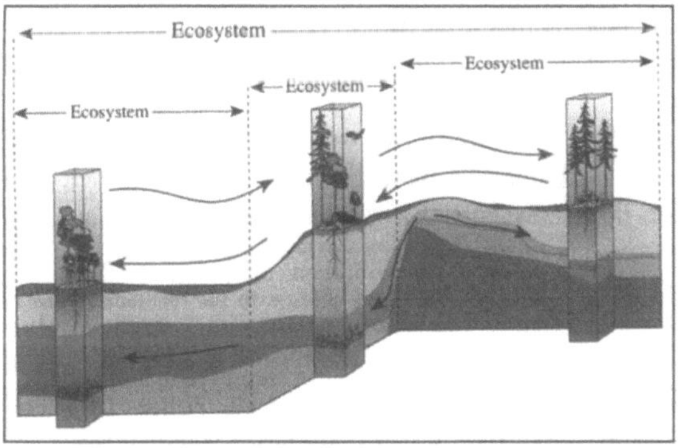


**Fig. 6.6a:** Permeable landscape

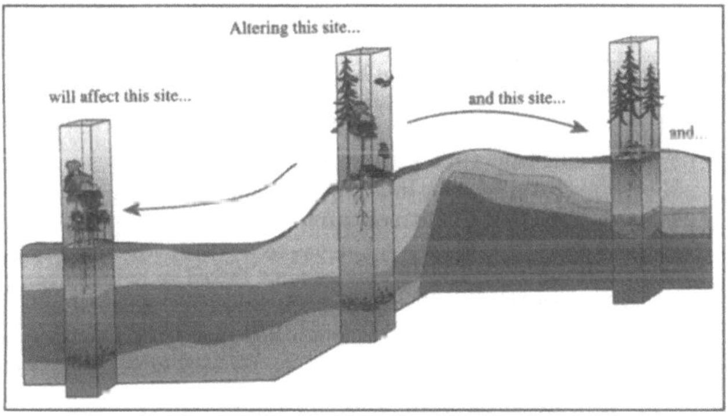


**Fig. 6.6b:** Sorce-sink effects boundaries (all figures from Bayley 1996)

### 6.2.4 Landscape heterogeneity and change

Spatial heterogeneity within ecosystems and among ecosystems arrayed on a landscape is critical to the functioning of individual ecosystems and of entire landscapes. For example, the patterns and distribution of plants within arid and semiarid ecosystems control patterns of nutrient cycling processes, with the highest accumulations of organic matter and the highest rates of nutrient cycling occurring under plants rather than in open spaces. Similarly, the configuration of ecosystems within watersheds and at even coarser scales determine the transfers and processes occurring at the landscape scale. For example, the adjacency of riparian systems to upland agricultural systems may prevent nitrate movement from the terrestrial watershed to downstream systems (Rau 1998, Steinhardt & Volk 2000). Furthermore, the degree to which a landscape is fragmented into smaller units determines variation in abundance and diversity of animals. All of the processes and mechanisms that operate in ecosystems have specific spatial dimensions. Knowledge of the physical, biological and ecological sources of this variation and of the resulting spatial patterns are essential in order to understand how both ecosystems and whole landscapes function. Additionally, it forms an important basis for predicting how ecosystems change temporally with natural and anthropogenic alterations.

A great variety of landscape metrics is available for describing and quantifying the degree of heterogeneity. Chap. 5 discussed how to apply these metrics and how to relate the metrics based on landscape structures to landscape processes - to use them as indicators for processes. The size, distribution and connectivity of the patches are quantifiable attributes of landscapes that provide a basis for evaluating change over time and space.

The concept of landscape, as used in ecology, considers a landscape as an ecological system comprising recognizable components such as managed forest patches, agricultural fields, human settlements and natural ecosystems. All landscapes can be thought of as mosaics, composed of discrete, bounded patches that have a distinct biotic structure or composition, and which in some cases are embedded in a predominant and more continuous cover-type matrix. Ecologists working at landscape scales are confronted with spatial heterogeneity both within and among patches of the landscape, and their research topics focus on both the interactions that take place among the units or patches on the landscape, and the behaviour and functioning of the landscape as a whole. Spatial heterogeneity on the landscape stems from both natural and human-caused disturbances, the successional status of vegetation communities, human land uses and land management, variations in state factors and environmental resources, and other anthropogenic influences such as the invasion of non-native species.

For instance, one concept for understanding the variation among soils and landscape is the "state factor" model by Jenny (1941), in which parent material, topography, climate, time and biota are all viewed as independently varying factors that exert control over soil development and ecosystem properties and processes. It is the interaction of these state factors with each other that underlies the formation of the very different types of landscapes existing today and have existed in the past. However, these state factors alone do not control the spatial array of the world's landscapes. Instead they determine the natural matrix upon which natural and human caused disturbances and land uses are overlain.

While natural disturbances have always been a force for spatial and temporal change in ecosystems, human activities have added another and increasingly overwhelming layer of change to the matrix of natural spatial variation. Half of the ice-free terrestrial surface has been transformed in some way by human activity (Turner & Gardner 1990). Human-dominated landscapes supply tremendous amounts of food, fibre and other landscape services to human populations. They have also changed the conditions of the landscape processes related to energy, nutrients and water, and have thereby left their mark on landscape-scale interactions and the earth system as a whole.

Concerning the example of intensive agriculture, it should be mentioned here that all of these land use changes result in significant alterations in the way ecosystems function, the way patches on the landscape affect each other, and the way landscapes as a whole function:

The majority of agricultural land use involves tillage as part of the management system, i.e. the ploughing of soil on a regular basis. During the last few decades, potential agricultural yield per unit of farmland has increased substantially, primarily through the development and use of high-yield crop varieties combined with industrially produced fertilizers, pesticides, herbicides and irrigation. This transition in cultivation methods represents a major shift in the way humans have traditionally practised agriculture. The combination of tillage, new cultivars and intensive inputs into agricultural systems has been largely responsible for keeping food production in step with the rapid human population growth of the last several decades. However, the practice of industrial agriculture carries significant consequences for local, regional and global environments (Matson & Boone 1984). Globally, organic matter stored in the soil profile represents a larger carbon pool than both the biota and the atmosphere together. Tillage disrupts the physical structure of the soil and exposes organic matter that is normally physically protected by its aggregation with soil minerals, thereby making it available to microbial decomposition. This physical change, along with the alteration of soil microclimate, results in faster decomposition and greater potential for erosion. Thus, carbon stored as soil organic matter decreases when subjected to tillage. Modern intensive agriculture also plays a significant role in the biogeochemical cycle of nitrogen. Because grain is harvested and removed from the fields yearly, agriculture depends on the regular input of nitrogen to maintain yields. Since 1945, industrial nitrogen fertilizers have largely supplanted organic nitrogen applied as animal manure or supplied by nitrogen fixation by organisms (such as leguminous plants) as a nitrogen source to crops. The increasing use of nitrogen fertilizers in all its forms has consequences for water and air quality as well as downwind and downstream ecosystems. Numerous local and regional studies have measured elevated nitrate concentrations in systems adjacent to intensive agricultural areas with consequences for both human health and for the functioning of ecosystems. The use of nitrogen fertilizers also influences atmospheric processes through the emission of a number of trace gases. Human-induced alteration in the landscape has become an overwhelming force of change. Land use changes are critical in terms of local scale consequences for biological diversity, air and water quality and other landscape services. At the regional and global scale, land use changes influence atmospheric composition and chemistry - and ultimately climate. At the same time, many land use changes are carried out to meet pressing human needs. These interconnecting causes and consequences call for integrative approaches to conservation, habitat protection, and land use planning that recognize the multiple and interacting roles of the landscape.

Spatial mosaics not only result in the variation and heterogeneity of the land surface cover and in landscape processes within the patch, but also determine the ways in which parts of the landscape mutually influence each other. There are three general pathways by which landscapes interact: via topographically controlled interactions (e.g. the topographically controlled redistribution of materials in water and via erosion); via transfers through the atmosphere (nitrogen and sulphur transfers, biomass combustion, dust transport); and via biotic transfers (movement of plants and animals). Some of these aspects will be discussed in detail in Chap. 7.

## 6.3 Scales and dimensions in landscape ecology

The whole universe can be considered as an organization consisting of a system hierarchy. Each higher level is constructed from systems of lower hierarchy levels. Lower components are relatively dependent on those above and vice versa any reverse influence by the lower components on the upper cannot be neglected either. All natural processes have their own scale domains, our observations are scale-specific, our interpretations depend on scales, and decisions relate to a certain time frame and spatial context. If we neglect scales, we draw wrong conclusions and take shortsighted decisions (Klijn 1995).

When ascending a scaling ladder, the loss of detailed information is compensated for by a gain in overview information about structures, relationships and interactions. Each scale level offers its own cognition facilities. Climbing up and down the scaling ladder reveals completely new insights into structural and functional phenomena which otherwise remain hidden. Even Troll (1939) said about archaeology, "On the ground we are standing in front of a clutter of lines that can only be arranged into a system from a height." (p. 250f) This view goes for modern landscape ecology, too: different landscape patterns and processes can be made out depending on whether the earth is observed from the ground, an aeroplane or a satellite.

Studies in landscape ecology, hydrology, meteorology and other related earth sciences have shown that different processes tend to dominate in distinctive, characteristic domains of scale in time and space. Thus, observations made on a single scale can, at best, only capture those patterns and processes pertinent to that scale of observation. The situation inevitably becomes complex when a description or explanation simultaneously invokes multiple levels of organization or domains of scale. While the issue of scaling has been widely recognized as essential in both basic and applied research, a general theory of scaling is still elusive due to the complex matter of scaling.

The discussion about the terms 'scale' and 'dimension' mentioned in Chap. 6.2.1. was accompanied - especially in Germany - by several proposals to indicate specific scale levels resp. dimensions (see overview in Leser 1991, 202pp). Unfortunately, some of the contributions only focus on a formal completion of the system of landscape units and there is no contribution on the development of methods. Table 6.2 gives an overview of some nomenclatura proposals and the related mapping scales and area of the adequate units. It should be mentioned here that this suggestion defining each scale level with a certain area size and mapping scale is ventured because the extent of the units also depends on the general structure (variety, diversity) of the landscape, which is why the extent of landscape units also differs between for example arid desert and temperate zone. Fig. 6.6 gives another example of landscape mapping at different scales. Bailey (1976, 1983, 1995) developed a technique for mapping ecoregions first for the United States that was subsequently expanded to include the rest of North America (Bailey and Cushwa 1981) and the world (Bailey 1989).

**Fig. 6.7:** Hierarchical landscape mapping on the example of Alaska (from Bailey 1995)

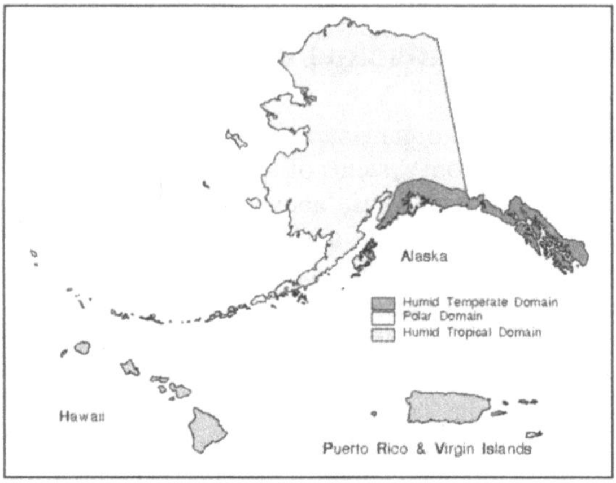


a) Ecosystem Domains

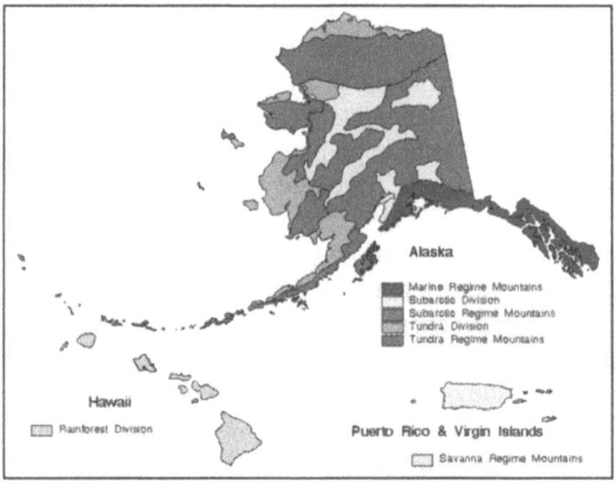


b) Ecosystem Divisions

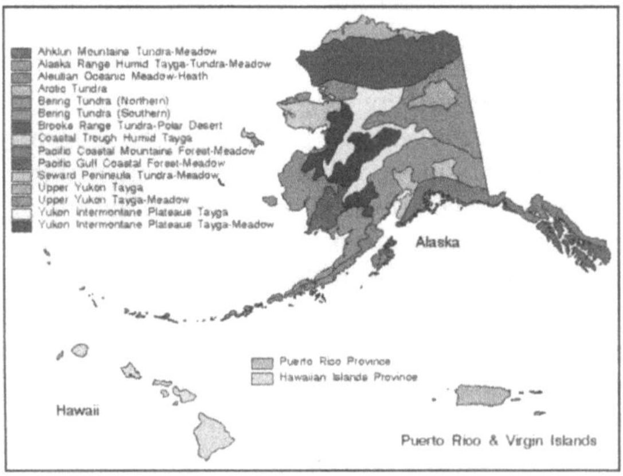


c) Ecosystem Provinces

**Table 6.2:** Different nomenclature proposals for hierarchical ecosystem classification

| Barsch 1975 | F. Klijn 1997 | indicative mapping scale | basic mapping unit |
|---|---|---|---|
| Zone | Ecozone | 1:50 Mio and smaller | >62,500 $km^2$ |
| Makroregion | Ecoprovince | 1:10 ... 50 Mio | 2,500 - 62,500 $km^2$ |
| Subregion | Ecoregion | 1:2 ... 10 Mio | 100 - 2,500 $km^2$ |
| Mikroregion | Ecodistrict | 1:500,000 ... 2 Mio | 625 - 10,000 ha |
| Mesochore | Ecosection | 1:100,000 ... 500,000 | 25 - 625 ha |
| Mikrochore | Ecoseries | 1:25,000 ... 100,000 | 1.5 - 25 ha |
| Nanochore | Ecotope | 1:5,000 ... 25,000 | 0.25 - 1.5 ha |
| Top | Eco-Element | 1:5,000 and larger | < 0.25 ha |

The regions delineated on the map were adopted for use in ecosystem management and are also used in the proposed National Interagency Ecoregion-based Ecological Assessments.

Each of these scale levels is characterized by specific temporal and spatial ranges and has to be investigated with specific methods (measuring, observation, mapping, modelling). Consequently the temporal and spatial resolution of all data collected or used has to match the scale to which it is to be applied. In reality, data are rarely available in the resolution required (Volk & Steinhardt 1998), which is why we have to tackle data homogenization and data or parameter transfer. Besides the problem of scale-appropriate data, another problem is scale-appropriate investigation methods. What methods of observation are appropriate to what scale level for data collection and what methods should be used to analyse and process them (Table 6.3) must be defined.

With respect to the above discussed problems of patterns and processes as well as scale-specific degrees of complexity, it must be emphasized that landscape ecology cannot be based exclusively on one type of hierarchy. According to Klijn (1995) we have to focus on:

**Table 6.3:** Complexity of ecological systems and hierarchy theory and appropriated methods

| Scales | Dimensions | Ranges of complexity | Methods of data gathering (selection) |
|---|---|---|---|
| Macroscale | regionic - global | disorganized complexity (to be dealt with statistical methods) | remote sensing techniques |
| Mesoscale | chorological - regionic | organized complexity (quantitative methods are lacking) | combination of fine and coarse scale methods |
| Microscale | topological (local) - chorological | organized simplicity (to be dealt with analytical mathematics) | point measurements, field mapping |

- Process-functional hierarchies based upon flow directions (relative position of systems in flows of energy, matter and information; ranking according to dependence on other systems)
- Hierarchies in complexity or organizational properties
- Temporal and spatial hierarchies

## 6.4 Regionalization in landscape ecology

The discussions concerning scale and dimension in landscape ecology resulted in division into specific hierarchical levels, known as micro-, meso- and macroscale. However, this delimitation must not be allowed to lead to the splitting of landscapes into stand-alone hierarchical elements. Despite this hierarchical structuring, a landscape has to be considered as a coherent unit, and so there must be connectivity between all the specific scales. The concept of regionalization can help solve this problem. It can bridge the gap between the ideas of scale-specific and cross-scale approaches. All hierarchical components can be assembled in this way (rather like the pieces of a jigsaw puzzle) to form a consistent landscape. To scale up from a leaf to a continent and beyond, we must understand how information is transferred from a fine to a broad scale and vice versa. We must learn how to aggregate and simplify, retaining essential information without getting bogged down in unnecessary details. Steinhardt & Volk (1999) edited a book about regionalization in landscape ecology which includes papers presented at an interdisciplinary conference in Germany in 1998. It shows the general necessity of regionalization methods and gives an overview of the variety of landscape-oriented research. Yet it brings home the fact that we still seem to be far away from the development of standards (assuming they are possible in the first place).

Initial ideas for the realization of transitions from one scale to another have been developed in hydrology. These transitions have been termed "regionalization" (Kleeberg 1992, 1998). It should be pointed out that this term is not related to the above-mentioned regional dimension. This additional level of dimensions was not introduced into German landscape ecology until 1973 by Haase. 'Region' in general refers to widespread areas. Although the term 'regionalization' has since come to be used in several disciplines, each discipline has its own narrow understanding of the word, with some interpretations actually being contradictory.

Bach & Frede (1999) launched a new methodological discussion concerning a general definition of regionalization and the development of regionalization strategies. At first sight this discussion seems to be very theoretical, but the more or less formal approach meets the requirements of an interdisciplinary research approach. Thus from a theoretical point of view, all data are characterized by three attributes: object (e.g. soil, climate, vegetation), feature (e.g. grain size, field capacity, mean annual air temperature) and scale (micro-, meso-, macroscale). Usually data have to be transferred to other objects, features and scales. This procedure of data transmission is defined as 'regionalization'. Depending on which of the three attributes of an existing primary data set is to be regionalized, three fundamental operations of transmission can be distinguished:

| | |
|---|---|
| **Translocation**: | The same feature is transferred from one object to *other objects* of the same class of objects on the same scale. |
| **Transformation**: | From one or more features for one object, *other features* for the same object on the same scale are derived. |
| **Up- or downscaling**: | The same feature is transferred from the objects on one scale to the same object on *another scale*. |

Fig. 6.8 shows the three transmission operations graphically. Accordingly, regionalization means the change of data $D_{i,j,k}$ in one or more of its attributes. All the three operations can be performed separately or combined. For this purpose, different methods are available or have to be developed. All transmission rules have to describe the way in which the destination data are generated from the primary data.

The implementation of these operations is currently limited by the rules available in the specific fields of landscape ecology. This must be one of the main tasks to be solved by several geosciences over the next years. Some of the existing transmission rules are mentioned below.

To answer the question: "What about the spatial validity of a data set measured at one point (e.g. different climatic parameters)?", some geostatistical approaches such as Thiessen polygons, kriging or the construction of isobars, isotherms, etc. are already available (Burrough 1986, Oliver 1990, Fohrer et al. 1999). Hence *translocation* is a resolvable problem.

The problem of *transformation* seems to be more difficult. Indicators and transfer functions have to enable new properties to be derived from measured/mapped data. Examples in this field include sediment ratio delivery (SDR) in the field of geomorphology (Hairston 1995), the unit hydrograph in hydrology (Sherman 1932) and pedo-transfer functions in soil sciences (Tietje & Tapkenhinrichs 1993).

The change of scale (*up-/downscaling*) is related to problems of aggregation or disaggregation of data. In this regard, the transmission of runoff data measured at the outlet of a watershed up to the whole watershed is an example of upscaling procedures (Fohrer et al. 1999). Sometimes it is necessary to downscale statistical data ascertained for an administrative unit (e.g. federal state) to the lower district level.

Regionalization is the key concept for reaching a compromise solution between scale-specific and cross-scale investigations. Scale-specific investigations have to be applied in the core areas of the different scale levels (mentioned in Table 6.1), and for the transition zones between the specific levels a cross-scale approach is necessary. Thus, a connection between the separate hierarchical levels and hence an uninterrupted systematical reflection can be implemented.

The problem of scale transfer was not realized a few years ago. The reason is that it occurs mainly in the face of heterogeneity (Bierkens et al. 2000). Despite (or maybe even because of) its recent emergence, research into scale transfer in environmental science has led to an enormous amount of different methods and approaches to upscaling and downscaling information. This makes it difficult for a practitioner to see where transfer occurs in the various steps of a research project scale, and what methods of scale transfer are available and should preferably be used for these cases. Bierkens et al. (2000) describe a number of available methods (Table 6.4) integrated into a decision support system for practitioners.

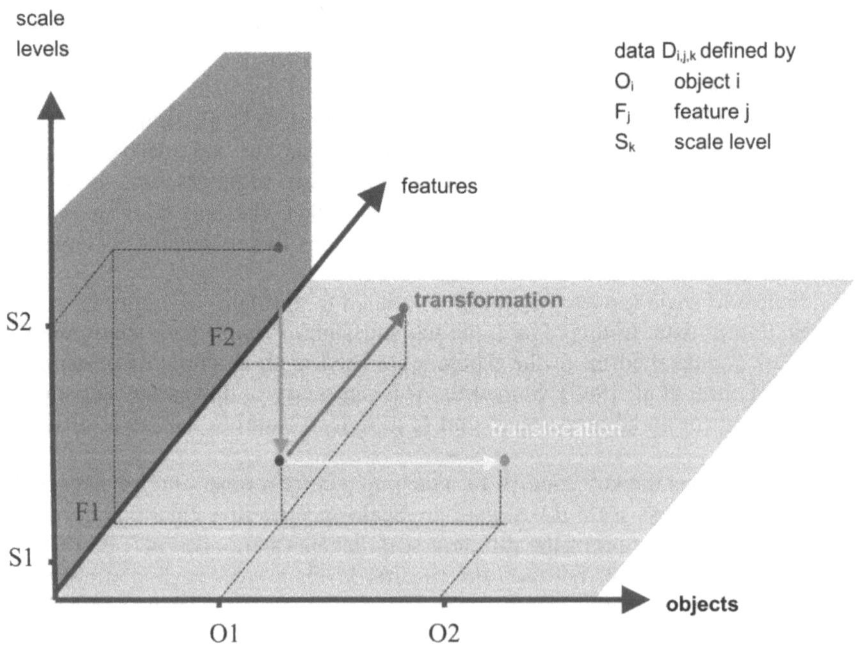


**Fig. 6.8:** Fundamental operations of regionalization

**Table 6.4:** Classification of upscaling and downscaling methods (after Bierkens et al. 2000)

| | | |
|---|---|---|
| UPSCALING | averaging of observations or output variables | - Exhaustive information (e.g. upscaling of measured daily precipitation to average precipitation over a decade)<br>- Design-based methods (e.g. averaging model output parameters to larger map units)<br>- Geostatistical prediction (e.g. block kriging)<br>- Deterministic functions (e.g. delineation of influence zones around sample locations by Thiessen polygons)<br>- Combinations and auxiliary information (e.g. stratified block kriging) |
| UPSCALING | finding representative parameters or input variables | - Exhaustive information (e.g. finding representative hydraulic conductivity for numerical block models)<br>- Deterministic functions (e.g. spline interpolation, inverse squared distance weighting)<br>- Indirect stochastic methods (e.g. estimating the statistics of the stochastic function from a limited number of observations)<br>- Direct stochastic methods (e.g. estimating statistical properties (e.g. mean, covariance function) from the observation)<br>- Inverse modelling (e.g. finding representative parameters) |
| UPSCALING | averaging of model equations | - Deterministic: temporal or volume averaging (e.g. estimating the uptake of the whole root system by averaging the uptake simulated for one root)<br>- Stochastic: ensemble averaging (e.g. one-dimensional steady-state groundwater flow in a heterogeneous porous medium) |
| UPSCALING | model simplification | - Lumped conceptual modelling (no standard solutions exist)<br>- Meta-modelling (e.g. calibrating parameters for a black-box model through regression) |
| DOWNSCALING | empirical functions | - Deterministic functions (e.g. splines, linear functions, general additive models)<br>- Conditional stochastic functions (e.g. using stochastic wavelets)<br>- Unconditional stochastic functions |
| DOWNSCALING | mechanistic models | - Deterministic functions (e.g. adjusting parameter values or boundary conditions of mechanistic models)<br>- Conditional stochastic functions (e.g. constructing equally probable realizations of a stochastic function by adding a noise component)<br>- Unconditional stochastic functions |
| DOWNSCALING | fine scale auxilary information | - Deterministic functions (e.g. determining the fine-scale variability of water storage in a sloping landscape using fine-scale topographic data and a value for the over-all water storage)<br>- Conditional stochastic functions (incorporating the ensemble of equiprobable functions instead of only one deterministic function)<br>- Unconditional stochastic functions (e.g. determining the probability density function at the detailed scale directly from the coarser scale) |

## 6.5 Examples for cross-scale and scale-specific investigations

Watersheds are affected by uncertain and complex interactive environmental and socio-poltical trends, some of local and many of regional or even global origin. Our ability to sustainably manage our natural resources is presently still constrained by our lack of knowledge about the hydrologic cycle and its relationship to the geosphere and the biosphere. However, water serves - at least in the temperate climates - as the most important carrying medium of all transport processes (cf. Chap. 7). Hence the watershed approach to the long-term research and monitoring of areas characterized by a high intensity of land use provides important data on ecosystem processes and interactions for detecting both spatial and temporal change in management practices as well as in environmental conditions. Watershed ecosystem studies are based on the collection of long-term data sets of the ecosystem conditions. At this spatial level, research and monitoring contribute to the accumulation of important baseline information on deposition, meteorology, hydrology, ecosystem functioning and land use. The data collections allow the partitioning of cause and effect relationships of ecological and management changes within watersheds. Currently, the investigations are focused on developing, testing and implementing state-of-the-art methods and procedures for application to improve water and land resource management at both the local and regional levels.

The quantification of the hydrologic cycle and chemical fluxes are the major objectives of the watershed programme. Such measurements, when combined with other geographic resources data (e.g. geology, land use, topography, historic and prehistoric records), permit a better understanding of ecosystem-level processes and how watershed ecosystems respond to various natural and human-induced stimuli. During the initial years a core set of variables to be monitored was defined, and sampling and database methods were established. Variables included precipitation, climate, vegetation, soils, hydrology and management practices.

Our integrated approach combines research, inventory, and monitoring within a focused programme for the collection of these data needed to test hypotheses regarding the contribution of human-induced stress to long-term ecological change within agricultural landscapes. To document the relationships between ecosystem effects and anthropogenic influences, long-term monitoring and research are essential. The existence of sites with a commitment to gathering long-term ecosystem level data permits research activities aimed at testing hypotheses relevant to ecosystem processes and structure.

*Combination of top-down and bottom-up approaches*

Combining 'top-down' and 'bottom-up' approaches in addition to coupled GIS-model applications and traditional classification and assessment methods appears to provide a way of investigating landscape-ecological structures and processes. Top-down approaches include the use of the inquiry function of the GIS to detect areas that can be defined as potential risk zones with vertical/horizontal material (and nutrient) leaching from agricultural areas. Assessment on this scale level is a rough filter that provides background information and identifies the properties for subsequent analysis. GIS can be used as a powerful tool to provide a process-based landscape typification. The outcome is units with similar conditions, characterized not only by their specific combination of structural components (grain size, slope angle, biotope type, etc.) but by the dominant processes (overland flow, macropore fluxes, percolation, interception, etc.). These process-based units can further indicate the dominant process direction (lateral - vertical) and determine the neighbourhood effects (either the adjacent or the upper/lower layer).

Going in the other direction, in a bottom-up approach the risk areas identified have to be investigated in detail - with other models and a database with a higher spatiotemporal resolution. For this purpose the scales have to be changed. Vertical and lateral water, material and energy fluxes from the designated risk areas can be qualified and quantified. By using a nested approach in small test areas, indicators for sustainable land use systems can subsequently be identified that can be applied to larger areas afterwards.

Examples of specific applications of 'top-down' (balancing - modeling - typifying) and 'bottom-up' (measuring - mapping - modelling) are discussed in Chap. 7.

## 6.6 Discussion and conclusion

Based on the fundamental theory of scales and dimensions in landscape ecology, our research components will integrate a series of analyses and assessments designed to create a rigorous context for decision making. We will apply quantitative tools and information systems as well as traditional methods to enable critical interpretation of the uncertainty associated with decisions about future alternatives. We will try to employ and combine three major research approaches and combined steps:

(1) Characterizing landscape status and changes (Chap. 3, 4 and 7);
(2) Identifying and understanding critical processes (Chap. 7); and
(3) Evaluating outcomes (Chap. 8, 9 and 10).

*(1) Characterizing status and change*

Research will assess trajectories of change from now through alternative future scenarios. If possible, the investigations should be extended to historical situations (retrospectively). The approach should be guided by initial assessments which will influence future research and environmental management. Assessment approaches should:

- Describe historical change;
- Describe current condition and function;
- Identify biophysical and socioeconomic processes and functions that constrain possible future ecosystem trajectories;
- Characterize the level of rigour and the uncertainty and unknowns in the assessment.

*(2) Identifying and understanding critical processes*

We try to select and apply a set of conceptual, quantitative and evaluative models to identify and analyse critical anthropogenic and non-anthropogenic processes related to ecosystems. This phase of research will:

- Identify critical ecological (biotic), environmental (physical and chemical) and socioeconomic (individual, domestic and institutional) influences on ecosystem structure and function;
- Select indicators in each process category that quantify the magnitude of response to these influences on ecosystems;
- Integrate quantitative tools, information systems and qualitative understanding to describe system responses both within and across process categories.

*(3) Evaluating outcomes*

We will evaluate and illustrate the social and ecological consequences of potential management practices for future landscape conditions. Furthermore, we will describe sources and levels of uncertainty and define likely boundaries of ecosystem and socioeconomic trajectories. This phase of work will entail:

- Creating alternative futures that illustrate the major strategic choices and explicitly identify the likelihood and advantage of relevant choices;
- Evaluating the consequences of alternative futures on critical anthropogenic and non-anthropogenic processes including the characterization of risks, technical limitations, scientific uncertainty and public response to alternative futures;
- Using different forms to present the results of these evaluations.

## 6.7 Outlook

Despite all the advantages of hierarchical theory, as well as of its application in landscape-related sciences and spatial planning, there are still many unresolved issues. Therefore, future research must focus on two directions. Firstly, a general theory of hierarchy that is acknowledged through all disciplines of geosciences or landscape-related research must be established. This includes a common use and understanding of terms related to scale and hierarchy, as well as the consideration and application of this theoretical background to the problem to be solved. Secondly, the great (technical) progress made in the development of tools for landscape analysis (GIS, remote sensing, modelling) must be critically examined. All these tools are often applied without respect to scale-related questions.

Bearing these problems, the following questions have to be solved:

- What properties of physical and human systems are invariant with respect to scale?
- What kinds of transformation of scale are available to aggregate or disaggregate data in ways that are logical, rigorous, and well-defined in theory?
- Is it possible to implement methods which assess the impact of scale through measures of information loss or gain, for example?

- How is the observation of processes affected by changes of scale, and how can we measure the degree to which processes are manifested at different scales?
- How is scale represented in the parametrization of process models, and how are models affected by the use of data from inappropriate scales?
- What is the potential for integrated tools to support multiscale databases, and associated modelling and analysis?
- What are the problems that must be resolved when integrating data from different scales?

Especially with respect to the inseparable coherence between nature and society, the following questions need to be answered:

- How can we *measure changes* in ecosystems across scales from individual sites to large basins or regions?
- What is the *current state* of ecological resources within a given region?
- How do *natural* patterns and processes of landscapes or ecosystems *interact with anthropogenic* patterns and processes?
- What *types of interactions* are consistent and what types are contradictory?
- What critical yardsticks are there for comparing and contrasting various alternative *future scenarios* (e.g. biological/ecological, economic, climate/hydrologic, demographic)?
- What *indicators of climatic/hydrologic/geomorphic processes* or components are most useful, meaningful and tractable for describing the historical condition, current status, and alternative futures across multiple spatial scales in a certain area?
- What *indicators of demographic and economic processes* or components are most useful, meaningful and tractable for describing the historical condition, current status and alternative futures across multiple spatial scales in a certain area?
- What *environmental management options* are available to alter future ecosystem conditions across a range of spatial scales?
- How can natural processes and human programmes be used to *maintain or restore ecosystem* processes and patterns?
- What *fundamental limits* govern the achievement of ecosystem management objectives?
- How can human efforts be designed to *enhance natural processes that restore ecosystems* and recognize the ecological benefits of future disturbances?

All the above tasks have to be solved with respect to spatial planning which is also organized hierarchically (Table 6.5., Chap. 9).

**Table 6.5.** Hierarchies in landscape and spatial planning in Germany (after Kiemstedt et al. 1997)

| Scale level | Planning level | Spatial planning | Landscape planning |
|---|---|---|---|
| Macroscale | Country | Spatial development policy | - |
| Mesoscale | Federal state | *Raumordnungsprogramm* | *Landschaftsprogramm* |
| | Region | *Regionalplan* | *Landschaftsrahmenplan* |
| Microscale | Municipality (town, village) | *Flächennutzungsplan* | *Landschaftsplan* |
| | Parts of municipalities | *Bebauungsplan* | *Grünordnungsplan* |

## 6.8 References


Bach M, Frede HG (1999) Regionalisierung als methodische Aufgabe. In: Steinhardt U, Volk M (Hrsg) Regionalisierung in der Landschaftsökologie - Forschung, Planung Praxis. Teubner, Leipzig-Stuttgart

Baldocchi DD (1993) Scaling water vapor and carbon dioxide exchange from leaves to a canopy: Rules and tools. In: Ehleringer JR, Field CB (eds): Scaling Physiological Processes: Leaf to Globe. Academic Press, San Diego, pp. 77-114

Barsch H (1975) Zur Kennzeichnung der Erdhülle und ihrer räumlichen Gliederung in der landschaftskundlichen Terminologie. Petermanns Geogr. Mitt. 119:

Barsch H, Billwitz K, Reuter B (1988) Einführung in die Landschaftsökologie. (Manuskript) Potsdam

Bailey RG (1976) Ecoregions of the United States (map1:7,500,000). Ogden, Utah: USDA Forest Service, Intermountain Region

Bailey RG (1983) Delineation of ecosystem regions. Environmental Management 7: 365-373

Bailey RG (1989) Explanatory supplement to ecoregions map of the continents (with separate map at 1:30,000,000). Environmental Conservation 16: 307-309

Bailey RG (1995) Description of the Ecoregions of the United States. http://www.fs.fed.us/land/ecosysmgmt/ecoreg1_home.html (01.11.2000)

Bailey RG (1996) Ecosystem Geography. Springer, New York

Bailey RG, Cushwa CT (1981) Ecoregions of North America (map 1:12,000,000). U.S. Fish and Wildlife Service, Washington, DC

Bierkens, MFP, Finke PA, de Willigen P (eds) (2000) Upscaling and Downscaling. Methods for Environmental Research. Kluwer Academic Publishers, Dordrecht

Burns TP, Pattan BC, Higashi H (1991) Hierarchical Evolution in Ecological Networks. In: Higashi, H. & T.P. Burns (eds): Theoretical Studies of Ecosystems: The Network Perspective. - New York

Burrough, P (1986) Principles of Geographical Information Systems for Land Ressources Assessment. Oxford University Press, New York

Di Castri F, Hadley M (1988) Enhancing the credibility of ecology: interacting along and across hierarchical scales. GeoJournal 17: 5-35

Dyck S (1983) Angewandte Hydrologie. Verlag für Bauwesen, Berlin

Egler FE (1942) Vegetation as an object of study. Philos. Sci. 9: 245-260

Fohrer N, Göbel S, Haverkamp P, Bastian P, Frede HG (1999): Regionalisierungsansätze bei der GIS-gestützten Modellierung des Landschaftswasserhaushaltes. In: Steinhardt U, Volk M (Hrsg) Regionalisierung in der Landschaftsökologie - Forschung, Planung Praxis. Teubner, Leipzig-Stuttgart

Goodchild MF, Quattrochi DA (1997) Scale, Multiscaling, Remote Sensing, and GIS. In: Quattrochi D A, Goodchild M F (eds) Scale in Remote Sensing and GIS. Lewis Publishers Boca Raton New York London Tokyo, pp 1 -11

Haase G (1973) Zur Ausgliederung von Raumeinheiten in der chorischen und regionischen Dimension - dargestellt an Beispielen aus der Bodengeographie. Petermanns Geogr. Mitt. 117:81-90

Hairston JA (1995) Soil Management to Protect Water Quality. Estimating Soil Erosion Losses And Sediment Delivery Ratios, Alabama Cooperative Extension Service, Auburn University, Alabama (http://hermes.ecn.purdue.edu/cgi/convertwq?7778, 09.11.2000)

Helming K, Frielinghaus M (1999) Skalenaspekte der Bodenerosion. In: Steinhardt U, Volk M (Hrsg) Regionalisierung in der Landschaftsökologie. Forschung - Planung - Praxis, Teubner, Stuttgart-Leipzig, pp 79-95

Herz K (1973) Beitrag zur Theorie der landschaftsanalytischen Maßstabsbereiche. Petermanns Geogr. Mitt. 117: 1-96

Herz K (1984) Theoretische Grundlagen der Arealstrukturanalyse. Wiss. Z. Päd. Hochsch., Dresden

http://www.geog.uni-hannover.de/phygeo/trianet/Grafik/Dimensionen.html (01.11.2000)

Jenny H (1941) Factors of Soil Formation. McGraw-Hill, New York.

Kiemstedt H, von Haaren C, Mönnecke M, Ott S (1997) Landschaftsplanung - Inhalte und Verfahrensweisen. Der Bundesminister für Umweltschutz und Reaktorsicherheit, 39 S.

Kleeberg HB (Hrsg) (1992) Regionalisierung in der Hydrologie. Deutsche Forschungsgemeinschaft, Mitteilung XI der Senatskommission für Wasserforschung, Weinheim, Basel

Kleeberg HB (Hrsg) (1998) Regionalisierung in der Hydrologie. Abschlußbericht des DFG-Schwerpunktprogramms. Wiley-VCH.

Klijn JA (1995) Hierarchical concepts in landscape ecology and its underlying disciplines. DLO Winand Staring Centre Report 100, Wageningen

Klijn F (1997) A hierarchical approach to ecosystems and ist implications for ecological land classification; with examples of ecoregions, ecodistricts and ecoseries of the Netherlands. Theses Leiden University

Kolasa J, Pickett STA (eds) (1992) Ecological Heterogeinity. Springer, New York Berlin Heidelberg

Koestler A (1967) The Ghost in the Machine. Random House, New York

Leser H (1991, 1997): Landschaftsökologie. Ulmer, Stuttgart

Matson PA, Boone RD (1984) Natural disturbance and nitrogen mineralization: wave-form dieback of Mountain Hemlock in the Oregon Cascades. Ecology 65(5):1511-1516

Moss M (1983) Landscape Synthesis, Landscape Processes and Land Classification, some theoretical and methodological issues. GeoJournal 7.2: 145-153

Neef E (1963) Dimensionen geographischer Betrachtungen. Forschungen und Fortschritte 37: 361-363

Neef E (1967) Die theoretischen Grundlagen der Landschaftslehre. Haack, Gotha

Oliver MA (1990) Kriging: A Method of Interpolation for Geographical Information Systems. International Journal of GIS Vol.4 No.4: 313-332

O'Neill RV (1988) Hierarchy theory and global change. In: Rosswall T, Woodmansee RG, Risser PG (eds) Scales and Global Change. John Wiley & Sons, New York, pp. 29 45

O'Neill RV (1989) Perspectives in hierarchy and scale. In: Roughgarden J, May RM, Levin SA (eds) Perspectives in Ecological Theory, Princeton University Press, Princeton, pp. 140-156

O'Neill RV et al. (1986) A hierarchical concept of Ecosystems. - Prrinceton University Press, Princeton, N.J.

Rau S (1999) Der Einfluß von Gewässerrandstreifen auf Stoffflüsse in Landschaften. Einsatz eines mobilen Pen-Computers zur Kartierung im Einzugsgebiet der Parthe. Diplomarbeit, unveröff., Universität Potsdam

Rowe JS (1961) The level-of-integration concept and ecology. Ecology 42/2: 420-427

Salthe SN (1991) Two forms of hierarchy theory in western discourses. International Journal of General Systems 18: 251-264

Schultz J (1995) The Ecozones of the World. Springer, Berlin, Heidelberg, New York

Sherman LK (1932) Streamflow from rainfall by unit-hydrograph-method. Eng. News-Rec. 108: 501-505

Simon HA (1973) The organizationof complex systems. In: Pattee HH (ed.) Hierarchy Theory: The Challange of Complex Systems, George Braziller, New York, pp1-27

Smuts JC (1926) Holism and Evolution (2nd printing, 1971). Viking press, New York

Steinhardt U (1999) Die Theorie der geographischen Dimensionen in der Angewandten Landschaftsökologie. In: Schneider-Sliwa R, Schaub D, Gerold G (eds) Angewandte Landschaftsökologie. Grundlagen und Methoden. Springer, Berlin-Heidelberg-New York, pp. 47-64

Steinhardt U, Volk M (eds) (1999): Regionalisierung in der Landschaftsökologie. Teubner, Leipzig-Stuttgart

Steinhardt U, Volk M (2000) Von der Makropore zum Flußeinzugsgebiet - Hierarchische Ansätze zum Verständnis des landschaftlichen Wasser- und Stoffhaushaltes. Petermann's Geogr. Mitt. 2/2000: 80-91

Tietje O, Tapkenhinrichs M (1993) Evaluation of Pedo-Transfer-Functions. Soil Sc.Soc.Am.J.57: 1088-1095

Troll C (1939) Luftbildplan und ökologische Bodenforschung. Zeitschrift der Gesellschaft für Erdkunde zu Berlin: 241: 298

Turner M, Gardner R (1990) Quantitative Methods in Landscape Ecology. The Analysis and Interpretation of Landscape Heterogeneity. Springer, New York Belrin Heidelberg

Volk M (1999) Interactions between landscape balance and land use in the Dessau region, eastern Germany. A hierarchical approach. - In: Hlavinkova P, Munzar J (eds) Regional Prosperity and Sustainability. Proceedings of the 3rd Moravian Geographical Conference CONGEO'99, pp 201-209

Volk M, Steinhardt U (1998) Integration unterschiedlich erhobener Datenebenen in GIS für landschaftsökologische Bewertungen im mitteldeutschen Raum. Photogrammetrie, Fernerkundung, Geoinformation 6/1998: 349-362

Wilmking M (1998) Von der Tundra zum Salzsee. Landschaftsökologische Differenzierungen im westlichen Uvs - Nuur - Becken, Mongolei. (Tundra to Salt Lake. Landscape Ecological Differentiation in the Western Part of the Uvs Nuur Hollow, Mongolia). Diplomarbeit, unveröff., Universität Potsdam

Wu J (1999) Hierarchy and Scaling. Extrapolating Information Along a Scaling Ladder. Canadian Journal of Remote Sensing 25(4): 367-380

# 7 Landscape balance

Martin Volk, Uta Steinhardt

## 7.1 Introduction

The characteristic distribution of the landscape's components land use, land cover, soil, morphology, hydrology, climate, geology, etc., forms the landscape structure. These components are interrelated by fluxes of water, material, energy and information (landscape-ecological processes), which result in the 'landscape balance'. This term is based on the German concept of the *Landschaftshaushalt*, which describes the associations between the geoecofactors in a geoecosystem due to the laws of nature (Leser 1997, Marks et al. 1992, Troll 1939, Zepp & Müller 1999)[1]. The geoecosystem is regarded as an 'open system' characterized by an equilibrium of flows, with input and output interactions with the landscape balances of the adjacent geoecosystems (the environment). Despite the dimension of the landscape ecosystem, a model of the landscape balance can be created for any order of magnitude. In doing so, methodological extensions or limitations arise for the different dimension steps. Several problems have emerged with the development of scale-specific methods, the improvement of knowledge about the interactions between landscape structure and landscape-ecological processes and the processual interactions and changes within the landscape ecosystem itself at different dimensions and scales. These questions become even more important when considering the impact of land use and its changes on the landscape balance and its assessment as a basis for a sustainable development.

Human impacts - such as land use - affect the interactions within a landscape ecosystem by changing the landscape structure and thus altering conditions for landscape-ecological processes. The human factor 'land use' within the complex ecosystem has a strong impact on the adaptability, regeneration and regulation capability of the landscape balance. It should be mentioned in this context that it is still a problem to assess the adaptability and dynamics of the landscape balance as a reaction to human impacts (feedbacks) within landscape analysis owing to the lack of knowledge about these interactions (Fig. 7.1 and Fig. 7.2). As most of the relevant processes in the landscape depend mainly on the mobile agent water, they have influences ranging from small to large scales. However, understanding of

[1] Schmithüsen (1973) transferred the theories and considerations of thermodynamics and synergism into geography with the term 'geosynergetic landscape research', which describes the totality of *all* interactions within a landscape (cf. also Müller 1999). Neef (e.g. 1973) also largely developed the system theories for 'his' landscape research on the basis of such knowledge.

these processes - especially on large scales - is still insufficient, as most of the processes take place on small scales. Concepts for sustainable development have to consider the implementation of information about the landscape balance on all scale levels. Special attention should paid to larger scales because most of the environmental conflicts and changes become apparent on the landscape scale. However, most of the useful methods for the analysis and assessment of landscape-ecological processes and parameters are limited to scales up to 1:25,000, and the importance of the parameters - and the parameters themselves - are limited to changes in a hierarchical spatio-temporal way.

To solve these problems, the following questions should be asked:

- *How does the importance of parameters (as well as the parameters themselves) of their landscape balance components (morphology, soil, hydrology, soil, hydrology, land use and cover and climate) change on different scales?*
- *How does the impact of changes to the landscape structure (especially land use) affect the water, material, energy and information fluxes (horizontal and vertical) on different scales?*
- *How does the land use influence the quality and quantity of soil and water?*

This also requires characterizing the processes concerning extension, duration, intensity and continuity - and improvingknowledge about possible feedback. The complex interactions of the landscape balance within the landscape system and the problems of its investigation and assessment are shown in Fig. 7.1. Fig. 7.2. contains an example of positive feedback.

In this paper, several national and international approaches and models for these investigations are presented. Finally, our hierarchical approach is described for mesoscale application. In addition, suggestions are made for the verification of large-scale calculations. For integrated landscape analysis, we aim to combine both 'top-down' and 'bottom-up' approaches with GIS-coupled model applications and traditional methods (e.g. mapping, measuring, etc). Using traditional methods is an essential part of verifying modelling results, as well as for improving knowledge of how landscape ecosystems function.

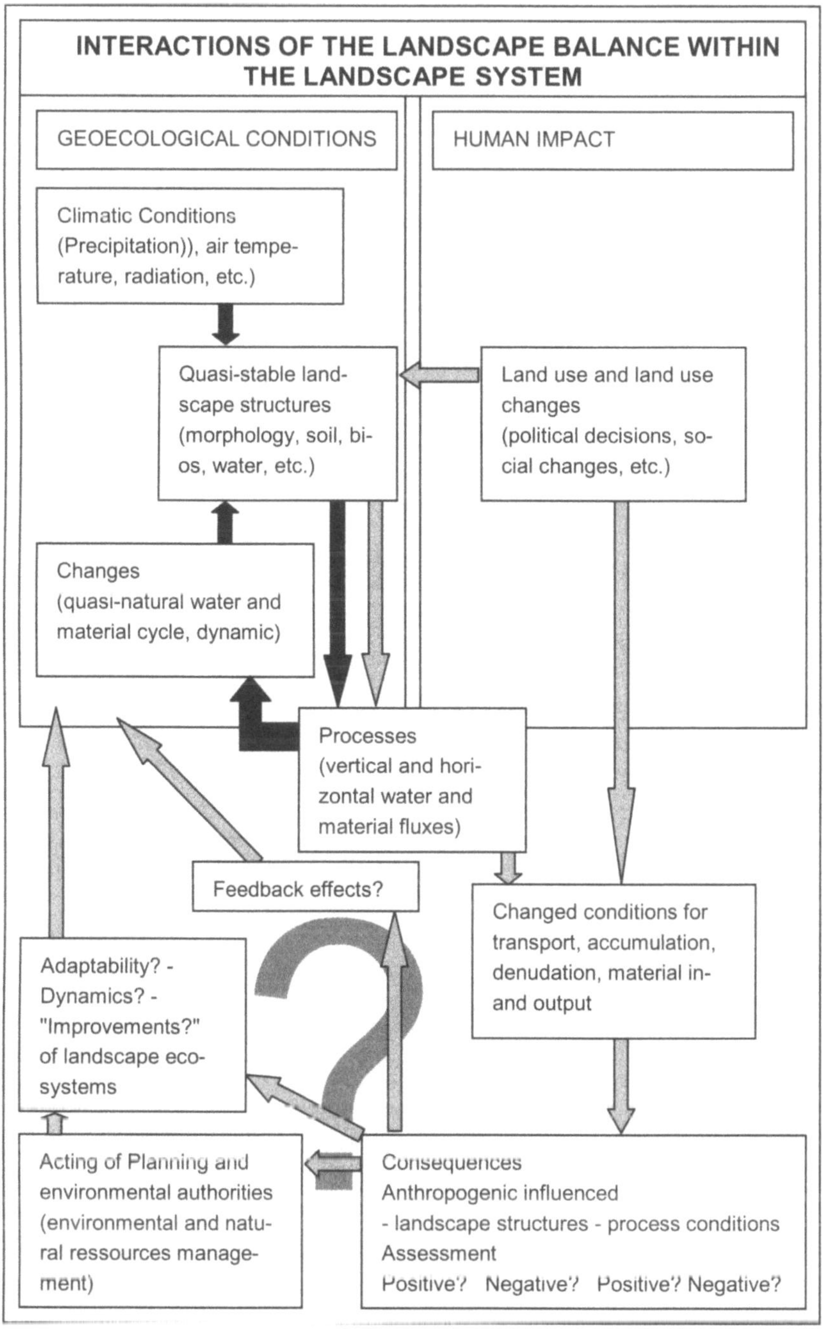


**Fig. 7.1:** Interactions of landscape balance within the landscape system and the problems of its investigation and the assessment of human impacts. This is even more difficult considering the fact that the intensity of the interactions and influences depend on time and the scales concerned.

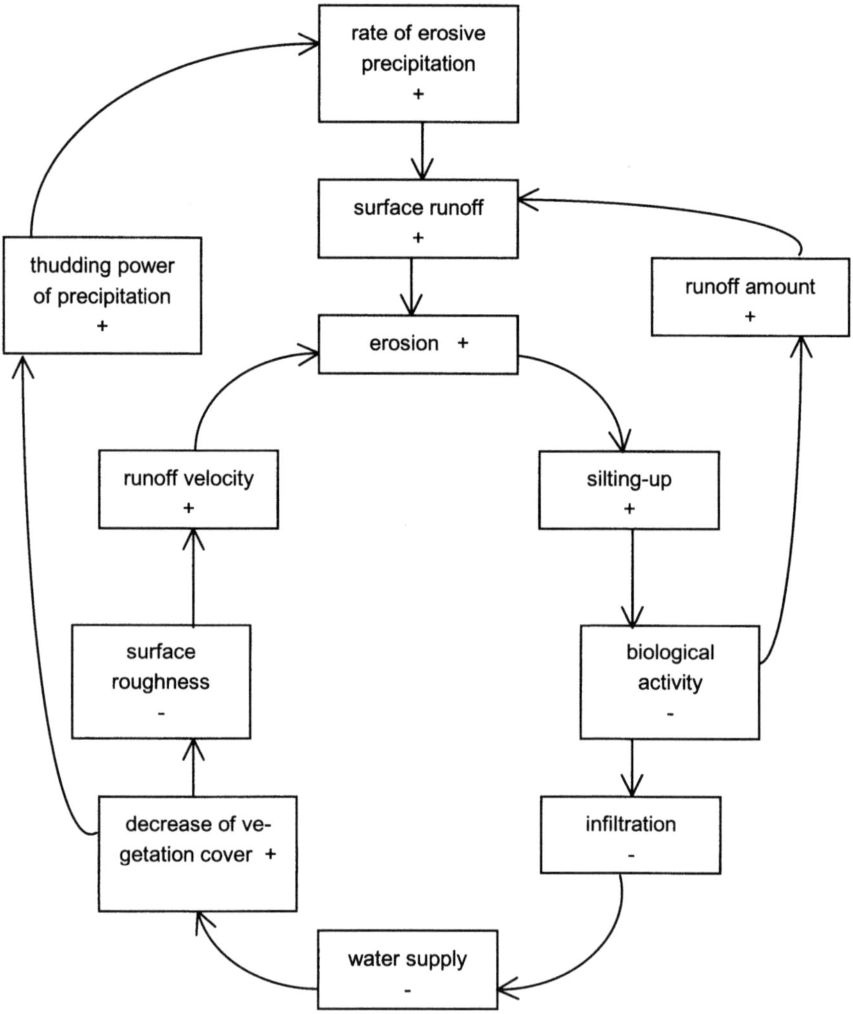


**Fig. 7.2:** Example for positive feedback: increased erosion and surface runoff caused by a decrease of the vegetation cover (+: increase; -: decrease) (after Rohdenburg 1989).

## 7.2 Interactions between landscape structures and processes

Due to its importance, some of the earliest works in ecology focused on understanding the relationships between patterns and processes at landscape scales (Cowles 1899, Cooper 1923, Gardner & O'Neill 1991). Klug & Lang (1983) described the investigation of the structure, functioning and dynamics of natural systems and their anthropogenic-technogenic transformation as a main task of geosystems research (see also Turba-Jurcyk 1990).

As mentioned in the introduction, the landscape structure causes different process conditions. "Despite the wealth of empirical and conceptual investigations that have been carried out since these early studies, the problem of predicting ecological processes at broad scales remains largely unsolved. This lack of resolution is due, in part, to the complexity of the problem and an intellectual tradition that has assumed that detailed measurements of fine-scale processes are necessary to predict broad-scale patterns." (Gardner & O'Neill 1991). This statement highlights the connection between landscape structure processes and the problem of scales in landscape ecology. In relation to this topic, special attention is given to the narrow, linear transition zones from one ecosystem to another, which are known as 'ecotones' (i.e. Jedicke 1994). These zones are characterized by a rapid change of the environmental conditions and site factors within a small area. Besides their importance for biodiversity (i.e. the "edge effect", Jedicke 1994), the main functions of ecotones consist in the protection of adjacent ecosystems, i.e. protection against unwanted material, nutrient and water fluxes (barrier effect, buffer function; Bastian & Schreiber 1999). Negative tendencies of land use development can reduce these transition zones and lead to drastic impacts on the living conditions for plants and animals, as well to completely changed fluxes of material, nutrient and water - which influences our natural resources such as water and soil, and in turn the mentioned living conditions for plants and animals (Plachter 1991). Due to the importance of ecotones for both biodiversity and as transition and buffer zone for processes (e.g. the influence of ecotones on flows of energy, material, organisms and water) within the landscape (structure), several publications deal with these topics (Hansen & di Castri 1992, Hansen et al. 1992). Delcourt & Delcourt (1992) tackles ecotone dynamics in time and space, while Weinstein (1992) suggests methods and models for monitoring ecotones to detect global change at different scales (see also Navch & Liebermann 1994).

One of the most important factors in relation to the landscape structure is georelief. In the landscape balance, georelief is a regulation factor and a structural area in or on which landscape ecological processes act (Leser 1999). Problems arise in connection with the fact that most of the process factors on the landscape scale - e.g. soil erosion - are mainly or partly derived from structural information such as soil maps etc. However, these methods do not enable the deduction of information about the processual and structural transformations at the structural boundaries between different landscape types. In the authors' opinion, due to the morphological conditions of a landscape (and thus with its changing conditions of soil, vegetation, micro- and mesoclimate, etc.), process-structural transformation zones (vertical to horizontal or reverse, interflow, etc.) of different widths have to be defined. This means that these transformation zones are narrow in landscapes with - for example - alpine conditions with high morphological energy, and wide in gently sloping or flat areas with low morphological energy. On the other hand, 'sharp' boundaries can often be observed in for example relatively small fluvial plains with a mainly ice-age glacial genesis. In these areas, both the heterogeneous distribution of the substrate and small differences in the morphological conditions of the surface can cause these boundaries.

A useful step to improve knowledge about the complex interactions between landscape structures, biodiversity and relevant processes appears to be to assess landscape structures with landscape metrics derived from satellite images (Turner & Gardner 1991, Antonova 1998, Boyce 1998, Richards 1998, Wallin 1998, Wallin & Boyce 1998, Lausch 1999, Lausch 2000)[2], which also allows the monitoring and documentation of land use changes and their impact on the landscape structure during time steps. Coupled with 'context-related' methods of digital image processing (improving land use classification with phenological and DEM-based morphological indicators, 'hydrological remote sensing', etc.), these studies seem to allow a connection with the more process-oriented studies. Finally, this method should allow the transferability and applicability of integrated models to the specific natural conditions of an investigated landscape to be examined. This is important in view of the fact that most of the models and algorithms applied are developed for specific research fields and special study areas with very different conditions in comparison to their own study areas. This approach is pursued by the authors and described in Chap.7.5.

[2] Even Troll (1939) mentioned the importance of interpreting aerial photographs for integrated landscape ecological analysis.

## 7.3 Fluxes of matter, energy and information in landscapes

Due to the complex processes of matter and energy transformation in landscapes, special attention is paid to the water as an essential element and a mobile agent which is the main transport medium in temperate climates. At first our investigations will concentrate on abiotic components of the landscape balance. Keeping its varied interactions with the bios in mind, these processes cannot be understand as purely abiotic (Finke 1994, Wohlrab et al. 1999). This is being tackled by other groups from our department (Lausch 1999, Lausch 2000; see also Chaps. 4 and 5).

### 7.3.1 Vertical and horizontal fluxes and processes

One of the most important and 'hottest' topics in landscape ecology is differentiation between vertical and horizontal fluxes and processes in landscapes. Most of the process-oriented investigations are concentrated on very small locations, which have all resulted in an improvement in the understanding of the horizontal and especially the vertical processes on the microscale. On this scale, the process system can be characterized as mostly vertical, whereas on the mesoscale horizontal processes are at the focus of consideration (Leser 1997). Schmidt (1978) pointed out that the biggest problem results in the transformation of the rare information about the horizontal processes into natural areas recorded with static methods. Bearing in mind that Schmidt's statement is related to the microscale, this remains valid nowadays, despite several works dealing with theoretical aspects, the improvement in field analysis and 'scale-transferring' techniques concerning this problem.

In an attempt to solve these problems, Menz & Kempel-Eggenberger (1999) combined two landscape-ecological methods on two different scale levels. On the microscale, they followed the concept of the landscape ecological complex analysis, with time-dynamic measurements of the landscape's water and material fluxes under climate-, bio- and hydroecological aspects (Leser 1991). The main step of these investigations is complex local analysis with the conceptual model 'local site regulation cycle' (*Standortregelkreis*, Chorley & Kennedy 1971, Mosimann 1978). This theoretical model includes different spatio-temporal dimensions for the measurements in the study area and should be the basis for upscaling the material fluxes and transformations. This means that material fluxes passing from one spatial dimension to the next higher level do not remain in one dimension. This results not only in a change of the transport direction from vertical to lateral, but also in an overstep to another spatio-temporal transformation network (e.g. the local fluxes are concentrated on direct surface runoff, after a short time they reach the drainage channel of the catchment, and hence a larger transformation network - 'positive feedback'[3]). On the other hand, the lapse of the local fluxes to the subsurface is defined by them as 'negative feedback' Because of the problems of

[3] The definition of 'positive feedback' has a different meaning here compared to Rohdenburg (1989).

transferring local process, information on larger areas - whose complexity is caused by heterogenization (Herz 1994) - they suggest the development of hypothetical key factors or connecting links between the different scale levels. Therefore, the coincident method used in this approach is the development of a digital geoecological risk analysis for a micro- to mesoscale application ("method of processual geoecotop and geoecochoric[4] segmentation"). This method is based on a typological classification of homogeneous assessment units considering processual characteristics as described by Leser & Klink (1988). The data basis of this integrated analysis comprises field data and maps (i.e. substrate, soil type, pH, soil depth, vegetation, land use and land cover, climate, etc.), and morphological parameters derived from a DEM. From the processes, it is possible to derive processual areas (material and nutrient balance, water balance, aerial balance, radiation balance). These process-based models are re-classified and form the basis for deriving ecological indicators (i.e. soil denudation, low soil depth, soil depth, pH, frost risk, etc.). The combination of structural and processual parameters and the application of classification and assessment methods (e.g. Marks et al. 1992) allows the designation of ecological risk zones (sensitivity of landscape to natural and anthropogenic impacts). Depending on the appropriate scales, the derivation of different information is possible. Transfer to larger areas (regions) is possible by modifying the given classification and assessment methods. Besides the problem that there is less information about the process dynamics and process behaviour in these structure-oriented studies, most of the given assessment methods are only valid for scales up to 1:25,000. Nevertheless, Menz & Kempel-Eggenberger (1999) suggest combining these two methods as a basis for the definition of connecting links between the dimensions that allow a scale-specific characterization of the process transformations (although the links are not defined in the publication).

Considering the above-mentioned studies, the following facts should be pointed out. The importance of ecological - and also socioeconomic - parameters changes depending on the spatio-temporal scale level concerned. For instance, within the material transformation process, a change of the scale level can happen - not only as 'downscaling' (top-down), but also as 'upscaling' (bottom-up), if material fluxes overstep under self-intensification into the next higher or lower dimension ('interflow-network'). The present focus on 'upscaling' methods is caused by small-scale concentrated research over decades. By combining both 'top-down' and 'bottom-up' approaches, it seems possible to link both methods. By doing so, suggest the authors, a contribution to the solution of problems related to the questions of continuous scale level transitions and the knowledge of horizontal processes on different scales can be expected (Mosimann 1999, Steinhardt & Volk 2000). It should be pointed out here that such approaches are very important for the progress of scale-related landscape-ecological research, considering questions about system behaviour, adaptability, feedback mechanisms, hierarchies, synergy, etc. The remaining problem is still the definition of the links between the different scale levels. Another question is the degree to which these 'philosophical', very difficult and complex system approaches have to be simplified for application to, for instance, environmental planning. Here, a combination with more practical ap-

[4] The concepts of topes and chores is explained in Chap. 6 of this book.

proaches - as is suggested in Bierkens et al. (2000) - is conceivable and should be aimed at.

### 7.3.2 Scale specific and cross scale investigations

The consideration and transferability of processes, local conditions and assessment methods to different scale levels plays a central role in landscape-ecological science as well as in planning practice. The main problem here is - as also in many other fields of landscape ecology (Müller & Volk 1998) - the lack of general theories allowing the derivation of rules for regionalization (Richter et al. 1997, Steinhardt & Volk 1999). Thus, two fundamentally different opinions can be pointed out:

- The need for methods that allow a transfer of locally valid (small-scale) information and results to larger areas through suitable indicators or transfer functions (Scheinost 1995, Tietje & Tapkenhinrichs 1993);

or

- The need for special methods on all scale levels.

The two cross-scale approaches mentioned are special studies in the field of soil sciences, with limited possibilities for transfer to other larger areas and fields of landscape ecology. Due to the current lack of cross-scale methods, scale-specific methods should be preferred. Currently, most of the studies in landscape ecology are concentrated on 'bottom up' approaches (King 1991, Meyer 1997, Gerold 1999, Diekkrüger 1999), which are defined by the translation or extrapolation of information from small scales to larger landscape or regional scales. King (1991) describes several methods for these 'scaling up' approaches. The inverse approach of 'top down' or 'scaling down' approaches, which can be defined as the translation of information from larger to smaller scales, is less commonly practised in research (it is mainly confined to climatological topics, such as predicting landscape response to climatic change by means of climate models, i.e. Gates 1985). Bierkens et al. (2000) suggest a theoretical framework for the large number of upscaling and downscaling methods used in environmental science. These methods are designed to help the practitioner assess whether scale transfer occurs in the research project, and if so, exactly where, and to descide what upscaling or downscaling methods are suitable for performing these instances of scale transfer. Their book includes a CD ROM with a simple Decision Support System (DSS), which helps choose the most appropriate upscaling or downscaling method depending on the 'research chain' of a project. The book is designed for applied research with a practical approach, and does not deal with more 'philosophical' approaches to scale, such as hierarchy, organization and synergy, etc. Steinhardt & Volk (Chap. 6) passed general comment on the above-mentioned regionalization problems. In the authors' opinion, both approaches - 'top down' and 'bottom up' - (along with their specific features) - are necessary and have to be combined in order to achieve an integrated landscape analysis for all spatial scales (cf. the following Chaps., see also Wrbka et al. 1999). Hence scale-specific approaches have to be applied to the "core zones" of each scale, and cross-scale investigations are used in the "transition zones" between the scales due to the loose coupling (Wu 1999) between the scales to guarantee a holistic consideration of landscape.

## 7.4 Water-carried fluxes of nutrients and pollutants

The outwash and transport of material, nutrients and pesticides is mostly linked to an amount of water flowing out of a region. This results in an input of this material into the groundwater and surface water. The investigation of these processes is often concentrated on phosphate (particle-bound transport through erosion: lateral processes) and nitrate (soluble transport through seepage: vertical processes). Because of the huge amount of different pesticides and the resulting complex chemical analysis, work needs to concentrate on estimating the spatio-temporal input behaviour of selected relevant pesticides (i.e. Grunewald et al. 1999).

### 7.4.1 Investigation methods

A comprehensive description of methods for landscape-ecological analysis applied in Germany is given by Bastian & Schreiber (1999) and Zepp & Müller (2000). The most highly developed investigation methods are for small-scale studies, including several proposals recommended for methodological standards in mapping, measuring and assessing up to a scale of 1:25,000 . There is still a lack of homogenous approaches for the various investigation methods for mesoscale and macroscale integrated landscape analysis (Lenz 1999).

### 7.4.2 GIS-coupled modelling on different scales

Most of the nutrient load of surface waters originate from non-point sources. To analyse these processes, the application of distributed parameter models in combination with geographical information systems (GISs) seems to be a useful method. Special attention has to be paid to the spatial variability of the landscape characteristics and their influence on the transport of water and nutrients within a given area. At present, many of the physically based approaches with a high spatio-temporal resolution cannot be effectively applied to medium-sized watersheds (Grayson et al. 1992), for example, because of the huge amount of input parameters required. Despite the much greater effort needed to parameterize, validate and run physically based models, simulated results often provide only slightly better or sometimes even worse correspondence with measured values than lumped-parameter models (Seyfried & Wilcox 1995). In this context, it should be mentioned that most of the common empirical models employed by environmental and planning offices and authorities rarely use more than three parameters (Hauhs et al. 2000).

Bearing these problems in mind, several models have been tested for their scale-specific applicability with respect to the time schedule and topics of research projects (Krysanova et al. 1996).[5] As most of the models were developed within research projects carried out in specific study areas, the possibility of transferring these methods to other regions needs to be tested. Table 7.1 gives an overview of several models, their capacities and operations, and their scale-specific applicability. These models have been selected by the authors to test their scale-specific applicability. More models are listed and described by Bork & Schröder (1996) and Grunwald (1997).

All input data used for model applications have to be prepared and modified depending on the specific calculation characteristics of the models (cp. Petry et al. 2000, Volk & Steinhardt 1998). This is also important for deriving indicators for environmental conflicts, land use, water balance and morphology interactions in catchment areas. One main problem of large-scale investigations is verifying the results. As measured data are mostly unavailable, the investigations have to be hierarchically linked to studies on smaller scales (sampling and analysis at representative locations, mapping, measuring, application of small-scale models, see above). Nevertheless, the application of these traditional methods is essential not only for verifying the modelling results, but also for improving basic knowledge about how the landscape ecosystem functions (Hauhs et al. 2000).

[5] Before applying a model, the algorithms used have to be checked. For example, most of the models that have an erosion component are based on different versions of the USLE (Bork & Schröder 1996). It seems important to be able to adapt the model algorithms to the specific conditions of a study area.

**Table. 7.1:** Selected models and their scale-specific applicability.

| Model system | Scales | Objectives, operations and capacities |
|---|---|---|
| SWAT (cp. Arnold et al. 1993, Srinivasan & Arnold 1993) | Large river basins, subbasins (up to several thousend square miles) | • Predict the effect of management decisions on water, sediment, nutrient and pesticide yields with reasonable accuracy on large, ungaged river basins.<br>• Daily time step to long term simulations<br>• Groundwater flow model<br>• Basins subdivided to account for differences in soils, land use, crops, topography, weather, etc.<br>• SWAT accepts output from EPIC (see below)<br>• SWAT accepts measured data & point sources<br>• Soil profile can be divided into ten layers<br>• Water can be transferred from channels and reservoirs<br>• Basin subdivided into subbasins or grid cells<br>• Nutrients and pesticide input/output<br>• Reach routing command language to route and add flows<br>• Windows/ArcView Interface<br>• Hundreds of cells/subbasins can be simulated in spatially displayed outputs |
| ABIMO (cp. Glugla & Fürtig 1997, Rachimov 1996) | Meso- to macro-scale | • Description of the basic elements of the water balance on the landscape scale (long-term values of runoff and evapotranspiration).<br>• "Mean runoff" is defined here as the difference between long-term mean annual precipitation and real evapotranspiration. This difference is equivalent is to the total runoff. In the case of a solely vertical seeping of the water this value corresponds with the groundwater recharge.<br>• Thus, the value must be understand as the sum only indifferent of both surface and subsurface runoff. Therefore, the results have be modified with a runoff quotient (based on slope inclination and groundwater level) after Röder (1998), which allows an estimation of the surface runoff and interflow. |
| AGNPS-WaSim-ETH Young et al. 1987, Schulla 1997 | Mesoscale watersheds | • System of computer models developed to predict non point source pollutant loadings within agricultural watersheds. It contains a continuous simulation, surface runoff model designed for risk and cost/benefit analysis. |

continued

| | | |
|---|---|---|
| | | • The set of computer programs consist of:<br>• input generation & editing as well as associated data bases;<br>• the "annualized" science & technology pollutant loading model (AnnAGNPS);<br>• output reformatting analysis;<br>• the integration of more comprehensive routines (CONCEPTS) for the instream processes;<br>• an instream water temperature model (SNTEMP);<br>• several related salmonid models (SIDO, Fry Emergence, Salmonid Total Life Stage, & Salmonid Economics).<br>• The application of AGNPS (Agricultural Nonpoint Source) can be used for the caculation of soil erosion, sediment transport and nutrient yield (N and P).<br>• The computed runoff and peak flowof the hydrological model WaSim-ETH will be used as input for the linked model AGNPS. |
| CANDY (cp. Franko et al. 1997) | Lower mesoscale, Farm level, fields | • Analysis of vertical carbon-, nitrogen and water fluxes between crops, soil, and groundwater; daily time step, consideration of land management practices |
| EPIC | Farm level, fields | • Capable of simulating the relevant biophysical processes simultaneously, as well as realistically, using readily available inputs and, where possible, accepted methodologies;<br>• Capable of simulating cropping systems for hundreds of years because erosion can be a relatively slow process;<br>• Applicable to a wide range of soils, climates and crops;<br>• Efficient, convenient to use, and capable of simulating the particular effects of management on soil erosion and productivity in specific environments.<br>• The model uses a daily time step to simulate weather, hydrology, soil temperature, erosion-sedimentation, nutrient cycling, tillage, crop management and growth, pesticide and nutrient movement with water and sediment, and field-scale costs and returns. |
| E3D (cp. Schmidt 1991, Von Werner 1995) | Subbasins, farms, fields | • Simulation of erosion processes during single strong rain events as well as for the calculation of annual or several year values.<br>• Based on a physical approach, characterized by a high spatio-temporal resolution and short calculation times |

## 7.5 Theory: A scale-specific hierarchical approach

The terms 'sustainable development' and 'sustainability' as well as 'ecological' are buzzwords which are being increasingly used and misused throughout society. These terms are frequently not understood in the sense of their original meaning as conceptual ideas, as was presented in the Brundlandt Report (Goodland 1992, Gore 1992). According to these ideas, the concept of sustainability is related to the whole earth. However, the realization of this concept requires a hierarchical approach with solutions for all spatial scale levels - from local to global. This must be seen in connection with the fact that human impacts - like land use - affect the landscape balance over a broad range of spatial scales (i.e. use of natural resources, input of agrochemicals, etc.). Several papers from different scientific disciplines deal with the hierarchical organization of nature (Burns et al. 1991, O'Neill et al. 1986). The hierarchical concept was introduced into German landscape ecology by Neef (1967) and continued by several other landscape ecologists (Leser 1997). An overview of hierarchical concepts in landscape ecology is given by Klijn (1995). These concepts are mainly focused on the hypothesis that each of these scale levels (micro-, meso- and macroscale) is characterized by specific temporal and spatial ranges. Due to the size and internal differentiation of the spatial reference level, different proceeds of cognition are related (spatio-temporal hierarchies of landscape-ecological processes). Consequently, each scale level needs layers (and indicators) with suitable spatio-temporal resolution and specific investigation methods, and provides specific information (Steinhardt 1999a, Steinhardt & Volk 2000). Regarding the increasing application of geographical information systems (GISs), this is often limited to consideration of the spatial resolution of the data layer. Technical problems caused by the huge amount of data used and the computer memory required are often the limiting factors for the application of high-resolution data. Despite all the advantages of these systems, GIS users often run the risk of applying a number of formal procedures which often do not help improve knowledge about the process behaviour in landscapes. The derivation of slope curvatures from digital elevation models (DEMs) with different spatial resolutions may be an example of those doubtful applications (Steinhardt & Volk 1998). Therefore consideration of scale-specific methods needs to be underlined. It is at this point that we touch upon the difficult field of regionalization. On the one hand, although small-scale investigations are useful and important for improving our knowledge of processes, they depend on complex, laborious methods such as detailed measurement or mapping. Many studies deal with the translation and extrapolation ('scaling-up') of these findings to larger scales ('bottom up'). By contrast, mesoscale and macroscale approaches enable landscape ecological interactions in large areas such as watersheds and regions to be detected while focusing on relevant areas ('scaling-down', 'top down'). This "step in the scale or consideration level" can be illustrated using the example of data layers based on remote sensing data gained from different recording platforms. As the survey altitude above the earth's surface increases (aerial photographs, satellite images, etc.), the visible of the whole landscape area grows but is accompanied by a loss of detail. Thus, more abstract information has to be used due to the increasing size of a given study area (e.g. the Normalized Difference Vegetation Index).

Regarding the investigation of vertical and horizontal material, nutrient and energy fluxes, we pursue the following hypothesis:

- *The basic components for vertical and horizontal material and energy fluxes - morphology, soil, hydrology, land use/management, and climate - are similar at all scale levels. It is only the importance of the parameters (and the parameters themselves) of these components which change for each scale (Helming & Frielinghaus 1999, Klijn 1995, Steinhardt & Volk 2000, Volk 1999).*

Main topic is here the definition of a linkage between the different landscape scales, as mentioned in the Chaps. before. To give an example for morphology and erosion: On the local scale, surface roughness is one of the main factors that will affect erosion deposition, whereas for larger scales (up to river catchments) the factors slope inclination, slope length, slope exposition up the shapes of streamlet, -net, -order and direction of flow are responsible for erosion processes. With the GIS-coupled combination of both “Top-down” and “Bottom-up”-approaches and more traditional methods, the investigation of the complex interactions between landscape structures and processes, as well as an assessment of the impact of land use changes on the landscape balance, seems to be enabled. This method is prefered by the authors. In the following Chaps., our approach will be presented with some examples. Fig. 7.3 gives an overview about our method.

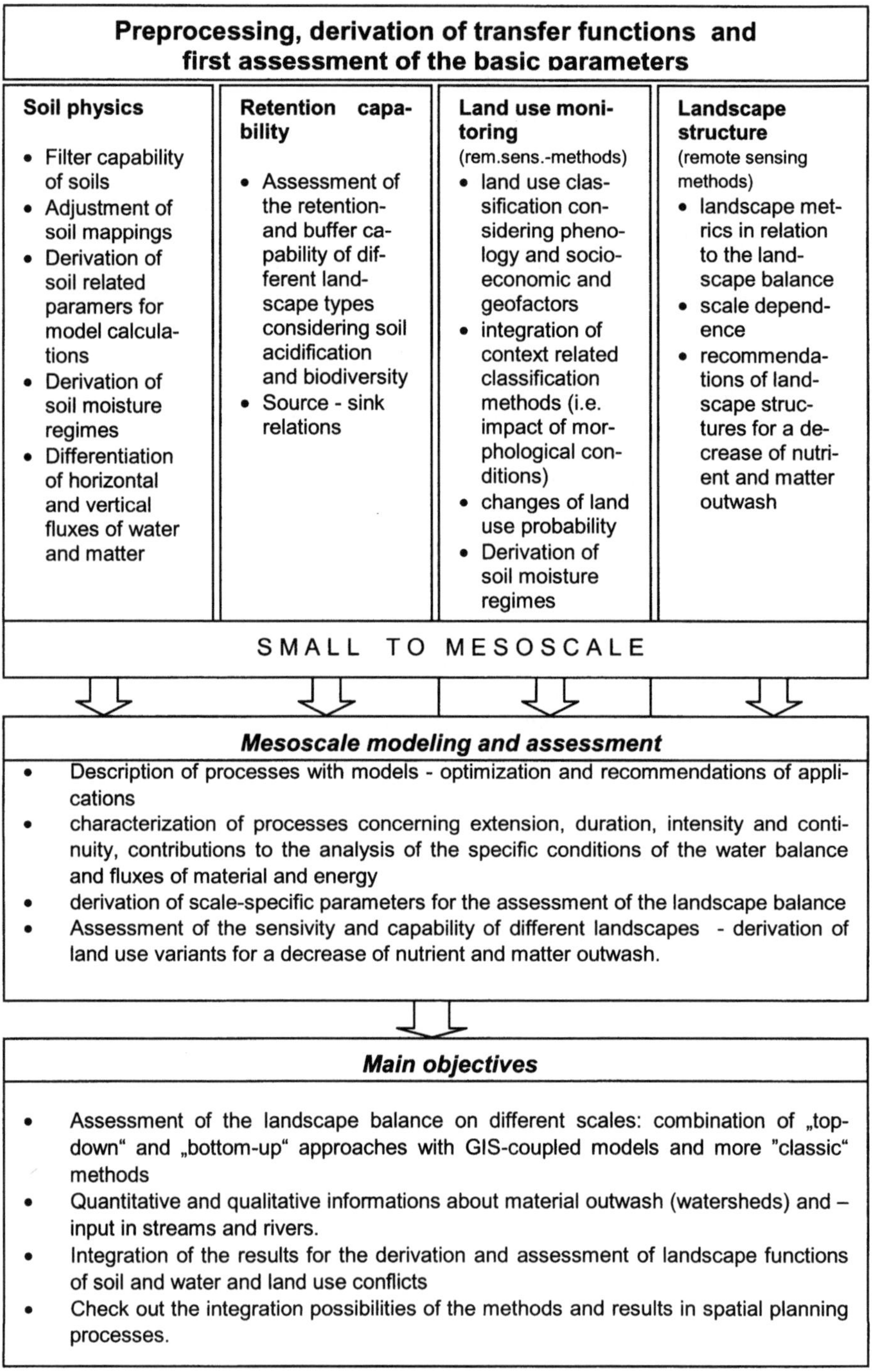


**Fig. 7.3:** Methodological steps for a hierachical, scale-specific landscape analysis of the landscape balance.

## 7.6 The study areas

The hierarchically nested approach presented here is tested on different scales in various areas of the states of Saxony and Saxony-Anhalt in eastern Germany with different natural and socioeconomic conditions. Therefore, the applicability and transferability of the method has to be checked. Below, the application of the approach is shown using the examples of the Dessau district (administrative unit) in the east of Saxony-Anhalt and on the River Parthe watershed in Saxony (cf. Fig. 7.4).

These two different types of investigation units are used due to the different objectives of the projects. The natural boundaries of *watersheds* (which can be described as 'quasi-closed systems') and their hierarchical organization form an appropriate structure for process-orientated environmental impact analysis. Administrative units should be used when the project's objective is to provide planning authorities with recommendations for land use and land management.

Generally speaking, it makes sense to consider both types of investigation units to ensure that information about the landscape balance is contributed to the planning processes in order to achieve sustainable development. The studies presented show a good way of combining these two approaches.

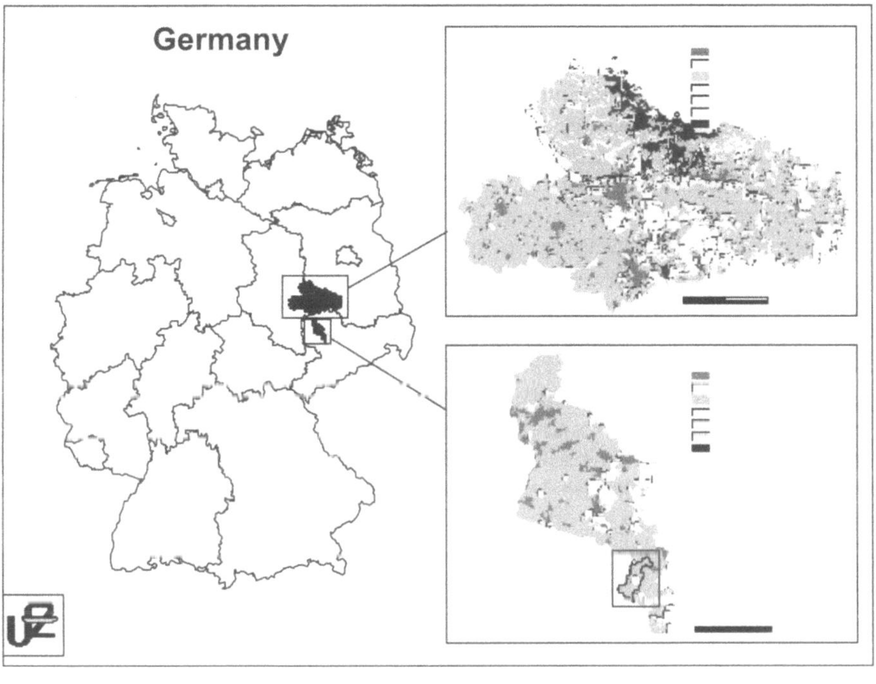


**Fig.7.4:** Location of the study areas in Germany.

### 7.6.1 The Dessau district

The district covers an area of about 4,300 $km^2$ in Saxony-Anhalt divided by the River Elbe. The study area is composed of various landscapes with very different conditions, ranging from Holocene floodplains and old moraine landscapes to very fertile loess plains. The mean annual precipitation varies between <500 mm in the western parts up to around 650 mm in the northern and southern district. The investigation area is one of the driest regions in Germany. Owing to the widespread fertile soils (chernozems) and lignite resources, agriculture, industry and other human activities have determined the main features of the region. On the other hand, there are almost undisturbed areas such as the riparian zone and the Elbe floodplain, which have been designated a biosphere reserve by UNESCO. The investigations are being carried out within the project "Landscape development, landscape balance and multifunctional land use in the Dessau district". This project is designed to derive strategies for sustainable development on the basis of investigations into the landscape balance (Petry & Krönert 1998, Volk 1999). Special attention is paid to restoring the landscape's multifunctionality by avoiding land use conflicts.

### 7.6.2 The Parthe river watershed

The River Parthe watershed (400 $km^2$) is a subbasin of the Elbe watershed and located in southeast Leipzig. It can be characterized as a representative part of the northwestern Pleistocene landscape with very different properties ranging from Permian porphyry hills and old moraine landscapes to fertile sandy loess plains. The mean annual precipitation is about 570 mm. The watershed is characterized by strong impacts of land use on the landscape balance resulting from the extraction of groundwater, sand, gravel and porphyry, lignite-mining, and the expansion of built-up areas on the outskirts, especially since 1990. The studies are part of an interdisciplinary project investigating the landscape water balance and the fluxes of material, nutrients and energy within the loess areas of the Elbe watershed. The interactions of the various material and energy fluxes are to be described here using a hierarchical network of various GIS model couplings. The results are to be used to calculate land use scenarios (impact of land use changes on the landscape balance) as a basis for the conclusion of sustainable land use variants.

Both study areas are dominated by agricultural land use. At present, the contrary development of agricultural land use is becoming increasingly dynamic and leading to landscape changes; while the loess-covered parts face further intensification, with marginalization becoming a widespread new phenomenon in the sandy Pleistocene areas.

## 7.7 'Top down': balancing - modelling - typifying

During the past few years, demands have increased for information about the landscape's water, material and energy balance to be integrated into planning processes. Consequently, the assessment of the impact of land use changes (i.e. groundwater abstraction, afforestation, etc.) on the landscape balance can be expected. In contrast to the increased application of GIS for environmental and planning surveys, a problem still arises with the availability of the relevant data layer for large areas. Moreover, most of the environmental parameters are only gathered and measured for short periods and in small areas. To solve this problem, investigating the landscape water balance on a regional scale (1:50,000 and higher) is proposed. The results of this investigation provide a basis for the designation of potential risk zones for vertical and horizontal material and nutrient outwash. The majority of the huge amount of data required was obtained by sharing data as part of our collaboration with geological and meteorological surveys, as well as planning and environmental authorities. One advantage of this exchange of data is that it strengthens communication and cooperation between research institutions and the relevant authorities responsible for landscape planning. On the other hand, this diverse information about soil, land use and land cover, groundwater and surface water, etc., is gathered using very different methods and for different projects and aims. As a result, the data are often inadequate for interdisciplinary landscape-ecological applications owing to their spatio-temporal resolution and their quality. Solving this problem requires standards and guidelines for the objective generalization and aggregation of the data layer (Volk & Steinhardt 1998, Petry et al. 2000).

Nevertheless, the calculations allow regional assessment and comparisons between areas of higher and lower *groundwater recharge and runoff* in relation to the prevailing natural conditions and the land use types. The water balances were calculated for both areas using the runoff simulation model ABIMO. Fig. 7.7 shows the groundwater recharge values using the example of the Dessau district. In comparison with other regions in Germany, both study areas exhibit low precipitation and groundwater recharge rates, the highest values in both cases being recorded in the morainic parts of the study areas. The dry western parts of the Dessau district with prevailing chernozems and cohesive substrate (highly important for the function of groundwater protection) are characterized by very low values. Within the existing priority areas for groundwater extraction in these western parts, only the extraction wells are protected. However, these sites are not necessarily the places where groundwater recharge and potential contamination occur. It is self-evident that the calculations can be used for the better designation of priority areas for groundwater extraction.

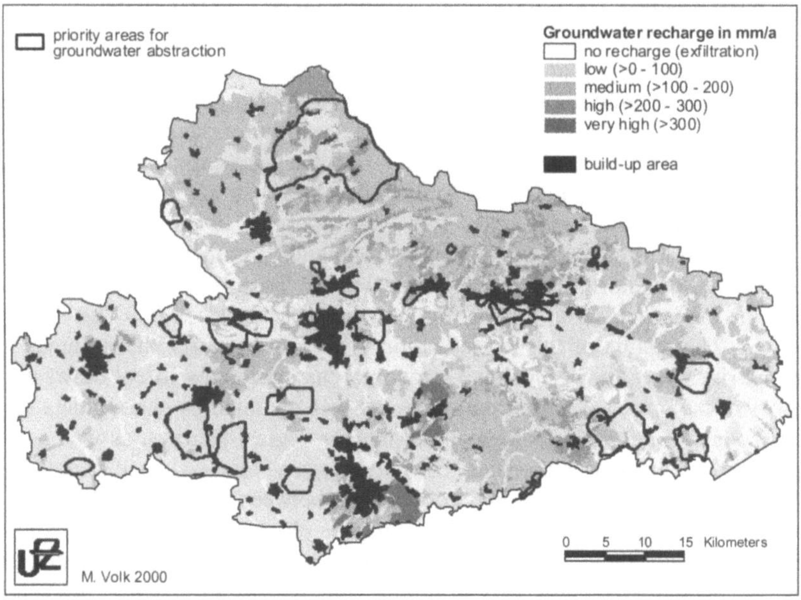


**Fig. 7.5:** Groundwater recharge in the Dessau district

Besides regional analysis, *scenarios* were calculated regarding the impact of land use changes on the water balance for smaller test areas within the Dessau district (Volk & Bannholzer 1999). As the database for the whole district has a relatively coarse spatial resolution, useful calculation of the scenarios is limited to areas >100 km². Fig. 7.6 shows an example of a test area. The conflicts in this part of the region stem from groundwater contamination by agrochemicals (e.g. nutrients and pesticides) and the overlap with priority areas for groundwater extraction and forestry. The calculations with different land use variants show for instance that afforestation in this area only slightly affects the groundwater recharge rate, but could improve the groundwater protection by a decrease in the agricultural areas. Although the potential lowering of the groundwater table caused by increased water extraction would only entail minor changes, its main ecological impact would be experienced by forestland (owing to dryness effects). These scenarios can only give rough indications of the impact of land use changes on the water balance for relatively large areas (average values). More detailed studies concerning the conditions within these areas require the application of other models and a database with a higher spatio-temporal resolution, as is shown in the following sections. Nevertheless, the results presented have been made available to the local water management and regional planning authorities to assist environmental planning decisions.

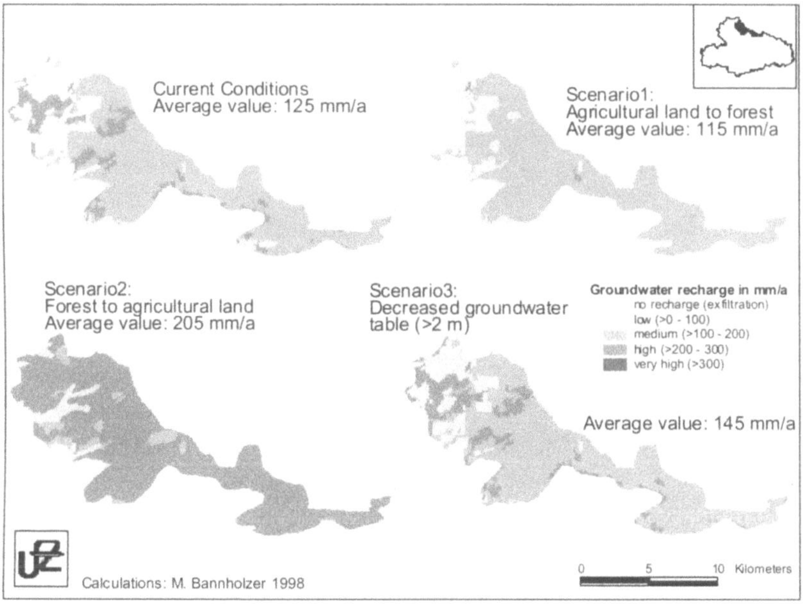


**Fig. 7.6:** Land use scenarios in a test area: Imapct of land use changes on the groundwater recharge.

In addition to the above studies, estimating *vertical and horizontal material and energy fluxes* on the regional scale is important in this part of our approach. These water-borne fluxes essentially depend on morphological conditions and surface cover. The percolation rates and an estimation of surface runoff and interflow were obtained by modifying the modelling results calculated with ABIMO after Röder (1998). Relevant geomorphologic parameters (e.g. slope angle and slope exposition) are coupled with the data layers of land use, soil conditions, the modelling results and climate in a GIS (ArcInfo). The soil data were classified by permeability, erosion disposition, etc. using the given assessment methods (AG Boden 1994).

In the first step, areas were identified using the query function of the GIS (ArcView), characterized by "arable land use", "percolation rate >180mm/a" and "slope inclination 0-2°". These areas are defined as potential risk areas ('hot spots') with vertical material leaching (e.g. nutrients, pesticides) from agricultural areas. According to the calculation results, the main risk areas are situated in the northern, eastern and partly the southern part of the Parthe watershed as well as in the northern and partly the southern part of the Dessau district, with permeable substrates.

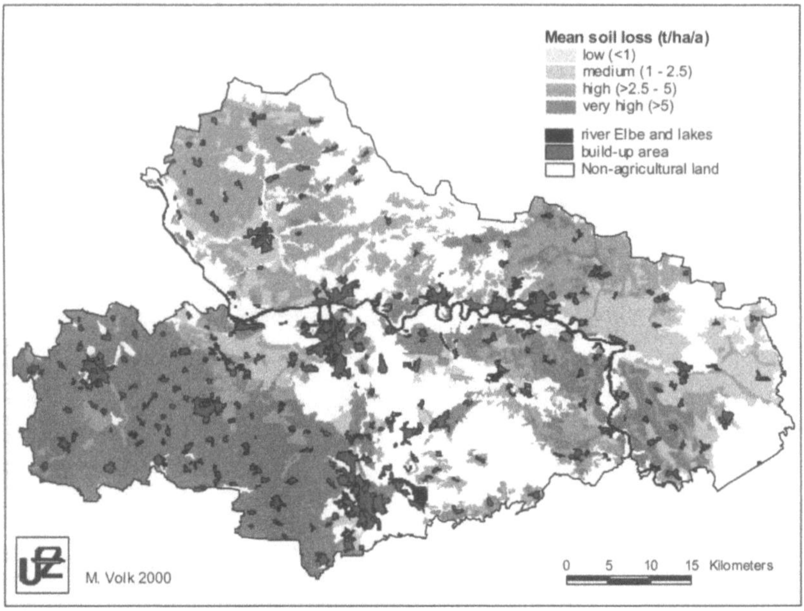


**Fig. 7.7:** Large scale calculation of mean soil loss of the Dessau distict.

For an initial estimate of the mean soil loss of the whole region, the modified Universal Soil Loss Equation ($R * L * S * K_B$) proposed by BGR (1994) was used. The equation factors were modified and reduced (due to the size of the study area) as follows: the R-factors (precipitation and surface runoff factor) were adapted to the conditions of the region (after Sauerborn 1994), the slope length factor L was equalized to 2.0 (slope factor S remains), and the factor $K_B$ determines the substrate-dependent rate of the erodibility factor K (after Schwertmann et al. 1990). The results of these calculations are shown in Fig. 7.7.

In combination with these results, an initial indication of potential risk zones with horizontal material (and nutrient) leaching from agricultural areas is obtained by selecting areas characterized by "arable land use", "cohesive substrate" and "slope inclination >1°" (relatively high soil loss) and "medium to high surface runoff". As the Dessau region is mostly flat, only a few risk zones for horizontal material (and nutrient) flow exist. The map in Fig. 7.8 shows both potential risk zones with vertical and horizontal material (and nutrient) leaching.

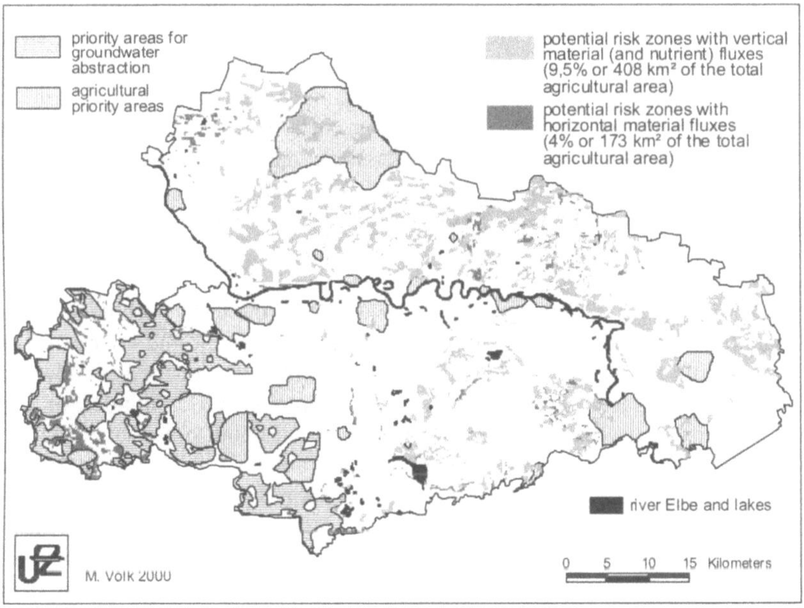


**Fig. 7.8:** Dessau district: Potential risk areas with vertical and lateral material (and nutrient) leaching

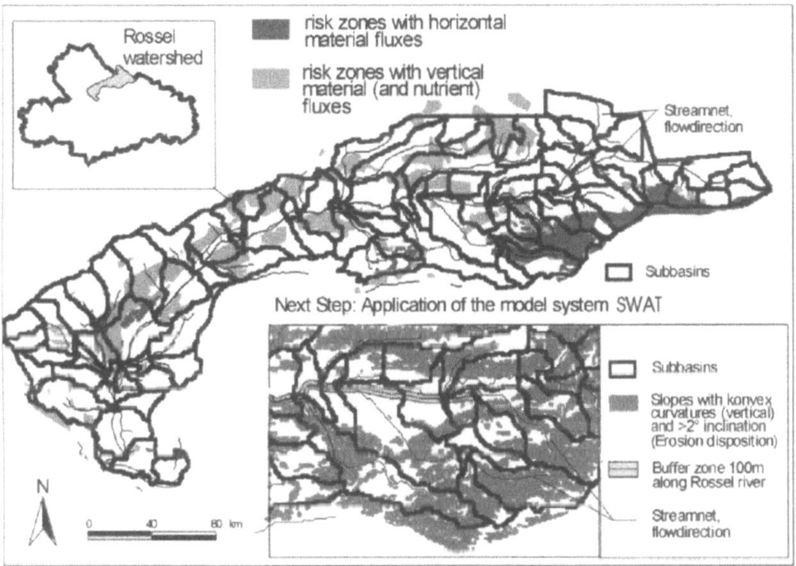


**Fig. 7.9:** Detailed investigations in the Rossel watershed.

The next step is to qualify and quantify the vertical and horizontal water, material and energy fluxes from the designated risk zones. These more detailed process-oriented studies require using watersheds as investigation units and a change of scale level (Steinhardt 1999b). The application of a database with a higher spatio-temporal resolution enables the derivation of hydro-morphological parameters such as flow direction, stream-order, watersheds, etc., which gives an impression of the transport conditions in the watershed. This analysis is carried out with hydrological functions in the Grid Module in ArcInfo or with the watershed module for ArcView, which is coupled with the HEC-HMS Hydrological Modelling System[6] (see also Olivera et al. 1998).

In addition, using different model systems allows the derivation of quantitative and qualitative information on the water, material, nutrient and energy fluxes in catchments and subbasins. As well as renewed calculation of the groundwater recharge and the surface runoff, this enables the improved differentiation of the risk zones. Moreover, useful information can be derived about potential risk zones in streams and rivers. Concluding the topographic factor LS in ArcInfo (Grid Module, Hickey et al. 1994) and determining the factors $K_B$ (substrate-dependent rate of the erodibility factor K) and R (precipitation and surface runoff factor, modified after Sauerborn 1994) and combining the two enables the modified usage of the Universal Soil Loss Equation after BGR (1994). The results of these more detailed calculations are shown in Fig. 7.9 using the examples of the Rossel watershed in the north of the Dessau district.

[6] A hydromorphological module is also integrated into the ArcView SWAT model.

## 7.8 'Bottom-up': measuring - mapping - modeling

To verify the mesoscale results and to improve knowledge about the process behaviour, we now attempt to combine the 'top-down' and the 'bottom-up' approaches. This will be done using the example of the Parthe watershed. For a detailed investigation of the landscape balance, four representative test sites were initially selected (Figs. 7.10, 7.11):

- Glasten - source area, slightly anthropogenically influenced
- Naunhof - middle course, impact through groundwater extraction
- Thekla - final gauge, partly urban influenced
- Schnellbach - subbasin, intensive agricultural impact

In November 1998, the installation of a network of measuring and survey stations to investigate the surface water was commenced (Figs. 7.10, 7.11).

**Fig. 7.10:** Gauge Glasten, V-weir (right);
Gauge Thekla, water sampler, rain-gauge (left)

**Table 7.2:** Selected precipitation values (daily totals) of the Parthe area, June 1999

| *Station* | *Precipitation values (mm)* | | | |
|---|---|---|---|---|
| | *2 July* | *18 July* | *19 July* | *20 July* |
| Glasten | 8.4 | 0.4 | 27.5 | 5.9 |
| Naunhof | 9.5 | 0.3 | 18.6 | 6.5 |
| Thekla | 5.0 | 0.3 | 2.6 | 1.7 |
| Schnellbach | 7.1 | 13.8 | 18.1 | 0.0 |

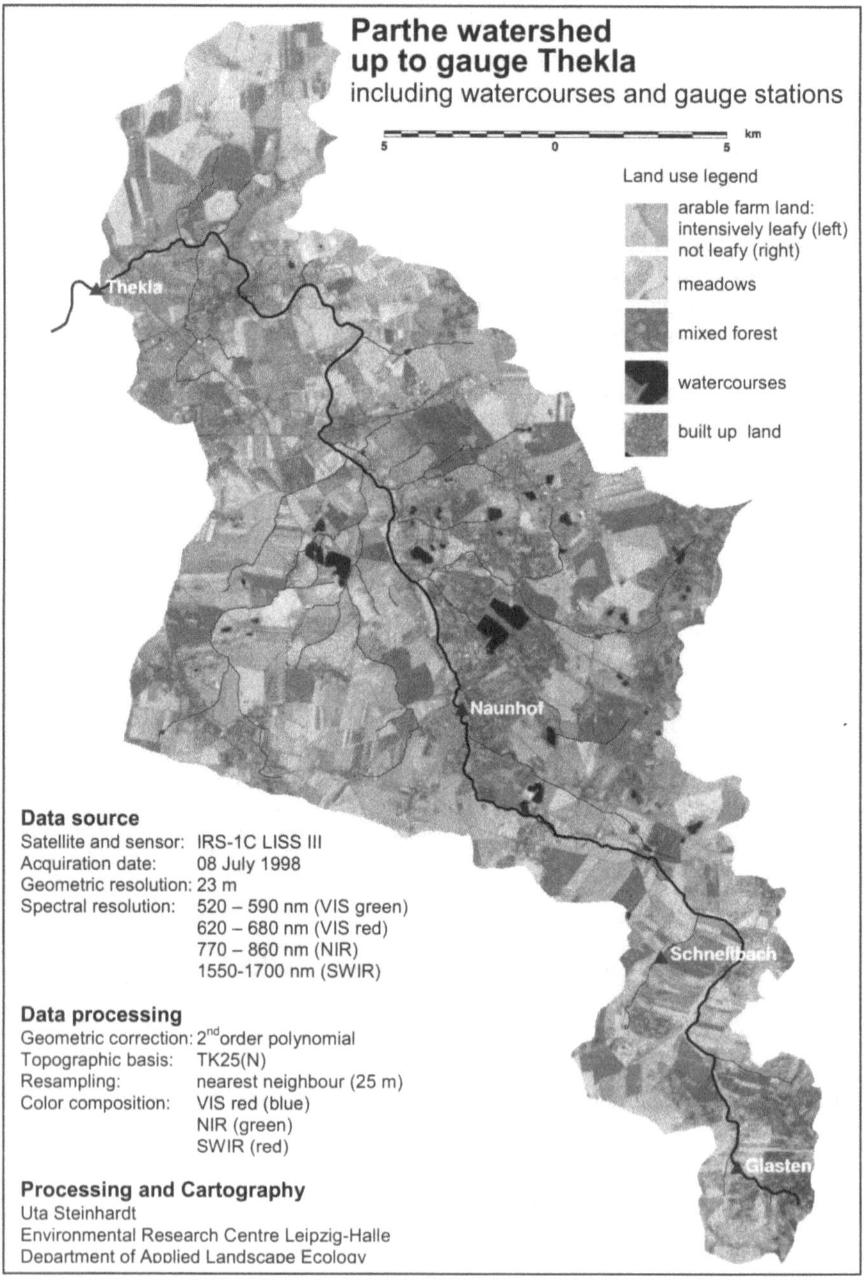


**Fig. 7.11:** Parthe watershed with its drainage network and location of the gauging stations.

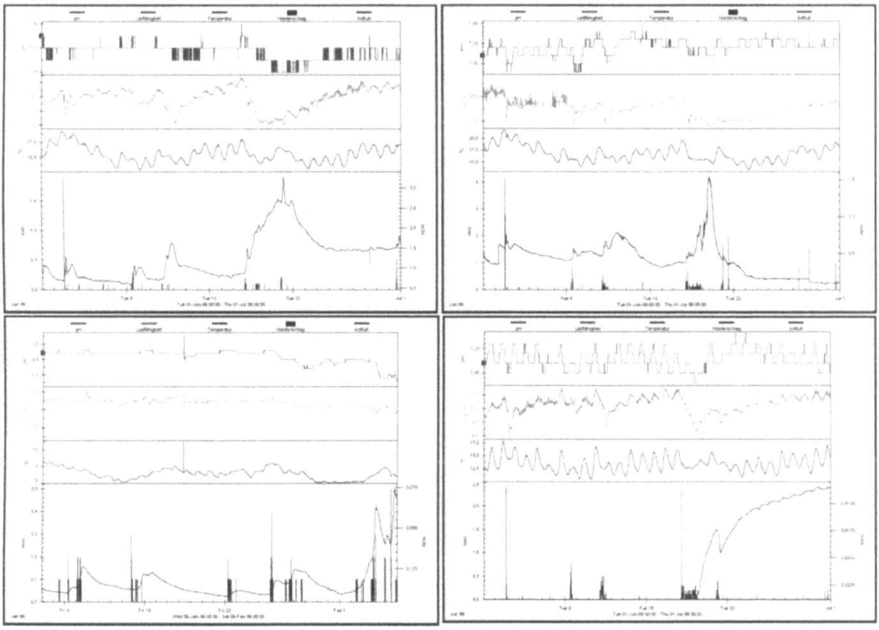

**Fig. 7.12:** Surface water parameters (pH, conductivity, temperature, precipitation and discharge) of the four gauges along the River Parthe for June 1999: Thekla (upper left), Naunhof (upper right), Glasten (lower left), Schnellbach (lower right).

The parameters discharge, pH, water temperature and conductivity and precipitation are measured at each station with a five-minute resolution. Additionally, the water is automatically sampled. Daily sampling (mixed samples) allows the derivation of information about the base load. The samples are analysed for their content of nitrogen and phosphorus components in the laboratory. At the same time, event-based sampling takes place during automatic sampling if a set flow rate value is exceeded. This enables the acquisition of the material components during the drain peaks. Fig. 7.12 shows the initial results using parameters recorded in June 1999.

The results show the occurrence of extreme precipitation events (short-term heavy rain, long-term light precipitation) throughout the whole watershed simultaneously, as well as those only locally surveyed or with a delay following events (Table 7.2).

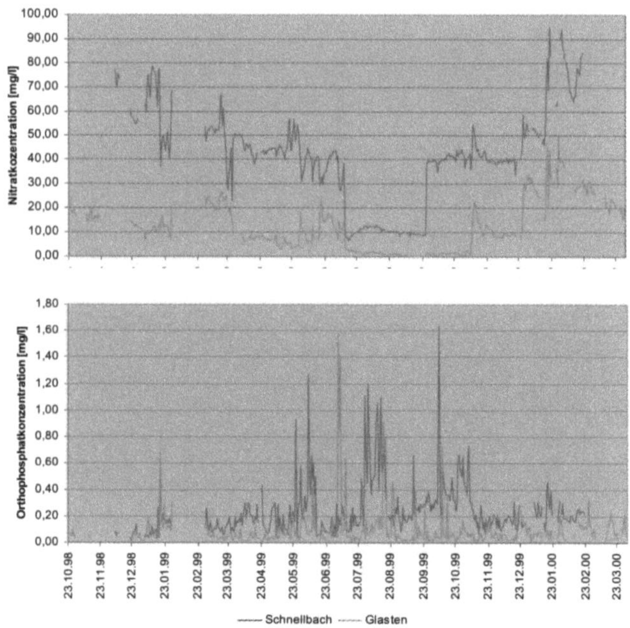


**Fig. 7.13:** Concentration of orthophosphate (bottom) and nitrate (top) in surface water.

The response of the receiving water to such precipitation events obviously depends more on the size of the watershed than on its natural conditions and land use and land cover situation. Although the size of the subbasin at the Glasten gauge is comparable to the size of the Schnellbach subbasin, both areas differ in their land use structure: nearly the whole subbasin at the Glasten gauge is used by forestry, whereas the Schnellbach subbasin is dominated by intensive agricultural use (Fig. 7.11). No retention influence by forest on drainage can be observed. Immediately after the precipitation event, the receiving water responds with a sharp rise in discharge, but also descends just as fast. At the gauges located downstream, such as Naunhof and to a much higher extent Thekla, the drainage curve rises with a delay after the precipitation event respectively with a stay away of the local precipitation. A decay can also be observed at the descent of the drainage curve.

The increased flow rate and its related dilution effect is naturally accompanied by a decrease in conductivity. The initial results of the investigations into the quality of the surface water indicate seasonal and temporal differences. All the samples were analysed for their orthophoshate and nitrate levels (Fig. 7.13).

These field measurements are planned to be the beginning of a long-term environmental monitoring. The results of the measurements help our understanding of the spatio-temporal distribution and organization of water-borne fluxes of material to be detected and improved, and also serve as input data for model applications. As the mesoscale processing of cultivation-related nutrient input into rivers by surface runoff entails the mathematical description of the fluxes of water, material and energy, the usage of simulation systems is essential.

The qualification and quantification of the vertical and horizontal fluxes of water, material and energy fluxes from the designated risk zones requires a change in the scale level, as mentioned in the previous section. A database with a higher spatio-temporal resolution allows detailed hydromorphological analysis (calculation of subbasins, flow direction, stream net and stream order as potential material transport courses, etc.) and is used as input data for various models. Due to the scale-specific applicability of the models tested, data with different spatio-temporal resolutions are needed. With the common consideration of both 'top-down' and 'bottom-up' approaches, the two methods converge at the 'subbasin scale', which is dealt with below.

As shown above, the application of GIS-coupled model applications also enables the identification of potential risk zones with horizontal material (and nutrients) outwash. However, these 'rough' identifications do not provide any indication of the potential input of the material concerned into the receiving water, which is related to two problems.

GIS-coupled model applications for describing the precipitation/runoff/drainage processes assume a 'depressionless' ('filled') DEM. The relevant GIS routines are only applicable if all drainless depressions of the DEM are filled (this is also true for the natural depressions!) and thus linked to the receiving water. This situation does not correspond to reality. The investigations by Fritsch (1998), for instance, show that in a study area in northeastern Germany, only 10% of the watershed area is linked to the receiving water. The situation differs due to the different natural conditions of landscapes, of course, but even in our own study areas the 100% drainage of the watershed is not guaranteed. Even with a hypothetical assumption of such a situation, the entire surface runoff does not reach the receiving water. This must also be considered in relation to the structure of the near-stream land. Near-stream land can exert a retention influence on the receiving water which is related to the dissolved substances in the water (nitrogen components), as well as the material fixed to the particles transported (i.e. phosphorus). This is the subject of several detailed investigations (e.g. Haycock & Burt 1993).

In our investigations, our goal was to produce a simple method to assess the effectiveness of the near-stream land along the 50 km Parthe riverbanks as a potential nutrient retention zone. This assessment necessitates measuring the relevant qualitative and quantitative parameters of the near-stream land. Therefore, the theoretical fundamentals of the nitrogen and phosphorus cycles had to be studied intensively. This enables the structures and processes to be identified which are relevant for the outwash of these materials out of the landscape and input into the water. This in turn formed the basis for the following development of a special mapping method. In doing so, various suggestions for the recording and assessment of the capability of the near-stream land (LUNRW 1993, Raderschall 1995, SMU 1995, LAWA 1998, DVWK 1997) were taken into account and combined with the above-mentioned aspects and adapted to the conditions of the study area (Fig. 7.14, Rau 1999).

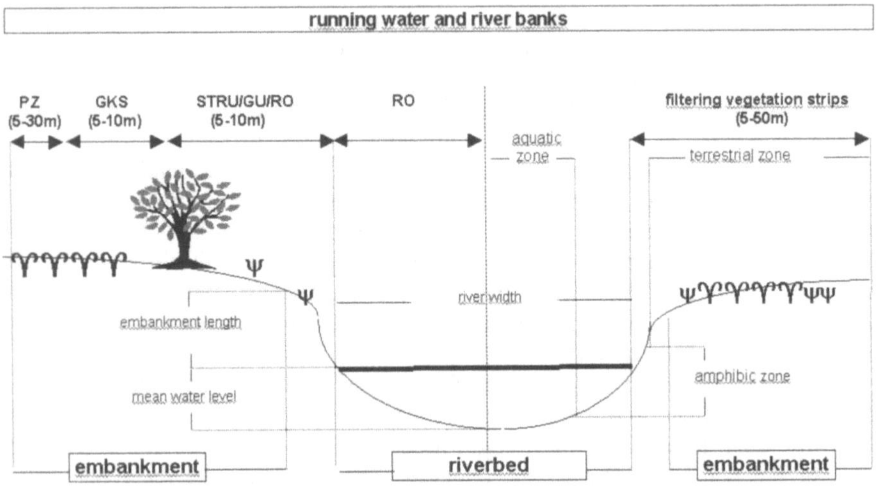


**Fig. 7.14:** Profile of a stream and structured retention zones (after Rau 1999)

As basic assessment units for recording and assessing, 250m-sections of the river are used. The size of this assessment unit is adapted to the length of the receiving water and suitable for the derivation of information for a watershed of this size. The method was developed using digital mapping and the application of a pen computer and GISPAD software (CON TERRA 1998). The assessment of the retention capability based on the mapping results was carried out using a three-step chart. In view of the amount of work involved, the time required and the related cost, the investigations showed that using the described mapping for a river with a length of 50 km is hardly useful - especially for application by environmental authorities. Nevertheless, the resulting assessments were coupled with other data layers in the GIS. The combination of defined buffer zones with low potential nutrient retention with areas showing high erosion risk, for instance, allows the identification of areas with a high risk of non-point source pollution.

All these investigations can be considered as an important preliminary stage and an addition to the following scale-specific application of models for the mathematical description of the transport processes and the measurement of nutrient input and output. In the Schnellbach subbasin of the River Parthe watershed, for instance, the small-scale model system E3D and the runoff simulation model WaSim-ETH were tested for their scale-specific applicability (Fig. 7.15), as well as to verify the large-scale calculations with ABIMO and later on SWAT. The simulations are based on data with a higher spatial, temporal and thematic resolution: For example, short-term individual rain events are considered as well as field-specific management (e.g. tillage, irrigation, drainage and fertilization). The combination of the investigation methods from different scales allows their scale-specific applicability to be checked. Thus, it should be possible to derive rules for transmission from one scale level to another.

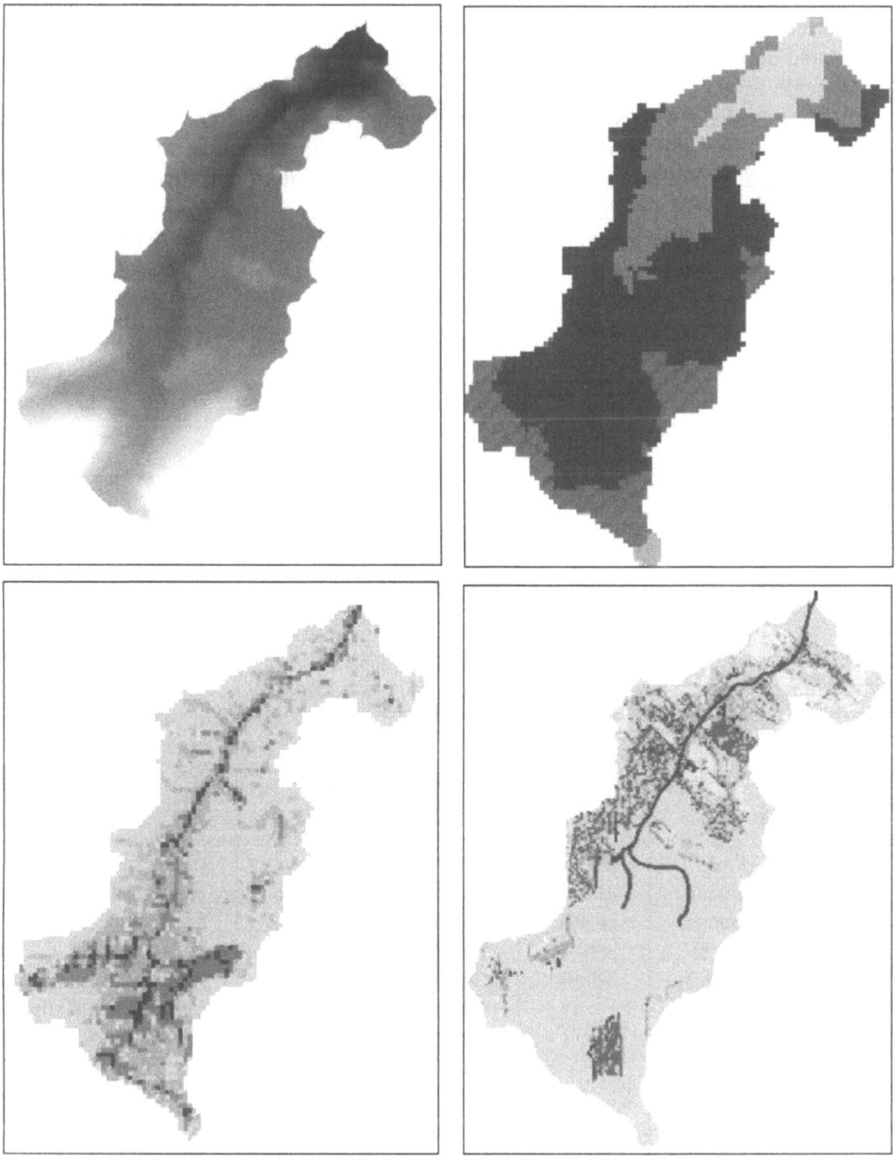

**Fig. 7.15:** Modeling event based soil loss with E3D and water balance (calculated with the runoff simulation model WaSim-ETH) within the Schnellbach subbasin (DTM (upper left), drainage time (upper right), discharge (lower left), soil erosion/accumulation (upper right))

## 7.9 Discussion and conclusions

The hierarchical approaches for the investigation and description of landscapes presented here ought to enable the optimization and regulation of process systems at all scale levels. Besides the reduction of material and nutrient outwash and the regulation of water flows, this also includes consideration of the internal interactions, as well as consideration of interactions with adjacent landscape ecosystems. The investigation of the landscape balance and the calculation of different land use scenarios allows an estimation of the impact of land use changes on the landscape's water, material and nutrient fluxes at the different scale levels (Table 7.3).

**Table 7.3:** A scale-specific approach for the landscape balance investigation and evaluation. The table includes also suggestions for the integration of the results in landscape planning processes.

| Scale level | Spatial planning level | Assessment Units | Data base (different spatial resolutions!) | Model applications | Derived information and application possibilities |
|---|---|---|---|---|---|
| 1:50.000 and larger<br><br>(areas >$10^3$ km²) | Regional level<br><br>(First identifications, coarse classifications) | • Land scape<br>• Units<br>• large river catchments<br>• areas with similar conditions | • Soil<br>• Morphology<br>• (DEM100-250)<br>• Climate<br>• Water<br>• Watersheds<br>• Land Use<br>• Landscape Units<br>• Spatial Planning Targets | • ABIMO<br><br>Description of the fundamental elements of the water cycle (e.g. runoff) | • Water balance and land use scenarios for areas >100 km²<br>• (Coarse) Identifications of potential risk zones with (water) and material fluxes (combination of modeling results with assessments guidelines)<br>• Analysis of land use conflicts on the regional scale and recommendations for land use (environmental and resources management and conservation) |

continued

| | | | | | |
|---|---|---|---|---|---|
| 1:25.000 to 1:50.000<br><br>(areas 10 to $10^3$ km²) | **Regional level**<br><br>District level<br><br>(Quantitative and qualitative information and assessments) | • Watersheds<br>• subbasins<br>• conservation areas<br>• indicated danger zones (cp. above) | • Soil<br>• Morphology<br>• (DEM40-100)<br>• Climate<br>• Water<br>• Watersheds<br>• Subbasins<br>• Land Use<br>• Spatial Planning Targets | • SWAT<br>• ABIMO<br>• WaSIM-ETH<br>• AGNPS<br>• (CANDY)<br><br>Investigation of vertical and horizontal matter and energy fluxes (Nitrogen, phosphorus and pesticides transport, Erosion) | • Identification of subbasins, streamnet and -order (flow of matter and nutrients)<br>• Indication of rivers and streams affected by matter and nutrient input<br>• Modeling of water balance and material and energy fluxes (qualitative and quantitative information) within the risk zones<br>• Combination of modeling results with assessment methods: Recommendations for land use variants for decreased material/ nutrient output related to agricultural areas. |
| 1:10.000 to 1:25.000<br><br>(areas 100m² to 10 km²) | District level<br><br>Community level<br><br>(Detailed quantitative and qualitative assessments) | • Fields<br>• biotopes<br>• river sections | (incl. mapping/ measuring)<br>• Soil<br>• Morphology<br>• (DEM<40)<br>• Climate<br>• Water<br>• Subbsins<br>• Land Use<br>• Spatial Planning Targets | • **Physical and empirical Models**<br><br>(WEPP, AGNPS, CANDY) | • Material and nutrient output (outwash) related to fields<br><br>• Polyfunctional landscape assessment and land use optimization |

On the basis of the results, recommendations can be concluded for land use variants with positive effects for environmental and natural resource protection. Considering the reduction of material and nutrient outwash out of agricultural areas (e.g. in conservation areas), this could be an important argument within the planning processes of the relevant agencies. Besides the scale-specific optimization of the model applications and landscape assessment methods, we are working on a hierarchical parameter indicator system, which allows both the assessment of other landscape functions and the integration of the model calculations. The first results of these investigations for the upper mesoscale are presented in Chap. 9 of this book. The augmentation and combination of 'classical' methods such as measuring, mapping and assessment with innovative GIS-model applications ought to solve the problem of verifying mesoscale and macroscale model calculations of the landscape balance. Various methods are being developed and tested by the authors' working group. The next important step is to integrate socioeconomic components into the approach. Examples of such integrated ecological and socioeconomical assessments are contained in Chap. 10 as well as in Horsch, Ring & Herzog (2001) and Dabbert et al. (1999).

## 7.10 Outlook

Considering the 'state of the art' of the field of landscape ecology dealt with here (including our own approach), future research should be focused on the following topics:

- *Topic 1:*
- Improving our understanding of landscape-ecological processes: interactions between landscape structures and processes.
- *Possible solution:*

  Further development of models and scale-specific assessment methods (optimization and verification)

  → Development and application of 'context-related' remote sensing methods

- *Topic 2:*
- Less availability of the required database for large areas
- *Possible solution:*

  → Development of transfer functions

  → Further development of 'hydrological remote sensing'.

- *Topic 3:*
- A lack of knowledge concerning the 'natural' dynamics and adaptability of ecosystems (current 'ecologiocal' assessments are mostly structure-oriented, especially at larger scales). As a result, insufficient attention is paid tothese factors.
- *Possible solution:*

  → Enforce of basic science in the field of the behaviour and adaptability of ecosystems to human impacts (with special attention to land use)

In addition, future landscape-ecological research should be directed towards the greater consideration of the 'driving forces' of land use development (supra-regional, national and international tendencies). Subsequently, the calculation of scenarios of these probable land use changes will allow better forecasts of their impact on the landscape balance. As socioeconomic parameters are integrated, multicriteria assessment can be carried out to derive land use and water management strategies geared towards sustainable development.

Future models and methods developed in landscape ecological research must be clear and understandable in their structure, procedure and applicability. This entails further investigations that contribute to a better understanding of the interactions between landscape-ecological processes and structures, and the hierarchical concepts of landscape ecosystems (i.e. Klijn 1995). This will lead to these models and methods being increasingly accepted by environmental and planning authorities as important instruments for the simulation of human impacts on the landscape balance and as decision instruments for relevant planning problems. This is all the more important considering the rapid development of computer-based models and methods in landscape ecology.

## 7.11 References


AG Boden (1994) Bodenkundliche Kartieranleitung. Hannover

Antonova N (1998) Land Cover Change in Tooele County, Utah, and its Effect on Ferruginous Hawk Nesting Sites: Analysis Using Satellite Data and GIS. - http://www.ac.wwu.edu/~n9649652/research.htm

Arnold JG, Allen PM, Bernhardt G (1993) A comprehensive surface-groundwater flow model. J. Hydrology 142: 47-69

Bastian O, Schreiber K-F (Hrsg) (1999) Analyse und ökologische Bewertung der Landschaft. Spektrum Akademischer Verlag, Heidelberg, Berlin

Bierkens MFP, Finke PA & de Willigen P (eds) (2000) Upscaling and Downscaling Methods for Environmental Research. Kluwer Academic Publishers, Dordrecht-Boston- London

Bork H-R, Schröder A (1996) Quantifizierung des Bodenabtrags anhand von Modellen. In: Handbuch der Bodenkunde 1, Erg. Lfg. 12/96, Chap. 7.3.5: S. 1-44

Boyce A (1998) Using Satellite Data to Investigate Relationships Between Land Cover Change and Water Resources in the Nooksack Watershed, Washington: 1973-1995. http://www.ac.wwu.edu/~n9649652/research.htm

Bundesanstalt für Geologie und Rohstoffe (BGR) (Hrsg) (1994) Methodendokumentation Bodenkunde. Auswertungsmöglichkeiten zur Beurteilung der Empfindlichkeit und Belastbarkeit der Böden. Geol. Jahrb., Reihe F, 31

Burns TP, Pattan BC, Higashi H (1991) Hierarchical Evolution in Ecological Networks. In: Higashi M, Burns TP (eds) Theorerical Studies of Ecosystems: The Network Perspective, New York

Chorley RS, Kennedy B (1971) Physical Geography. A Systems Approach

CON TERRA (1998): GISPAD 2.0 Benutzerhandbuch. Münster

Cooper WS (1923) The recent ecological history of Glacier Bay Alaska: II. the present vegetation cycle. - Ecology 4: 223-247

Cowles HC (1899) The ecological relations of the vegetation on the sand dunes of Lake Michigan: I. geographical relations of the dune floras. Botanical Gazette 27: 95-117, 167-202, 281-308, 361-391

Dabbert S, Herrmann S, Kaule G, Sommer M (Hrsg) (1999) Landschaftsmodellierung für die Umweltplanung. Methodik, Anwendung und Übertragbarkeit am Beispiel von Agrarlandschaften. Springer, Berlin-Heidelberg-New York

Delcourt PA, Delcourt HR (1992) Ecotone dynamics in space and time. In: Hansen AJ, di Castri F (eds): Landscape boundaries. Springer, New York, pp. 19-54

Deutscher Verband für Wasserwirtschaft und Kulturbau e.V. (DVWK 1997) Uferstreifen an Fließgewässern - Funktion, Gestaltung und Pflege. DVWK-Merkblätter zur Wasserwirtschaft, Bonn

Diekkrüger B (1999) Regionalisierung von Wasserqualität und -quantität - Konzepte und Methoden. In: Steinhardt U, Volk M (eds): Regionalisierung in der Landschaftsökologie. Forschung, Planung, Praxis. Teubner, Stuttgart-Leipzig, pp 67-78

Finke L (1994) Landschaftsökologie. - Das Geographische Seminar. Westermann, Braunschweig

Franko U et al (1997) Einfluß von Standort und Bewirtschaftung auf den N-Austrag aus Agrarökosystemen. UFZ-Bericht 10/97, Leipzig

Fritsch U (1998) Zur Bestimmung potentieller Abflußbahnen aus einem digitalen Geländemodell am Beipiel einer Jungmoränenlandschaft. - Diploma thesis (unpub.), Univ. Potsdam

Gardner RH, O'Neill RV (1991) Pattern, Processes and Predictability: The Use of Neutral Models for Landscape Analysis. In: Turner MG, Gardner RG (eds) Quantitative Methods in Landscape Ecology. Ecological Studies 82, Springer, New York, pp. 289-307

Gates WL (1985) The use of general circulation models in the analysis of the ecosystem impacts of climatic change. Climatic Change 7: 267-284

Gerold G (1999) Regionalisierung und upscaling des Wasserumsatzes in Einzugsgebieten. In: Steinhardt U,Volk M (eds) Regionalisierung in der Landschaftsökologie. Forschung, Planung, Praxis. Teubner, Stuttgart-Leipzig, pp 79-95

Glugla G, Fürtig M (1997) Dokumentation zur Anwendung des Rechenprogrammes ABIMO. Bundesanstalt für Gewässerkunde, Berlin

Goodland R (1992) Nach dem Brundtland-Bericht: umweltverträgliche wirtschaftliche Entwicklung. - Deutsches Nationalkomitee für das UNESCO-Programm "Der Mensch und die Biospaere" (MAB) und Deutsche UNESCO-Kommission

Gore A (1992) Wege zum Gleichgewicht. Ein Marshallplan für die Erde. Fischer, Stuttgart

Grayson RB, Moore ID, McMahon TA (1992) Physically based hydrological modeling. 2. Is the concept realistic? Water Resources Research 26, 19: 2659-2667

Grunewald K, Mannsfeld K, Gebel M (1999) Raum-zeitliche Maßstabsprobleme und deren Ergebnisrelevanz - dargestellt am Beispiel der Quantifizierung diffuser Stoffeinträge in Oberflächengewässer. In: Steinhardt U, Volk M (eds) Regionalisierung in der Landschaftsökologie. Forschung, Planung, Praxis. Teubner, Stuttgart-Leipzig, pp 180-193

Grunwald S (1997) GIS-gestützte Modellierung des Landschaftswasser- und Stoffhaushaltes mit dem Modell AGNPSm. Boden und Landschaft. Schriftenreihe zur Bodenkunde, Landeskultur und Landschaftsökologie 14, Gießen

Hansen AJ, di Castri F (eds) (1992): Landscape boundaries. Springer, New York

Hansen AJ, Risser PG, di Castri F (1992) Epilogue: Biodiversity and ecological flows across ecotones. In: Hansen AJ, di Castri F (eds) (1992): Landscape boundaries. Springer, New York, pp. 423-438

Hauhs M, Lange H, Kastner-Maresch A (2000) Die Modellierung ökologischer Systeme - wissenschaftliche Computerspiele oder theoretische Alchemie? Petermann's Geogr. Mitt. 2000/2: 52-57

Haycock NE, Burt TP (1993) The sensivity to rivers to nitrate leaching: The effectiveness of near-stream land as a nutrient retention zone. In: Thomas DSG, Allison RJ (eds) Landscape Sensivity. Springer, New York, pp.261-272

Helming K, Frielinghaus M (1999) Skalenaspekte der Bodenerosion. In: Steinhardt U, Volk ; (eds): Regionalisierung in der Landschaftsökologie. Forschung, Planung, Praxis. Teubner, Stuttgart-Leipzig, pp. 221-232

Herz K (1994) Ein geographischer Landschaftsbegriff. - Wiss. Z. Techn. Univ. Dresden 43/5: 82-89

Hickey R, Smith A, Jankowski P (1994) Slope length calculation from a DEM within ArcInfo GRID. Comput. Environ. and Urban Systems 18: 365-380

Horsch H, Ring I, Herzog F (eds) (2001) Nachhaltige Wasserbewirtschaftung und Landnutzung. Methoden und Instrumente der Entscheidungsfindung und -umsetzung. - (in press)

Jedicke E (1994) Biotopverbund. Grundlagen und Maßnahmen einer neuen Naturschutzstrategie. Ulmer, Stuttgart

King AW (1991) Translating Models Across Scales in Landscape Ecology. In: Turner MG, Gardner RH (eds) Quantitative Methods in Landscape Ecology. Ecological Studies 82, Springer; New York, pp. 479- 517

Klijn JA (1995) Hierarchical concepts in landscape ecology and its underlying disciplines. Winand Staring Centre for Integrated Land, Soil and Water Research, Report 100, Wageningen

Klug H, Lang R (1983) Einführung in die Geosystemlehre. Wissenschaftliche Buchgesellschaft, Darmstadt

Krysanova V, Müller-Wohlfeil D-I, Becker A (1996) Integrated modelling of hydrology and water quality in mesoscale watersheds. PIK-Report 18, Potsdam

Landesumweltamt Nordrhein-Westfalen (LUNRW) (Hrsg) (1998): Gewässerstrukturgüte in Nordrhein-Westfalen. Kartieranleitung. Merkblätter 14

Lausch A (1999) Möglichkeiten und Grenzen der Einbeziehung von Fernerkundungsdaten zur Analyse von Indikatoren der Landschaftsstruktur - Beispielsregion Südraum Leipzig. - In: Steinhardt U, Volk M (eds) (1999): Regionalisierung in der Landschaftsökologie. Forschung, Planung, Praxis. Teubner, Stuttgart-Leipzig, pp 162-179

Lausch A (2000) Raum-zeitliches Monitoring von Landschaftsstrukturen in der Tagebauregion Südraum Leipzig mit Methoden der Fernerkundung und Geoinformation. UFZ-Bericht 12/2000, Leipzig

Länderarbeitsgemeinschaft Wasser (LAWA, 1998) Gewässerstrukturgütekartierung in der Bundesrepublik Deutschland. Verfahren kleine und mittelgroße Fließgewässer. Berlin

Lenz R (1999) Mittelmaßstäbige Raumbewertungsverfahren für die Angewandte Landschaftsökologie. In: Schneider-Sliwa R, Schaub D, Gerold G (eds) (1999): Angewandte Landschaftsökologie. Grundlagen und Methoden. Springer, Berlin-Heidelberg-New York, pp 151-161

Leser H (1997) Landschaftsökologie. Ulmer, Stuttgart

Leser H (1999) Georelief. In: Zepp, H, Müller MJ (Hrsg) (1999) Landschaftsökologische Erfassungsstandards. Forschungen zur deutschen Landeskunde 224, Flensburg, pp 29-49

Leser H, Klink HJ (1988) Handbuch und Kartieranleitung Geoökologische Karte 1:25.000, Forschungen z. Deutschen Landeskunde 228, Trier

Marks R et al (1992): Anleitung zur Bewertung des Leistungsvermögens des Landschaftshaushaltes. Forschungen z. Deutschen Landeskunde 229, Trier

Menz M, Kempel-Eggenberger C (1999) Gegenüberstellung von Stoffflußmessungen nach dem Konzept der landschaftsökologischen Komplexanalyse mit einer GIS-basierten ökologischen Risikokarte in topischer bis mesochorischer Dimension im schweizerischen Jura. In: Steinhardt U, Volk M (eds) (1999) Regionalisierung in der Landschaftsökologie. Forschung, Planung, Praxis. Teubner, Stuttgart-Leipzig, pp 109-121

Meyer B (1997) Landschaftsstrukturen und Regulationsfunktionen in Intensivagrarlandschaften im Raum Leipzig-Halle. Regionalisierte Umweltqualitätsziele - Funktionsbewertungen - Multikriterielle Landschaftsoptimierung. UFZ - Bericht 24/1997, Leipzig

Mosimann T (1978) Der Standort im landschaftlichen Ökosystem. Catena 5: 351-364

Mosimann T (1999) Angewandte Landschaftsökologie - Inhalte, Stellung, Perspektiven. In: Schneider-Sliwa R, Schaub D, Gerold G (eds): Angewandte Landschaftsökologie. Grundlagen und Methoden. Springer, berlin-Heidelberg-New York, pp 5-23

Müller E, Volk M (1998): Entwicklung, Stand und Perspektiven der Landschaftsbewertung. In: Grabaum R, Steinhardt U (eds) Landschaftsbewertungsverfahren unter Anwendung analytischer Verfahren und Fuzzy Logic. UFZ Bericht 6/1998, Leipzig, pp 10-27

Müller F (1999) Ökosystemare Modellvorstellungen und Ökosystemmodelle in der Angewandten Landschaftsökologie. In: Schneider-Sliwa R, Schaub D, Gerold G (eds): Angewandte Landschaftsökologie. Grundlagen und Methoden. Springer, berlin-Heidelberg-New York, pp 25-47

Naveh Z, Lieberman A (1994) Landscape Ecology. Springer, New York

Neef E (1967) Die theoretischen Grundlagen der Landschaftslehre. Haack, Gotha
Neef E (1973) Beiträge zur Klärung der Terminologie in der Landschaftsforschung. - Arbeitspapier zum Symposium vom 28.11. - 01.12.1973 in Smolenice, CSSR
Olivera F, Reed S, Maidment D (1998) HEC-PrePro v. 2.0: An ArcView Pre-Processor for HEC's Hydrological Modeling System. Contribution to the 1998 ESRI User's Conference July 25-31, 1998, San Diego, California http://internetcity.crwr.utexas.edu/gis/gishydro99/watchar/papers/esri98/p400.htm.
O'Neill RV et al. (1986) A hierarchical concept of Ecosystems. - Princetone University Press, Princeton
Petry D, Krönert R (1998) Towards the integration of land use and natural resources: contributions of landscape analysis to regional planning. In: Dover JW, Bunce RGH (eds) (1998) Keyconcepts in Landscape Ecology. IALE (UK), Preston, pp. 405-410
Petry D, Herzog F, Volk M, Steinhardt U, Erfurth S (2000) Auswirkungen unterschiedlicher Datengrundlagen auf mesoskalige Wasserhaushaltsmodellierungen: Beispiele aus dem mitteldeutschen Raum. Z. Kulturtechnik und Landentwicklung 1/2000: 19-27
Plachter H (1991): Naturschutz. Ulmer, Stuttgart
Rachimov C (1996) Algorithmus zum BAGROV-GLUGLA-Verfahren für die Berechnung langjähriger Mittelwerte des Wasserhaushaltes (Abflussbildungsmodell ABIMO, Version 2.1). - Programmbeschreibung, pro data consulting Claus Rachimow
Raderschall R (1995) Bisheriger Kenntnisstand zur Funktion und Bedeutung der Gewässerrandstreifen. In: Behrendt H, Raderschall R, Pagenkopf W, Frielinghaus M, Winnige R (1995) Studie zur Erarbeitung von Grundlagen für die Ausweisung von Gewässerrandstreifen im Auftrag des MUNR Brandenburg, pp 10-51
Rau S (1999) Der Einfluss von Gewässerrandstreifen auf Stoffflüsse in Landschaften. Einsatz eines mobilen Pen-Computers zur Kartierung im Einzugsgebiet der Parthe. Diploma thesis (unpub.), Univ. Potsdam
Richards B (1998) A Regional-Scale Analysis of Wildlife Dispersal Success in a Fragmented Landscape. http://www.ac.wwu.edu/~n9649652/research.htm
Richter O et al. (1997) Kopplung Geographischer Informationssysteme (GIS) mit ökologischen Modellen im Naturschutzmanagement. In: Kratz R, Suhling F (eds) (1997) GIS im Naturschutz: Forschung, Planung, Praxis, pp 5-29
Röder M (1998) Erfassung und Bewertung anthropogen bedingter Änderungen des Landschaftswasserhaushaltes - dargestellt an Beispielen der Westlausitz. PhD-thesis TU Dresden
Rohdenburg H (1989) Landschaftsökologie - Geomorphologie. - [Aus d. Ms. bearb. u. hrsg. von Margit Rohdenburg]. Catena
Sächsisches Staatsministerium für Umwelt und Landesentwicklung (SMU 1995) Materialien zur Wasserwirtschaft. Richtlinien für die naturnahe Gestaltung der fließgewässer in Sachsen
Sauerborn P (1994) Die Erosivität der Niederschläge in Deutschland - Ein Beitrag zur quantitativen Prognose der Bodenerosion durch Wasser in Mitteleuropa. Bonner Bodenkundliche Abhandlungen 13
Schmidt R (1978) Geoökologische und bodengeographische Einheiten der chorischen Dimension und ihre Bedeutung für Charakterisierung der Agrarstandorte der DDR. In: Beiträge zur Geographie 29/1, pp 81-157
Schmithüsen J (1973) Die ökologischen Aspekte der Landschaftsforschung. 3rd Int. Symposium "Content and Object of the Complex Landscape Research", Smolenice, CSSR.
Scheinost A (1995) Pedotransfer-Funktionen zum Wasser- und Stoffhaushalt einer Bodenlandschaft. Shaker, Aachen
Schulla J (1997) Hydrologische Modellierung von Flußgebieten zur Abschätzung der Folgen von Klimaänderungen. ETH-Diss. 12018, ETH Zürich
Schwertmann U, Vogl W & Kainz M (1990) Bodenerosion durch Wasser. Ulmer, Stuttgart
Seyfried MS, Wilcox BP (1995) Scale and the nature of spatial variability: Field examples having implications for hydrological modeling. Water Resources Research 31/1: 173-184
Srinivasan R, Arnold JG (1993) Basin scale water quiality modelling using GIS. Proceedings, Applications of Advanced Inform. Technologies for Managem. Of Nat. Res., June 17-19, Spokane, WA, USA, pp 475-484

Steinhardt U (1999a): Die Theorie der geographischen Dimensionen in der Angewandten Landschaftsökologie. - In: Schneider-Sliwa R, Schaub D, Gerold G (eds): Angewandte Landschaftsökologie. Grundlagen und Methoden. Springer, berlin-Heidelberg-New York, pp 47-67

Steinhardt U (1999b) Quantifying landscape ecological processes on the landscape scale. In: Proceedings of the 5th World Congress of the IALE: Landscape Ecology - The science and the action, Vol. II, Snowmass Village, Colorado, p 144

Steinhardt U, Volk M (1998) Wasser- und Stoffhaushalt auf mesoskaliger Ebene: Modellierung, Bewertung und Visualisierung. - GEOSYSTEMS-Fachtagung 1998: Modellierung und Visualisierung von räumlichen dynamischen Prozessen, CD-ROM.

Steinhardt U, Volk M (eds) (1999) Regionalisierung in der Landschaftsökologie. Forschung, Planung, Praxis. Teubner, Stuttgart-Leipzig

Steinhardt U, Volk M (2000) Von der Makropore zum Flußeinzugsgebiet - hierarchische Ansätze zum Verständnis des landschaftslichen Wasser- und Stoffaushaltes. Petermann's Geogr. Mitt. 2000/2: 80-91

Tietje O, Tapkenhinrichs M (1993) Evaluation of Pedo-Transfer-Functions. Soil Sc. Soc. Am. J. 57: 1088-1095

Troll C (1939) Luftbildplan und ökologische Bodenforschung. Z. Gesellschaft für Erdkunde zu Berlin 7/8: 241-298

Turba-Jurcyk B (1990) Geosystemforschung. Giessener Geogr. Schriften 67

Turner MG, Gardner RH (eds) (1991) Quantitative Methods in Landscape Ecology. Ecological Studies 82, Springer, New York

Volk M (1999) Interactions between landscape balance and land use in the Dessau region, eastern Germany. A hierarchical approach. In: Hlavinkova P, Munzar J (eds) (1999) Regional Prosperity and Sustainability. Proceedings of the 3rd Moravian Geographical Conference CONGEO'99, pp. 201-209

Volk M, Bannholzer M (1999) Auswirkungen von Landnutzungsänderungen auf den Gebietswasserhaushalt: Anwendungsmöglichkeiten des Modells ABIMO für regionale Szenarien. - Geoökodynamik 20: 193-210

Volk M, Steinhardt U (1998) Integration unterschiedlich erhobener Datenebenen in GIS für landschaftsökologische Bewertungen im mitteldeutschen Raum. Photogrammetrie-Fernerkundung-GIS 6/98: 349-362

Wallin D (1998) Using Remote Sensing to Model the Effects of Land Use on Carbon Flux. http://www.ac.wwu.edu/~n9649652/research.htm

Wallin D, Boyce A (1998) Analysis of Land-use Effects on Landscape Patterns and Biological Diversity in Pacific Northwest Forests: 1972-1992. http://www.ac.wwu.edu/~n9649652/research.htm

Weinstein DA (1992) Use of simulation models to evaluate the alteration of ecotones by global carbon dioxid increase. In: Hansen AJ, di Castri F (eds) Landscape boundaries. Springer, New York, pp. 379-393

Werner M v (1995) GIS-orientierte Methoden der digitalen Reliefanalyse zur Modellierung von Bodenerosion in kleinen Einzugsgebieten. Dissertation, Freie Univ. Berlin

Wohlrab B, Meuser A, Sokollek V (1999) Landschaftswasserhaushalt. - In: Schneider-Sliwa R, Schaub D, Gerold G (eds): Angewandte Landschaftsökologie. Grundlagen und Methoden. Springer, berlin-Heidelberg-New York, pp 277-302

Wrbka T, Reiter K, Szerencsits E, Beissmann H, Mandl P, Bartel A, Schneider W, Suppan F (1999) Landscape structure derived from satellite images as indicator for sustainable landuse. In: Nieuwenhuis GJA, Vaughan RA, Molenaar M (eds) Operational Remote Sensing for Sustainable Development. Proceedings of the 18th EARSeL Symposium on "Operational Remote Sensing for Sustainable Development", 11-14 May 1998, Enschede, A.A. Balkema, Rotterdam-Brookfield, pp. 119-127 and color plate

Wu J (1999) Hierarchy and Scaling. Extrapolating Information Along a Scaling Ladder. Canadian Journal of Remote Sensing 25(4): 367-380

Young RA, Onstad CA, Bosch DD & Anderson WP (1987) AGNPS - A nonpoint-source pollution model for evaluating agricultural watersheds. J. Soil and Water Conservation 44/2: 169-173

Zepp H, Müller MJ (Hrsg) (1999) Landschaftsökologische Erfassungsstandards. Ein Methodenbuch. Forschungen zur Deutschen Landeskunde 244, Flensburg

# 8 Landscape assessment

Burghard Meyer

## 8.1 Introduction

When elaborating concepts of landscape development designed to provide long-term environmental sustainability, the key issue at stake comprises competing land use claims to the area concerned (i.e. multifunctional land use). These conflicting aims compete for the area's natural resources and functions. The starting-point for researching landscape assessment is that every landscape must simultaneously fulfil regulation, production, carrier and information functions to differing extents, and this can lead to land use conflicts. By analyzing these conflicts, environmental problems can be identified. In the view of the Department of Applied Landscape Ecology at the UFZ Centre for Environmental Research Leipzig-Halle, special attention needs to be paid to maintaining and restoring regulation functions. Nevertheless, other functions must not be neglected. Indicators must be introduced so that functions can be measured and monitored. Therefore, interest is primarily directly towards methods which consider a number of different functions simultaneously.

According to Mosimann (1999), landscape assessment in applied landscape ecology deals with information at the level of both facts and values which distinguish landscapes from each other: "It compares and contrasts scientific findings and land use claims, conservation aims and development goals of society and its stakeholders." According to Naveh & Lieberman (1994), this necessitates adopting a holistic approach which also takes into account society and the economy. However, says Leser (1999), this cannot be modelled as a whole with a set of individual techniques; instead, an integrated approach must be applied based on the complex of geofactors, biofactors and human factors for the main systems by means of special methods. Examples of such approaches are provided by Schellnhuber et al. (1997), Lemly (1997), Dale et al. (1998), Lee et. al. (1999), Schmid & Schelske (1997), and Grabaum et al. (1999).

Preu & Leineweber (1996) differentiate between geomorphological, pedological, hydrological, bioecological and landscape-ecological approaches for assessing natural areas. Although the approach put forward in this article can (abiding by this classification) be based on the landscape-ecological assessment of natural areas, it also involves large parts of the other approaches mentioned. The method is based on the management of area-related data in geographic information systems. For an in-depth theoretical background, see the methodological and theoretical fundamentals contained in the following references: Naveh & Lieberman (1994), Preu & Leineweber (1996), Finke (1994), Marks et al. (1989), Leser (1997), de Groot (1992), Zepp & Müller (1999), Dollinger (1989), Leser & Klink (1988) Duttmann & Mosimann (1994), Neumeister (1989), Niemann (1977, 1982), Schreiber (1988), Grabaum (1996), Haase (1991), Neef (1965), Blaschke (1997), ANU (1996), Sukopp & Wittig (1993), Scott et al.(1998), Poschmann et al. (1998), and Farina (1998).

A system of landscape assessment based on landscape-ecological and specialist methods became established in planning practice in the mid-1980s. Nowadays, landscape assessment methods are an essential feature of key planning instruments in Germany and the EU (e.g. flora-fauna-habitat and environmental impact assessments, landscape and agricultural planning, and regional policy). Numerous manuals describe assessment methods with various objectives and purposes from both a practical and a more systematic angle, e.g. Bastian & Schreiber (1994), Haber, Riedel, & Theurer (1991), Hennings (1994), and Plachter (1994).

A distinction must be drawn in landscape assessment between two important trends:

- On the one hand, models and large quantities of data are used to evaluate a single function. Producing results which are very accurate and very complex, this trend can be summed up as "specialization" in monofunctional assessment methods. These assessment methods are being increasingly integrated into the neighbouring sciences of landscape ecology, where they are further developed.
- On the other hand, some or even many functions are being increasingly assessed simultaneously or consecutively and then systematically linked together with respect to sustainability. This trend can be described as "landscape-ecological integration" into multifunctional or polyfunctional assessment methods. The degree of complexity here needs to be reduced.

In the following, the author will restrict himself to the topic of landscape-ecological integration with examples of multifunctional and polyfunctional assessment. This assessment naturally builds upon the results of monofunctional assessment.

Multifunctional assessment methods cannot be applied blindly without awareness of the aims of assessment. The three examples and their aims are as follows:

(1) The assessment of topological landscape structures in order to conclude development aims in agricultural landscapes;
(2) The usage of GIS for multifunctional landscape assessment and multicriteria optimization in landscape development scenarios for a landscape or agricultural development plan;
(3) The polyfunctional assessment of landscape units in order to conclude the measures necessary to protect the regional environment and natural resources.

In the face of the recent flood of diverse digital data and the availability of geographic information systems, as well as the development of expert systems, too little attention has been paid to the structured combination of different types of information. However, this combination or aggregation of information is typical of the practical application of landscape assessment methods. The regional applicability and the scale-independence of assessment methods needs to be increasingly discussed nowadays, especially in connection with:

(1) The assessment of landscape structures using newly concluded indicators;
(2) The assessment of a number of landscape functions in the same reference period as a way of combining landscape-ecological information;
(3) Seeking compromises between conflicting assessment results by using multicriteria optimization as a technique for further, scale-independent assessment integration;
(4) Polyfunctional assessment as a method for aggregating assessment results for mesoscale assessment.

All the examples listed above can be used to conclude scenarios of possible future landscape developments within the framework of model conclusions for sustainable development. Functions can be used to measure sustainability as indicators. Whenever a model is implemented, just how close it comes to the intended goals can be measured with functional assessments by using these indicators. For more about the model method, see Wiegleb (1997)

The methodological approach of landscape assessments is shown by means of examples for different study areas in the Leipzig-Halle region. The study areas and the scales used are listed in Table 8.1 :

Chap. 8.2 shows how assessing landscape structures can be used to conclude development goals for agricultural landscapes at a topological level on the basis of the model of the "ideal structure of a rural landscape". For this purpose, bases for the assessment of biotic and abiotic landscape structures are discussed using the example application (see also Gustafson 1998).

Chap. 8.3 shows how multifactorial functional landscape assessments followed by landscape optimization can be put to practical use as an integrative concept for landscape development. The advantages and disadvantages of this GIS-based, highly formalized method for planning agricultural areas and landscapes are discussed. Developed to the level where it can be used in practice, this method refers to landscape elements.

Chap. 8.4 describes the method for the polyfunctional landscape assessment of landscape units in order to conclude the measures necessary for regional environmental and resource protection by using a mesoscale assessment method based on landscape units and landscape types.

**Table 8.1:** Study areas and the scales used

| Chapter: name and location of the study area | Size of study area (ha) | Scale | Related literature |
|---|---|---|---|
| 8.2. Jesewitz/north Saxony | 4,817 | 1:10,000 | Meyer (1997, 1998) |
| 8.2. Barnstädt/Saxony-Anhalt | 4,530 | 1:10,000 | Meyer (1997) |
| 8.2. Gimritz/Saxony-Anhalt | 3,680 | 1:10,000 | Meyer (1997) |
| 8.2. Nerchau/Saxony | 3,189 | 1:10,000 | Meyer (1997) |
| 8.3. Barnstädt/Saxony-Anhalt | 4,240 | 1:10,000 | Grabaum, Meyer, Mühle (1999) |
| 8.4. Halle-Leipzig-Bitterfeld region | 487,776 | 1:400,000 | Meyer & Krönert (1998) |

## 8.2 Assessment of landscape structures at the topological level for the conclusion of development goals in agricultural landscapes

The concept of landscape structure should be discussed in connection with other concepts describing the landscape such as “land use”, “biotope types”, “land use and structural types” and also “naturalness”, i.e. with the regionally and locally changing combination of geographical and landscape components. In the Central European cultural landscape, land use is closely related to human economic activity, and genuine natural landscapes no longer exist. Agricultural landscapes in central Germany have been altered several times since forest clearance by the changing cultivation systems and the usage of chemical and mechanical farming methods. This change culminated in the extensive clearance of landscapes in the 1960s and 1970s during the collectivization of East German agriculture.

Analyzing and assessing landscape structures is highly suitable for substantiating the reconstruction of a new cultural landscape. The methodological basis used is the model of the “ideal structure of a rural landscape” as described by Mander et al. (1998). Assessment covers not only the current landscape structure (as described by the secondary landscape structures), but also the primary landscape structures derived from site analyses. The latter substantiates a spatial pattern for land use adapted to the natural area. The extent to which the secondary landscape structure corresponds to the ideal of the primary landscape structure (as an agricultural model) is ascertained in landscape structure assessments (see below). Assessment results used as environmental quality objectives can be spatially quantified by means of a GIS and applied in a suitable combination for first deriving and then evaluating models.

### 8.2.1 Topological dimension and assessment

The term ‘landscape structure’ is used to refer to “the structure of a landscape, which comprises the appearance of the spatial pattern and the balance function of the spatial units, and which consists of formal and functional features” (Leser et al. 1984). According to Bastian & Schreiber (1994), landscape structure is restrictively defined as “The entirety of the landscape and geographical components and elements making up a landscape as well as their structure (spatial arrangement and connection), albeit with the causal loop being neglected.” The terms ‘spatial patterns’ and ‘landscape components’ each refer to anthropogenic influences on the landscape. The former definition has the advantage for landscape assessment that the landscape structure can be described with both formal and functional characteristics. Landscape structures are described in this section with dimensions derived from the formal structure.

The necessary basic data are supplied for example by terrain maps and laboratory investigations, as well as indirect methods such as the evaluation of aerial photographs, map analyses or satellite pictures. Leser (1991) uses in this respect the term ‘semi-quantitative data’, since they are not determined from one’s own field analyses as “genuine ecological parameters” for the quantitative determination of the process and turnover in landscape ecosystems. However, in the author's opinion, researchers' own field analyses are selective too, owing to the heterogeneity of the landscape. They cannot usually be generalized or transferred to other landscapes. Moreover, landscape structure parameters derived from public primary data (maps and aerial photographs) provide better reproducibility than basic landscape ecology data obtained from complicated measuring programs. Then again, the latter are indispensable for initial model formulation and validation. It should be pointed out that if followed consistently, the principle of restricting landscape ecology to "genuine ecological parameters" would mean that bioecological and statistical methods such as the compilation of "red lists" or biodiversity measures could not be used, since they are based on statistical information, selected considerations, and the analysis and further-processing of data sources already in existence. This is contradicted by the extensive application of assessment approaches in science and practice (Finke 1994).

For a definition of the spatial orientation of topological assessments, see Chap. 6. Assessments on any scale are designed to process functional or structural relationships as a basis for political action. Generally speaking, (natural) scientific facts are determined using a suitable method and scale-dependent precision, and then compiled in different classes. These are then combined with non-scientific quality ratings (good/bad, high/low, etc.), something which is frequently the subject of sharp criticism by scientists. Furthermore, when various objects of assessment are analysed and integrated, weighting is also often used. Weighting is a process in which objects of assessment that are difficult to compare are relativized by weighting them with the aid of additional sources (literature, surveys, expert discussions). For example, soil protection is weighted with a factor of 0.5, while groundwater protection is given a factor of 0.8. This last integration step in particular is usually rejected in disciplinarily orientated basic research - because, according to Marks et al. (1989) for instance, the interactions cannot be evaluated. These interactions must be regarded as an important field of research.

On a topological scale, the smallest unit of a landscape structure to be assessed is a landscape element, which in assessment practice can be equated with a biotope type or with a unit of a land use category. For the soil, the smallest unit assessed is an area with the same soil type (a 'pedotope') "which is determined by uniform pedogenetic and ecological processes" (Leser (1997). Sufficient homogeneity is assumed for both biotopes and pedotopes.

Landscape structure analyses and assessments are performed nowadays in order to contribute to the definition of environmental quality objectives for the model ("Leitbild") discussion (ANL 1994). Environmental quality objectives determine factually, spatially and temporally defined qualities of structures, functions etc. which are to be preserved or changed. The ideal structure of a rural landscape according to Mander et al. (1988) (Fig. 8.1) is regionally adapted to the landscape structure by means of assessments. Jedicke (1991) combines the landscape structure in his joint biotope concept with functional questions. In addition, Jedicke (1991) augments his concept with the necessity of large-scale material extensivation, in which the functional approach of landscape assessment is combined with the structurally orientated approaches of landscape ecology.

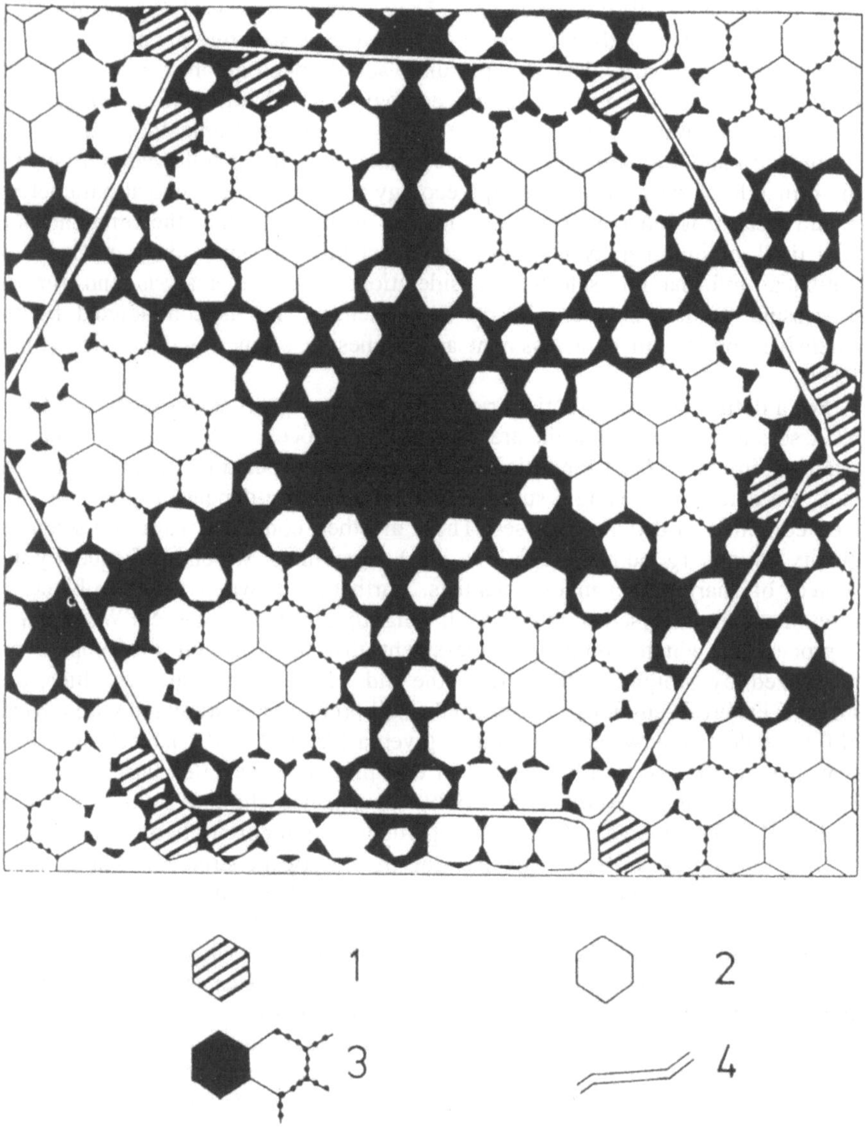


**Fig. 8.1:** The ideal structure of rural landscape: Ecological diversity is guaranteed by heterolevel structure of ecotones and network of compensative areas. 1 = urban areas, 2 = fields, 3 = compensative areas (forests, swamps, woodlots, buffer-strips etc.), 4 = main roads. (Mander et al 1988)

Mander et al. (1988) assume on the basis of ecotone theory that ecological landscape diversity is given by a multilevel structure of ecotones, areas subject to different forms of cultivation depending on local site factors, and a network of 'compensation areas'. Moreover, this landscape diversity should be interrupted as little as possible by roads and building development. Ascertaining and substantiating the landscape structures necessary for this is based on theoretically hexagonal areas. The formulation of this theoretical land use arangement is subject to enormous variation in the real landscape and must therefore be investigated more precisely at a topological level of consideration.

### 8.2.2 Primary and secondary landscape structure

In the following sections, landscape structure analyses are described with the help of regional case studies from four study areas in the Halle-Leipzig district, in different natural areas, and ranging in size between 36 and 47 km², so that related problems can be discussed. At a topological level, a distinction is drawn between the primary and secondary landscape structure. These two terms harbour the aspects of naturalness and land use, which ought to form the basis for all planning. By contrast, land use nowadays only conforms with naturalness to a limited extent.

The **primary landscape structure** ($D_b$) is the distribution of geoecofactors, illustrated in the example for simplicity's sake by the heterogeneity of the soil cover. It reflects the local conditions shaped by natural areas, landscape genesis and land use history. This primary landscape structure is derived from a soil map with a scale of 1:10,000. It can be measured by the soil diversity (Map 8.1).

Calculation of the primary landscape structure ($D_b$) incorporates the size of the study area (BS) and the lengths of the borders between the pedotopes (Bj) into the calculation. The calculation formula is as follows: Db = sum Bj/BS. By comparison, the more complicated method of calculation by Mander et al. (1988) using the calculation formula $D_b$ = e(M/N)BS * sum BJ (incorporating the total number of pedotopes (N) and the number of different pedotopes (M)) produced an identical evaluation of the study areas. $D_b$ is a measure of diversity quantified in m/ha, which can only be used for landscapes in order to take into account fundamental statistical principles. In order to preserve statistical significance, the number of landscape elements should not be less than n = 100, although a much larger figure is actually preferable.

The **secondary landscape structure** (D) comprises anthropogenic land use in the form of landscape elements (biotope types classified by main land use). It is measured by the indicator of landscape diversity and quantified via the exclusion of biotope types as secondary landscape elements on a scale of 1:10,000. Different secondary landscape structures can be determined for different time periods, e.g. for historical reference periods (Map 8.2). Similar to the primary landscape ($D_b$), all areal landscape elements are incorporated into the calculation of D. D is calculated on the basis of the size of the study area (S) and the length of the ecotones (Lj) using the formula D = sum Lj/S.

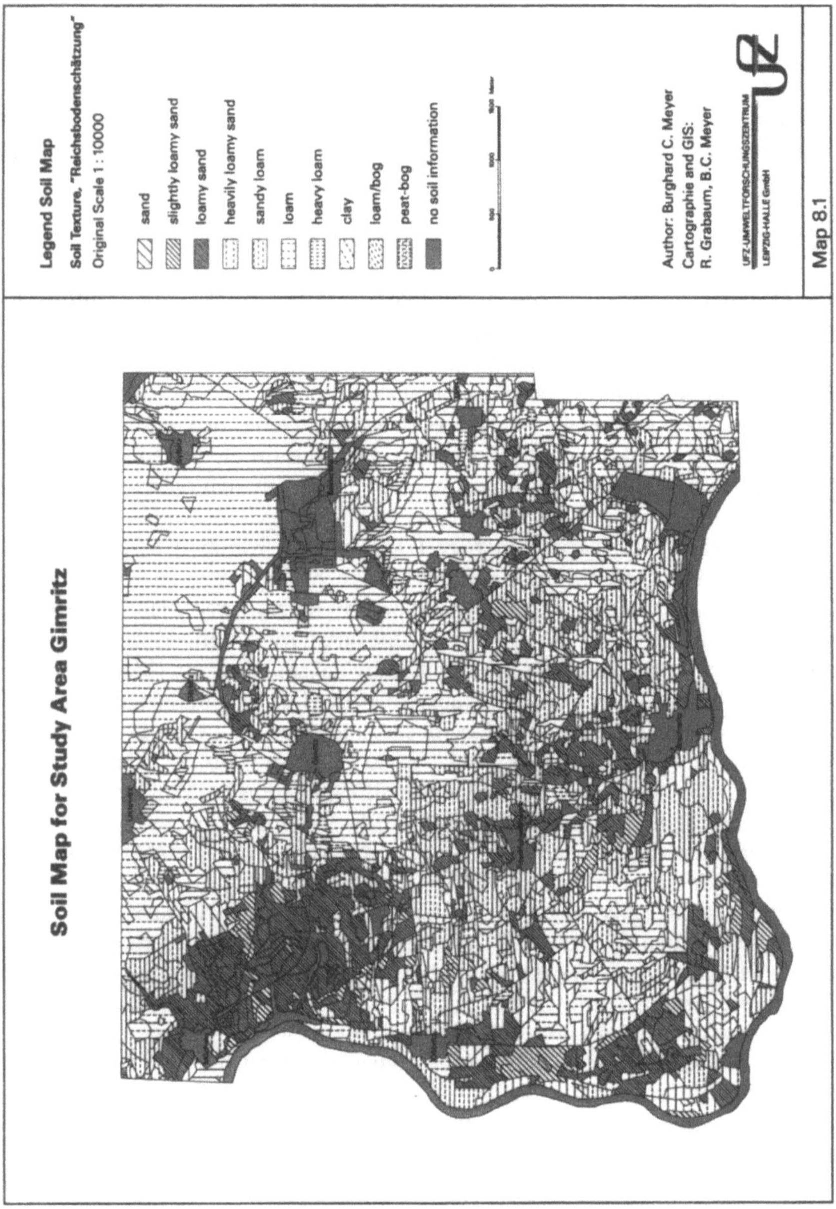


**Map 8.1:** The map shows an extract from the digitized soil appraisal map (1936) of the study area Gimritz at the original scale of 1:10,000

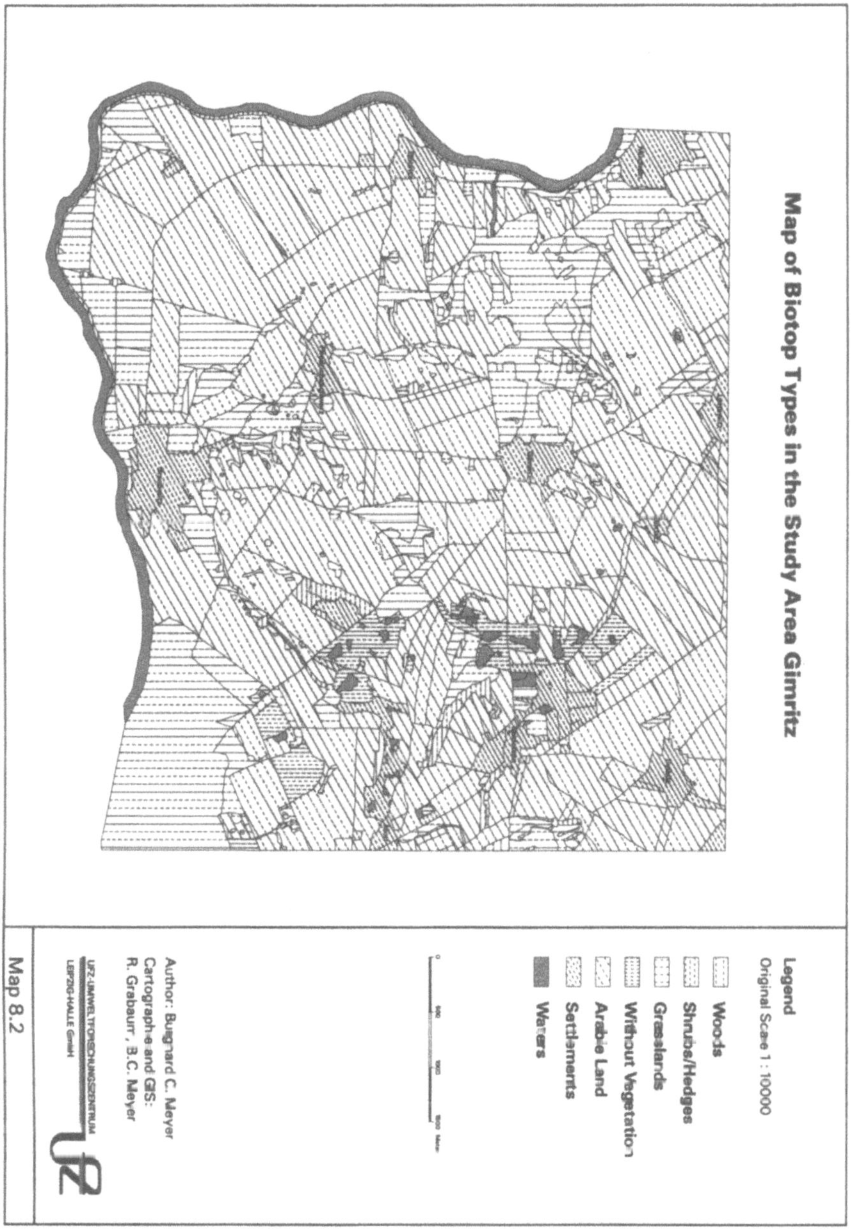


**Map 8.2** The map shows the distribution of biotope types based on our own mapping in 1994 of the study area Gimritz at the original scale of 1:10,000. Comparison of maps 8.1 and 8.2 reveals the soil heterogeneity of the locations used for agriculture and the expansion of residential and mining areas between 1936 and 1994.

The **primary landscape structure potential** quantifies the aim in rural areas of bringing D into line with $D_b$. The difference between the two values is the unexhausted landscape structure potential $D_{diff}$ (Table 8.2). In comparison to the two landscape diversity indicators, $D_{diff}$ is used to derive a regionally measured diversity appropriate for the location in agricultural landscapes. The absolute value of $D_{diff}$ indicates the length of the missing structures. The prioritization of measures for improving landscape diversity given in Table 8.3 varies greatly from one region to the next.

For the assessment, different (regional) values are to be prioritized in a practical, five-level scale. In the example, this takes into account both the degree of heterogeneity of an agricultural landscape and the general level of the heterogeneity index. For agricultural landscapes in the example areas, soil heterogeneity on large fields is crucial for future land use. However, this information is for example not shown on the mesoscale agricultural site map (MMK) owing to the chorological homogenization of the presentation of soil type communities on a scale of 1:100,000. For other regions, the prioritization in Table 8.3 can be normatively determined for specific regions, as D can also assume higher values than $D_b$.

**Table 8.2:** Balance of the diversity of the secondary (D) and primary landscape structure ($D_b$) in four study areas in the Halle-Leipzig district (in 100 m/ha) (Meyer 1997)

| Landscape | D (Secondary LS) | $D_b$ (Primary LS) | $D_{diff}$ = Primary landscape structure potential | Soil region |
|---|---|---|---|---|
| Jesewitz | 1.57 | 2.80 | -1.23 | Sandy loess |
| Nerchau | 1.73 | 2.31 | -0.58 | Loess |
| Barnstädt | 1.14 | 2.25 | -1.13 | Loessy black soil |
| Gimritz | 1.57 | 3.19 | -1.62 | Loess (eroded) |

**Table 8.3:** Classification of the regional assessment of the necessity of measures for landscape diversity improvement in agricultural landscapes (A) and agricultural landscapes with a small proportion of woodland (AW) (Meyer 1996)

| class | $D_{diff}$ | Classification |
|---|---|---|
| 1 | - 0 - 0.4 | Medium |
| 2 | -0.4 - 0.8 | Medium-high |
| 3 | -0.8 - 1.2 | High |
| 4 | -1.2 - 1.6 | Very high |
| 5 | -> 1.6 | Especially high |

### 8.2.3 Regional landscape structure assessments and environmental quality objectives

Working on the basis of the digital data levels listed in Table 8.4, the author performed further landscape structure analyses. Their quantification with GIS should similarly only be undertaken in sufficiently large, statistically meaningful areas (e.g. landscape units or natural area units). For a detailed description, see Meyer (1997).

Based on the data levels listed in Table. 8.4, regionalized environmental quality objectives were concluded and assessed. They are summarized below:

1. A proportion of at least 15% "conservation-relevant land" needs to be achieved in landscapes used for intensive farming in accordance with the normative demands of conservation, especially in view of the usually much smaller proportion of land devoted to this purpose.
2. A soil type/farming land ratio of 1:1 needs to be achieved for field size adapted to the landscape's locational features which can be sustainably used for agriculture in the long term, and which conforms with the landscape diversity shown in Section 4.
3. A potential field margin length needs to be achieved in agricultural landscapes which corresponds to the length of prominent soil borders in landscapes, and which provides a basis for the regionally different lengths of hedges and field margins to be laid out.
4. The stream and river length of the hydrographic network needs to be restored to the situation prevailing in 1935-36.
5. The further expansion of the road network needs to be curbed.

**Table 8.4:** Data levels used in the GIS in four study areas for regional landscape structure assessments (Meyer 1997)

| Jesewitz | Nerchau | Gimritz | Barnstädt |
|---|---|---|---|
| Biotope types<br>-Linear elements<br>-Areal elements | Biotope types<br>-Linear elements<br>-Areal elements | Biotope types<br>-Linear elements<br>-Areal elements | Biotope types<br>-Linear elements<br>-Areal elements |
| Soil types<br>-State stages<br>-Geol. origin<br>Soil form communities | Soil types<br>-State stages<br>-Geol. origin<br>Soil form communities | Soil types<br>-State stages<br>-Geol. origin | Soil types<br>-State stages<br>-Geol. origin |
| Hydrographic network 1936<br>Hydrographic network 1990 | Hydrographic network 1936<br>Hydrographic network 1990 | Hydrographic network 1936<br>Hydrographic network 1990 | Hydrographic network 1936<br>Hydrographic network 1990 |
| Roads and paths 1936<br>Roads and paths 1990 | Roads and paths 1936<br>Roads and paths 1990 | Roads and paths 1936<br>Roads and paths 1990 | Roads and paths 1936<br>Roads and paths 1990 |

This selection of environmental quality objectives can justify various linear and areal secondary landscape structures which would ameliorate the problems stemming from the landscape structure in agricultural landscapes. As their influence on the habitat functions for fauna and flora can only be determined to a limited extent, areal landscape-function assessment methods (pollutant balance, regulation functions) should also be included in order to achieve integrated agricultural landscape development. It should, however, be noted that they are related in different ways to landscape structures and are also largely moulded by the land use structure.

### 8.2.4 Advantages and disadvantages of landscape structure assessments

In all assessments used to classify the landscape, attention must be paid to numerous methodological problems. These start with selecting the appropriate method of assessment. The main problem areas can be grouped together under the categories of selectivity, functional meaningfulness and the frequently incomplete or absent stocks of data:

1. Although the selectivity of individual assessment methods (e.g. a biotope assessment during fruit orchard mapping) gives a good overview of a detailed structure within a landscape with respect for example to certain biotopes which need to be protected, the information's selectivity means it usually cannot be generalized for a whole region.
2. The functional effects of linear landscape structures such as hedges, field margins etc. are at present only indirectly quantifiable and meaningful. For instance, only the influence of an idealized hedge on soil erosion caused by the wind - which only acts in an indirectly functional manner - can be determined. The same applies to the effect of a hedge on the habitat function for pest-destroying species in an agricultural landscape, the estimation of which is highly complicated.
3. One problem is the lack of data stocks relevant to landscape structures, which are normally not publicly available (at least in digital form). However, the comprehensive biotope type mapping now carried out almost everywhere in Germany using CIR (Color Infrared) aerial photographs provides a good basis for the regional quantification of landscape structures in sufficiently large areas necessary for the statistically validated calculation of diversity indices. This necessitates a sufficient statistical stock of basic data.

These problems can basically be solved and must be offset against the major advantages, i.e. the new technical possibilities of quantification using GIS, regional data integration and comparability:

1. **Quantification with GIS:** The landscape structure quantifications shown in Fig. 8.1 and Table 8.3 can only be derived from spatial data allowing the quantification of structural spatial patterns. They concern orders of magnitude which are barely known (if at all), such as those in the lists of biotope types which are endangered or are worth preserving (e.g. Riecken et al. 1994) for specific questions - but not for general landscape structure considerations.

2. **Regional data integration:** These landscape structure quantifications enable new information about the primary landscape structure and the landscape diversity of large test areas or regions regarding both linear and areal landscape structures. This also enables the comparability and assessment of individual structural parameters (e.g. the proportion of grassland in the historical cultural landscape with the primary landscape structure). Only thus can environmental quality objectives be made regionally assessable.
3. **Comparability:** Explaining and comprehending interventions in the landscape structure is greatly improved by the comparability of quantitative regional and local data. Onē important advantage in this respect is for example enhancing the degree to which measures for the improvement in the secondary landscape structure (e.g. the information that 95% - compared to a quantifiable reference year - of a biotope type or a landscape structure has been destroyed, and that as an environmental quality aim at least 10% should be renaturalized) can be translated into practice.

The methodology of the multifactorial assessment of regional landscape structures is still in its infancy. GIS are indispensable for treating cleared landscapes used for intensive agriculture and concluding environmental quality objectives, as this is the only way in which landscape structure assessment can be carried out. Multifactorial landscape structure assessments can solve problems such as:

- The quantification of regionally adapted landscape diversity
- Calculating the length of necessary landscape structures
- Assessing a necessary proportion of grassland
- Concluding field sizes and shapes adapted to the landscape
- The establishment of mosaics of alternating forest and open land

Remote sensing in the area of landscape metrics and pattern identification contains a few methods which can be used to analyse landscape structure parameters (Forman 1985; Chap. 4). It should be noted that remote sensing is based on raster data. Area-specific assessment for planning is not yet possible using landscape metrics with the current state of the art, or at least has not yet been plausibly verified. Area-specific assessment for practical use would have to satisfy at least the following requirements:

- Clear spatial and scalar orientation
- The uniform application of methods to a uniform data basis to ensure the results are comparable;
- A solution to the problems of raster-vector data integration;
- The formulation and verification of the relationship between social norms and the measured structural dimensions (by means of research in the social sciences) as a basis for assessment.

The practical application of the landscape structure assessments presented in this chapter is doubtless more likely to be carried out using GIS analysis methods rather than remote sensing. However, it will be possible to integrate remote-sensing data and methods for individual data levels.

# 8.3 Using GIS for landscape ecological assessments and multicriteria optimization for the Barnstädt test area[1]

## 8.3.1 Introduction

Many ecological problems result from the prolonged monofunctional use of cultural landscapes. This is apparent from regulation disruption. As cultural landscapes perform a variety of functions in nature, accurate planning tools are required to ensure optimal landscape planning. However, the methods used are frequently inadequate because various factors, such as the complexity of landscape structures and reciprocal influence, often make it impossible to conclude an optimal land use concept for a certain region. The common approach of superimposing different assessment maps to generate 'conflict maps' does not fully meet our requirements. As it only highlights the incompatibility of different land use options when the conflict zones are obvious, it cannot produce the integrated view needed of a planning region.

Hence there is a need to develop and practise methods able to generate land use options taking into account a landscape's multifunctionality. One such method is the mathematical technique of multicriteria optimization. The goal functions needed for optimization can be derived from the multifunctional requirements of a landscape. Using this method calls for landscape-ecological assessments geared to goal functions.

Grabaum (1996) developed a computer-based method combining landscape-ecological assessment with optimization. This formed the basis of the "Method of Multicriteria Assessment and Optimization", designed for the low structured agrarian landscape near Querfurt (Saxony-Anhalt) and other test sites. (Grabaum & Meyer 1997, 1998, 1999; Map 8.3).

The possibility of linking up assessments based on landscape elements with the method of multicriteria optimization was described by Koch et al. (1989), and then put forward by Grabaum (1996) as a computer-based integrated method.

As discussed in Chap. 3, the main tool employed for processing a project is a GIS. It is used for data acquisition, the further processing of data, geoecological assessment, presenting scenarios and showing the findings of optimization. Editing functions can be used to integrate information about the cultural landscape and other aspects relevant to land use and implementation (such as site boundaries, pathways and lines of sight, etc.) into planning. Geoecological functions can subsequently be assessed. Direct comparison of the current state of the landscape and land use scenarios can hence be visualized (in maps) and shown statistically.

The interdisciplinary research project "Landscape Assessment and Optimization" (part of the joint project "Impact of land use on landscape balance and biodiversity in agricultural dominated landscapes") is addressing the landscape and its functions, the aim being to ascertain and evaluate the

[1] With the assistance of R. Grabaum and H. Mühle

landscape's current status and changes if land use is altered. This provides a basis for using landscape optimization to draw up proposals (scenarios) for an improved landscape structure. The procedure is as follows (Fig. 8.2):

- Definition of a provisional model for the study area;
- Landscape analysis, landscape assessment;
- Definition of scenarios;
- Ascertaining land use options based on geoecological functions and functions of cultural landscapes as well as functional assessments;
- The effects of implementing these options on landscape-ecological functions.

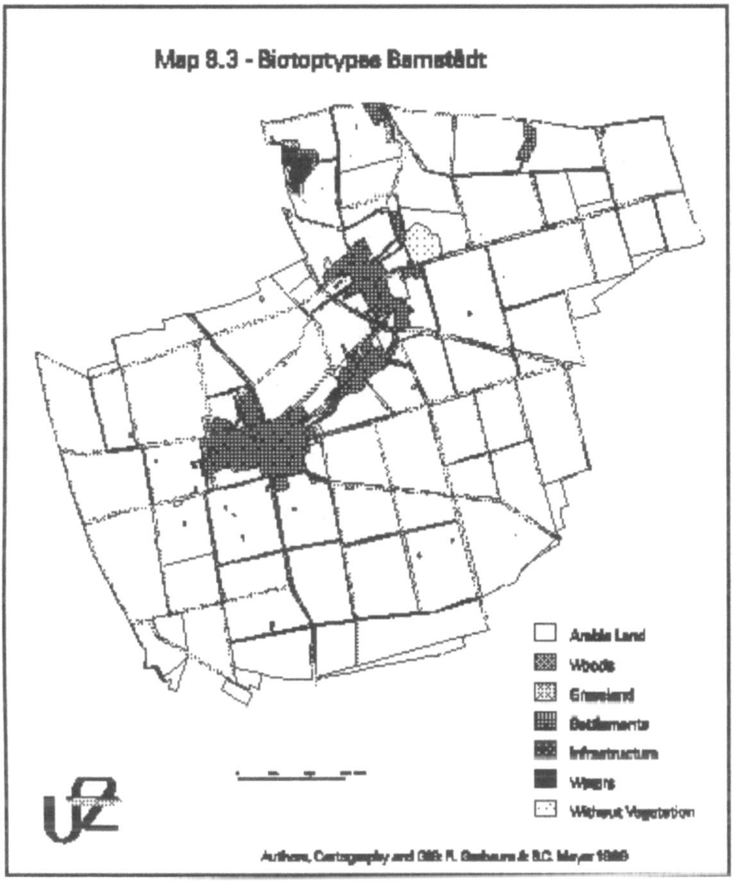


**Map 8.3:** Biotoptypes of the Barnstaedt test site

Cooperation with practical associates, i.e. the presentation and discussion of options as proposals for local planning, is essential both at the start of the proceedings and for the definition of the scenarios, as well as during the discussion of implementation.

### 8.3.2 Initial questions and functional assessment

Landscape-ecological assessments begin by seeking guidance criteria oriented towards regulation and other functions. As is standard practice in landscape planning, the survey region was first demarcated and basic data were compiled. The research examples used below refer to the test site Barnstädt west of Leipzig in an intensively used arable landscape. This test site covers an area of 4,240 ha. The nearly plain landscape is dominated by black soils. The arable land is mainly used to cultivate cereals (Table 8.4).

The models, targets and a selection of the geoecological and landscape functions to be taken into account were discussed in internal expert discussions at several workshops held during the project. Moreover, regional expertise could be used thanks to the project participants' many years of research and the practical partners' extensive experience in the study area. Another basis was provided by the existing plans (Regional Development Plan (Sachsenn-Anhalt 1994), Structural Preliminary Agricultural Planning (Landesgesellschaft Sachsen-Anhalt 1995), and the landscape plan of the administrative district of Wein-Weida Land 1995).

The selection of functions depends on the model chosen for the study area and the relevant issues, and takes place after an initial landscape analysis. The theoretical framework is formed by natural functions in accordance with de Groot (1992), who distinguishes between regulation, carrier, information and production functions. Although the integration of as many functional levels of consideration as possible theoretically best reflects the multi-functionality of the landscape, for practical reasons this approach should be avoided. It should be borne in mind that an excessively large number of functions will reduce the clarity and comprehensibility of the findings (due to overlapping by functions with a similar effect). The selection of geoecological functions (optimization goals) should therefore focus on the main ones.

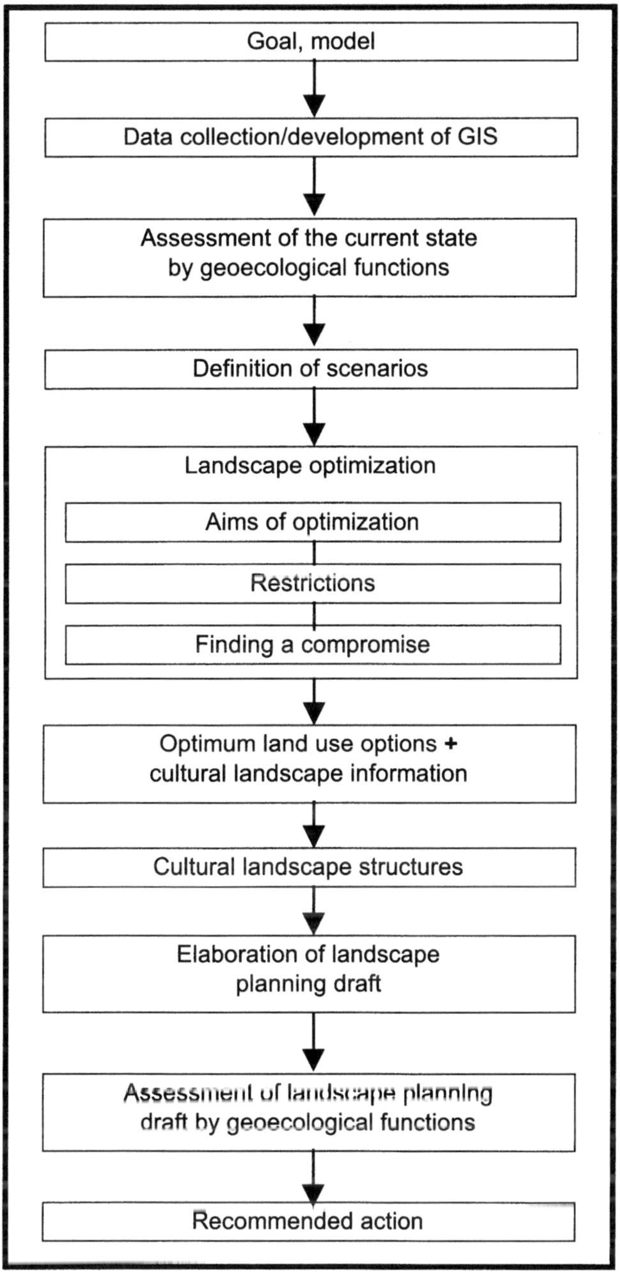


**Fig. 8.2:** Procedure for working out viable action recommendations on the basis of the landscape assessment and multicriteria landscape optimization

**Table 8.4:** Land use types in the study area

| Land use | Area (ha) | Percentage |
|---|---|---|
| Arable land | 3,629 | 85.6 |
| Hops | 24 | 0.6 |
| Built-up areas | 208 | 4.9 |
| Transport | 287 | 6.7 |
| Grassland in built-up areas | 13 | 0.3 |
| Herbaceous vegetation | 11 | 0.3 |
| Bushes/bushes | 18 | 0.4 |
| Land with no vegetation | 22 | 0.5 |
| Surface water | 28 | 0.7 |
| Total | 4,240 | 100.0 |

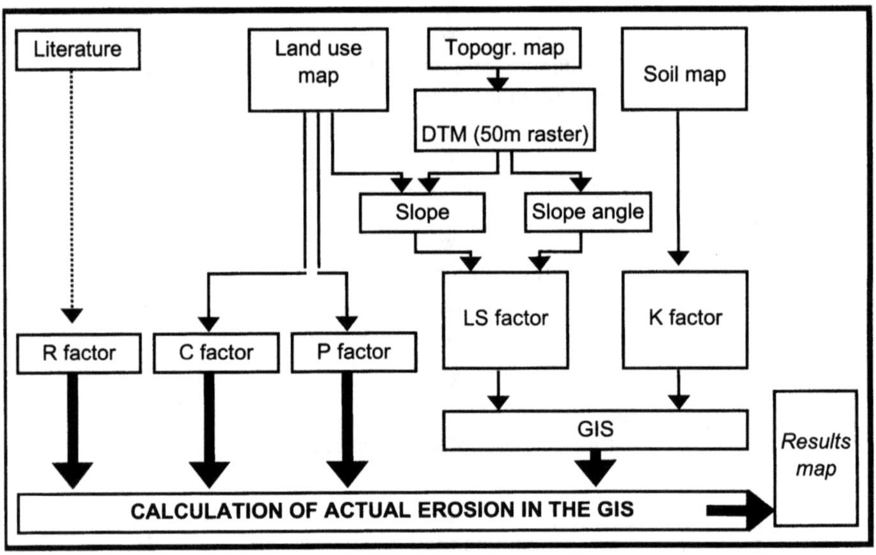


**Fig. 8.3:** Linking up primary data levels in the GIS to assess the danger of soil erosion according to Schwertmann et al. (1987).

### 8.3.3 Geoecological assessment of the current state of the landscape

The geoecological assessment applied is mainly based on techniques described in the "Instructions for the Assessment of the Capacity of the Landscape Balance" (Marks et al. 1989) and other validated methods in the literature:

- Soil erosion by water (Schwertmann et al. 1987)
- Soil erosion by wind (Smith et al. 1992)
- Runoff regulation (Marks et al. 1989)
- Production function (soil indices) in Scheffer & Schachtschabel (1984)

These can be plausibly used in the study area taking into account their scopes of validity, as they are partly based on tests which have been conducted for many years. Landscape-ecological assessments are carried out in the GIS by linking up the corresponding data levels with comprehensible rules. Fig. 8.3 shows by way of example how various (primary) data levels are interconnected in the assessment of medium soil erosion caused by water according to Schwertmann et al. (1987).

A detailed description of the assessment methods would extend beyond the bounds of this article. For more details, see in particular Grabaum et al. (1999). The findings of the assessments are shown in ordinal classes (frequently with 3 or 5 levels). This classification into classes is essential for the further use of the results in the optimization.

The main basis for such techniques is landscape analysis, in which the required basic data are recorded. This entails using the generation of data in the Geoecological Mapping Instructions (Leser et al. 1988) and Hennings (1994). In the method presented, the possibility of the direct further-processing of data using the GIS (e.g. as a guide for optimization) is of key importance.

### 8.3.4 Significance of scenarios within the methodology for land use options

A scenario is a hypothetical sequence of event (in public and industrial planning) fabricated to take account of causal relationships (Duden 1990). Scenarios demarcate the method's scope for action and provide information about development possibilities under certain defined assumptions. They are thus necessary to describe the provisional model more precisely. Scenarios are used in the methods at various points.

- During geoecological assessment, scenarios can provide an overview of the effects of changing individual parameters (basic assumptions) on the assessment results - and hence information about the laws to be taken into account for optimization restrictions (see also Meyer & Grabaum 1996).
- During landscape optimization, scenarios can be defined by changing the weighting of individual partial functions.
- When describing future land use options, the effects of changing land use proportions on the target functions can be formulated in scenarios.

In order to describe future land use when describing the model more precisely, as well as to present the land use changes and their effects on the (landscape-ecological) functions, various scenarios are defined and evaluated with respect to the objectives. Each of these scenarios contains a different scope of land use changes. When defining the objectives and during problem analysis, it is initially determined whether optimization takes place or whether the change in land use can be described with other methods. As far as optimization is concerned, this means that for each function which can be described in an areal manner, an objective is defined based on the evaluating analysis which is to be achieved with a change in land use (for example, the reduction in erosion by at least 30%).

If optimization is used to ascertain land use options, function-related goals need to be formulated. Achieving these aims entails defining restrictions. These include in particular future areal percentages of the land use elements under consideration. These areal percentages will not be exact, but will instead be defined within certain limits (for instance the growth of forestland may account for between 6% and 8% of the area to be optimized). The extent to which the goals are achieved can be reviewed after each optimization run. Should it not be sufficient, a new optimization run with different areal percentages can be started at any time. However, it should be noted that as soon as the aim is defined a certain areal percentage corresponding to the original model must be assumed (for example, a very high proportion of forest does not correspond to the model of an open agricultural landscape). Thus compromises are to be sought if the (geoecological) aims are to be achieved.

In addition, decisions will have to be made concerning the exclusion of areas or the exclusion of uses for certain areas. The scenario thus developed containing indications of the goals of future land use changes provides the framework for landscape optimization. The scenario hence corresponds to an initial, more precise specification of the model.

The three scenarios drafted during the project for the study area are outlined below:

- **Scenario 1**

The proportion of grassland and bushes/woodlands in the study area is increased to 15%. This value is based on Heydemann (1981, 1983), who proposed increasing conservation areas in Germany to a total of 15%. The scenario chosen is designed to examine the effects of these changes.

- **Scenario 2**

The proportion of grassland and bushes/woodlands is 7.5%. This value is half the percentage of quasi-natural structures mentioned by Heydemann (1981, 1983).

- **Scenario 3**

The proportion of grassland and bushes/woodlands is 30%. This value is double the percentage of quasi-natural structures mentioned by Heydemann (1981, 1983).

**Table 8.5:** Assessment of the current state of the landscape (function classification in %)

| Assessment Class \function | Soil erosion by water (5 classes) (arable land) | Soil erosion by wind (2 classes) (arable land) | Runoff regulation (5 classes) (current crop sequence) | Production function (5 classes) (arable land) |
|---|---|---|---|---|
| 1 | 0.5 | | 0.7 | 0.0 |
| 2 | 21.0 | 36.5 | 82.0 | 0.4 |
| 3 | 46.8 | | 5.4 | 1.6 |
| 4 | 9.2 | 63.5 | 0.05 | 14.8 |
| 5 | 8.9 | | 11.9 | 83.2 |

#### *8.3.4.1 Optimization aims for conservation goals*

After extensive assessments of geoecological functions had been carried out in the study area, hence enabling the degree of fulfilment of the provisional model to be estimated, multicriteria optimization is required in order to ascertain land use options for the nature conservation scenarios.

The results of the assessment of the current state of the landscape are shown in Table 8.5 (Assessment Class 1 means very low; Assessment Class 5 means very high). For a breakdown of the assessment methods, see Grabaum et al. 1999).

Landscape optimization is geared towards important area-related aims, for which (geoecological) assessments were therefore concluded. The aim is to secure or even improve the effectiveness of the (geoecological) functions, as required in the updated model. This model can be defined for the study area as follows:

The study area should preserve the character of an open agricultural landscape with extensive soil protection, an increase in the soil's retention capacity, and the maintenance of high productivity. Economic and social conditions are to be organized such that they are in tune with the preservation or improvement in biotic (biological diversity) and abiotic resources, and that the population can be ensured a sufficient income with high social acceptance of their work. The proportion of biotope structures for nature conservation is to be increased.

Starting from the current state, the ecologically substantiated main aims listed below for the implementation of this model can be formulated for all three nature conservation scenarios following on from Mühle (1998):

- Reduction in soil erosion by water as a contribution to soil protection (Function 1)
- Improvement in the retention capacity (Function 2)
- Continuation of production on soils with the highest soil indexes (Function 3)
- Reduction in soil erosion caused by the wind as a contribution to soil protection (Function 4)
- Increasing landscape and species diversity
- The creation of biotopes for species and nature conservation

The aims of increasing landscape and species diversity and the creation of biotopes for species and nature conservation can be achieved if the functions/aims 1-3 are met by changing land use. The aim of reducing wind erosion is incorporated after optimization when drawing up the landscape planning draft, as it is here that linear landscape elements are of crucial importance. Regionally specific environmental quality standards are defined by the formulation of very concrete normative suggestions:

**Water erosion**
The total erosion taking place on the arable land in the study area must be halved. To meet this goal, new landscape structures are to be placed directly on relatively steep or long inclines. This objective should be achieved for the most important areas by means of land use options. Methods of cultivation which are designed to reduce soil erosion should be applied on all areas.

**Runoff regulation**
The retention capacity should not be worsened anywhere, but should instead be improved by one class on 5% of the land. This can in particular be achieved by increasing the percentage of woodland.

All in all, water erosion is to be reduced and the retention capacity increased in those areas which are also subject to wind erosion.

**Production function**
The goal is to maintain production on soils with the highest soil indexes. Furthermore, at least half the land with a soil index-60 should continue to be used for agriculture.

**Wind erosion**
All in all, the amount of land exposed to wind erosion should be reduced by 10% (without buffers).

Following the introduction of buffers, the halving of the amount of land exposed to wind erosion would be desirable. (This goal is not regarded as a direct aim of optimization, but is instead to be achieved by increasing the proportion of grassland and bushes/woodland. Whether this aim has been achieved can only be examined by reassessment.)

#### 8.3.4.2 Restrictions

First of all the optimization area within the study area is defined. Although basically all the land evaluated can be covered by optimization, it makes little sense to include built-up areas, transport routes, surface water, bushes or existing grassland, as their current uses are to be preserved. Hence only cultivated areas should be included.

Starting from scenarios 1 to 3, the following areas are defined for optimization (Table 8.6).

As only the arable land is regarded as the optimization area, the percentage of land relevant for conservation purposes compared to the entire area is somewhat lower. However, this is made up for by the about 1% of the study area which

**Table 8.6:** Restrictions of elements for optimization (sections of the optimization area)

| Element | Scenario 1 | | Scenario 2 | | Scenario 3 | |
|---|---|---|---|---|---|---|
| | Proportion in ha | Proportion in % | Proportion in ha | Proportion in % | Proportion in ha | Proportion in % |
| Grassland | 183-366 | 5-10 | 109-254 | 3-7 | 362-1086 | 10-30 |
| bushes/ woodland | 183-366 | 5-10 | 109-254 | 3-7 | 362-1086 | 10-30 |
| Area of quasi-natural structures | 512-585 | 14-16 | 254-290 | 7-8 | 1,050-1,122 | 29-31 |
| Arable land | 3,070-3,143 | 84-86 | 3,331-3,367 | 92-93 | 2,498-2,570 | 69-71 |

already contains semi-natural structures, which are not included in optimization. As limiting use to a certain area is not planned for the optimization area, this eliminates the need for additional restrictions.

Complete assessment results already exist for the three target functions, i.e. for the elements arable land, grassland and bushes/woodland. In order to obtain the smallest common geometry in the GIS for the optimization area, the assessment results of the target functions are divided into separate areas. The result is that the study area (4,240 ha) is divided into 2,485 polygons, and the optimization area (3,621.1 ha) is divided into 1,871 polygons. However, since the smallest areas created below a size of 200 m² are not taken into account in the GIS, 1,676 polygons currently used as arable land (3,620.3 ha) remain for optimization.

#### 8.3.4.3 Multicriteria optimization - definition and solution

Optimization is a branch of numerical mathematics which includes the optimal stipulation of quantities, temporal courses and other properties of a system while simultaneously taking restrictions into account.

The mathematical method of multicriteria optimization achieves results which can be considered as optimal compromises between different goal functions. These goal functions may often be mutually conflicting. The assessment results (i.e. the results of each goal divided into classes) are used as coefficients for the goal functions. Hence, assessments have to be carried out for the entire set of variables. The number of variables is equal to the number of evaluated landscape elements multiplied by the number of polygons. Thus, each landscape element has to be considered on the level of each polygon. A landscape element may completely cover a polygon or share it with other elements (see Fig. 8.4).

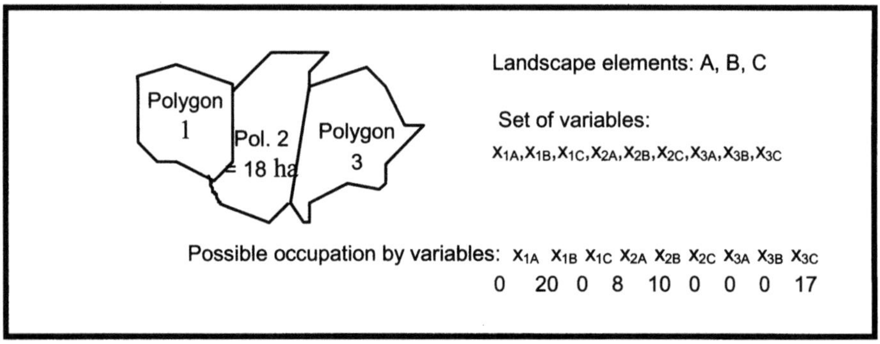


**Fig. 8.4:** Set of variables and a fictitious solution (assignment of variables)

Equality restrictions (area of polygons) or inequality restrictions (the whole size of the landscape elements) define the boundaries of optimization. Optimization is a linear problem. Such approaches are often called "linear programming" (e.g. Werner 1993). The optimum in this case is defined as follows: A solution (variable assignment) is optimal if a higher value cannot be achieved for one goal without decreasing the goal function value of at least one other goal. This case of optimality is denoted as "PARETO optimality" (Wierzbicki 1979, Dewess 1985). The values of goal functions are in turn used as a criterion to measure the optimum. These values can be calculated by adding the products of areas obtained as solutions (see areas with values greater than 0 in Fig. 8.4) and their corresponding goal function coefficients.

Many different techniques can be used to solve multicriteria optimization problems, including statistical methods (Changkong & Haimes 1983), methods using the Simplex algorithm (e.g. Falkenhagen 1989), and reference point methods (Wierzbicki 1979). The method used here is based on game theory and was elaborated by Dewess (1985). The optimum is calculated by minimizing the maximum difference of each goal value from its optimal value. This method involves calculating optimum values for each goal (without considering the other goal functions). This enables the risks of a monofunctional landscape to be identified. One peculiarity of this method is that the user can interactively calculate an arbitrary series of solutions by weighting each of the goal functions subjectively. The set of solutions is therefore infinite.

Optimization is carried out using the software LNOPT (Grabaum 1995). The software has to be adapted to the problem at hand by defining different values.

After definition, the optimization problem has the following structure:

n - number of variables
$n = np * ne$ (number of polygons np * number of landscape elements ne);
**x** - vector of variables; $\mathbf{x} = (x_1,...,x_n)$
m - number of goal functions
**Q** - Matrix of goal functions; $\mathbf{Q} = (\mathbf{q}^1,...,\mathbf{q}^m)^T$, where $\mathbf{q}^i = (q^i_1,...,q^i_n)$, i=1,..,m

$$\mathbf{Qx} = \text{Max!}$$

restrictions:

1. $\mathbf{x} >= \mathbf{0}$
2. $\sum_{j=1}^{np} x_{ij} <= b_i$ , i=1,..,ne (boundary of element i)
3. $\sum_{i=1}^{ne} x_{ij} = b_j$ , j=1,..,np (area of assessment polygon j)

or in short

$\mathbf{Ax} <= \mathbf{b}$
$\mathbf{x} <= \mathbf{0}$ , where **A** is a (0,1) matrix. The matrix only shows whether the variable in the restriction is activated (1) or not (0).

The restrictions of element size (the elements used in the example being arable land, grassland and forest) have to be fixed before optimization. This was done by setting the upper and lower boundaries (see Table 8.6). Alternatively, the optimization problem can be solved without defining an element size. Other restrictions can be integrated (for example specifying definite landscape elements on a restricted number of polygons).

#### 8.1.1.4 Maximization and compromises

During optimization, first of all the maximum values of the functions 'Reduction in water erosion', 'Improvement in the retention capacity' and 'Improvement in the production function' calculated. The function 'Reduction of soil erosion due to water' is a minimization function (minimizing potential erosion by the suitable choice of erosion-inhibiting landscape elements); the other two functions are maximization functions. As all the aims are maximized the LNOPT program, the minimization function 'soil erosion due to water' is converted into a maximization task by multiplying it by -1 ('Maximization of the resistance to soil erosion due to water')

The maximum values can be used to identify the problem areas for the individual geoecological functions and to improve them by means of land uses changes. The other functions are not considered. Therefore the realization of the results from the maximization of one single function is not recommended.

Consequently, a compromise then has to be found. For this purpose, the individual functions can be weighted so that in each case a different 'optimum' compromise can be determined. For the optimization area, three compromises were calculated for each scenario. They differ in terms of the weighting of the individual goals (Table. 8.7). First of all, all the objectives are weighted equally. Other possibilities include favouring two functions over the third function (Compromise 2) and the gradual preference of each function (Compromise 3).

The improvement in the individual functions can be gauged from the functional value. The functional value is the sum of the products of area size, containing element x, and its assessment. If for example Area 1 (Fig. 8.4) has for a Function 1 the following value vector (Element A:3, B:2 C:2), Area 2 the value vector (A:5, B:1,C:1) and Area 3 the value vector (A:4, B:2, C:2), the area occupation shown has the following (non-dimensional) functional value Z:

$$Z = 18 * 2 + 5 * 5 + 11 * 1 + 21 * 2 = 114.$$

This formula also allows the value of the current use situation to be calculated for all functions. The following tables (Tables. 8.8-10) show the optimization results (area occupations and target function values) for the conservation scenarios (1-3). By way of comparison, the current land use is always given in the first line. The optimization results are shown in Map 8.4.

The tables show the areas in hectares of the individual elements in the optimization area, indicating the degree to which the limits of the restrictions can be reached (Table. 8.6). For example, in the maximization of production, the largest possible arable area is completely used, whereas when erosion protection is maximized, the maximum areas of woodland and grassland is used. Moreover, the (target) functional values Z are shown. In each case, the maximum values and the compromises shown in Fig. 8.5 are highlighted in bold type. It can be seen that in scenarios 1-3, the current functional values are significantly less than the functional values of optimization (Compromises 1-3) for the two regulation functions ‘Erosion protection’ and ‘Retention’. By contrast, the current functional value for the production function is higher than the optimum values - for during reorganization, farming is ceased on some areas. The difference depends on the area envisaged for the new biotope structures.

In Fig. 8.5 it is apparent that the functional values for current use (blue line) are partly far below the optimum values of the compromise solutions (with the exception of the production function). Optimization thus brings about an improvement in the regulation functions (line for Compromise 1). In Fig. 8.5 it should be noted that the three axes are completely independent of each other and for graphic reasons their origin cannot be shown. The “Utopia Point” consists of the maximum values of the individual functions (bold figures in Tables. 8.8-10). It is actually a theoretical value, because the maximum values of all the functions included in the method can never be simultaneously reached owing to their opposing aims.

The choice of a correspondingly weighted compromise solution must be satisfactorily explained. Assuming the equal importance of all goals (Table. 8.7), which is defensible from an ecological viewpoint, in the following only Compromise 1 for Scenario 2 is considered (Map 8.4.)

**Table 8.7:** Weighting of the functions in compromise optimization

| Function | Compromise 1 | Compromise 2 | Compromise 3 |
|---|---|---|---|
| Water erosion | 1 | 101 | 75 |
| Retention capacity | 1 | 101 | 74 |
| Production function | 1 | 100 | 73 |

**Table 8.8:** Optimization results for scenario 1 (15% conservation area)

| | Arable land in ha | Grassland in ha | Woodland in ha | Functional value of water erosion | Functional value of retention | Functional value of production |
|---|---|---|---|---|---|---|
| Current use | 3620.80 | 0 | 0 | 123,889,537 | 142,426,629 | 296,280,021 |
| Maximization of water erosion protection | 3070.00 | 184.28 | 366.00 | **140,179,779** | 147,540,732 | 254,370,622 |
| Maximization of retention | 3071.28 | 183.00 | 366.00 | 136,331,583 | **148,471,196** | 256,592,817 |
| Maximization of production | 3143.00 | 183.00 | 294.28 | 132,545,553 | 146,540,761 | **267,099,797** |
| Compromise 1 | 3143.00 | 183.00 | 294.28 | 138,127,081 | 146,993,753 | 263,188,568 |
| Compromise 2 | 3135.19 | 183.00 | 302.09 | 138,734,346 | 147,011,491 | 261,728,372 |
| Compromise 3 | 3093.67 | 183.00 | 343.61 | 139,561,761 | 147,471,167 | 258,830,960 |

**Table 8.9:** Optimization results for scenario 2 (7.5 % conservation area)

| | Arable land in ha | Grassland in ha | Woodland in ha | Functional value of water erosion | Functional value of retention | Functional value of production |
|---|---|---|---|---|---|---|
| Current use | 3620.28 | 0 | 0 | 123,889,537 | 142,426,629 | 296,280,021 |
| Maximization of water erosion protection | 3331.00 | 109.00 | 180.28 | **134,579,723** | 145,417,426 | 273,505,728 |
| Maximization of retention | 3331.00 | 109.00 | 180.28 | 131,024,639 | **146,613,995** | 275,507,210 |
| Maximization of production | 3368.00 | 109.00 | 144.28 | 128,397,386 | 144,685,849 | **282,169,661** |
| Compromise 1 | **3368.00** | **109.00** | **144.28** | **132,783,811** | **145,027,518** | **278,404,220** |
| Compromise 2 | 3351.35 | 109.00 | 159.93 | 133,390,430 | 145,318,354 | 276,907,032 |
| Compromise 3 | 3331.00 | 109.00 | 180.28 | 134,475,360 | 145,445,273 | 274,432,156 |

**Table 8.10:** Optimization results for scenario 3 (30 % conservation area)

| | Arable land in ha | Grassland in ha | Woodland in ha | Functional value of water erosion | Functional value of retention | Functional value of production |
|---|---|---|---|---|---|---|
| Current use | 3620.28 | 0 | 0 | 123,889,537 | 142,426,629 | 296,280,021 |
| Maximization of water erosion protection | 2498.00 | 362.00 | 760.28 | **151,619,779** | 152,081,860 | 207,554,668 |
| Maximization of retention | 2498.00 | 362.00 | 760.28 | 147,023,054 | **152,413,995** | 212,151,185 |
| Maximization of production | 2570.00 | 362.00 | 688.28 | 142,997,202 | 150,749,305 | **222,528,647** |
| Compromise 1 | 2570.00 | 362.00 | 688.28 | 149,259,200 | 151,362,911 | 219,064,083 |
| Compromise 2 | 2570.00 | 478.36 | 571.92 | 149,579,122 | 150,362,649 | 217,360,024 |
| Compromise 3 | 2548.71 | 622.15 | 449.43 | 150,357,254 | 149,129,592 | 214,790,985 |

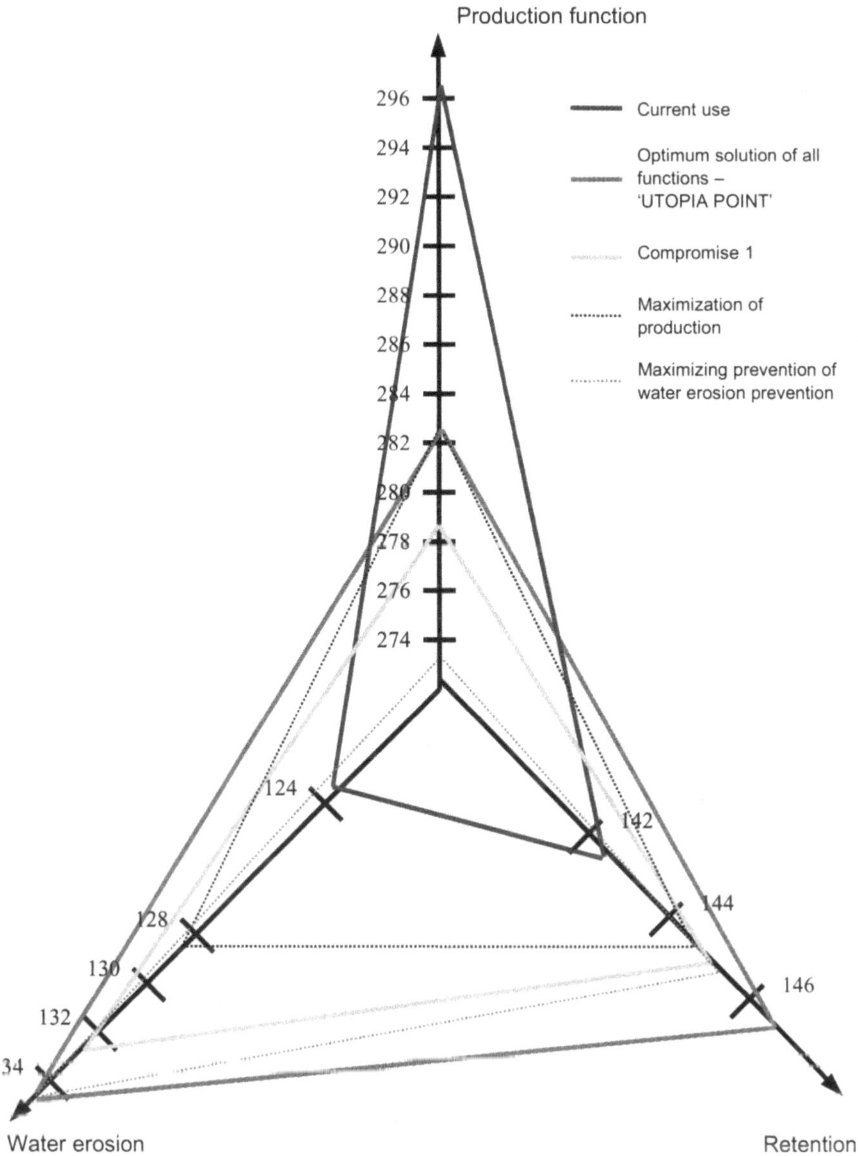


**Fig. 8.5:** Comparison of (target) function values (× $10^6$) of various optimal solutions with the current use for Scenario 2 (7.5% conservation areas)

#### 8.1.1.5 Landscape planning draft

To develop feasible action recommendations, the optimization results require further treatment. First of all, the right scenario must be chosen. The detailed model is very helpful for this purpose. The study area will continue in future to be mainly used for sustainable agriculture. Accordingly, implementing other land use forms over a large area is not sustainable, particular from an economic and social viewpoint. Therefore, the authors chose Scenario 2 (7.5% of arable land to be transformed into quasi-natural structures). Although this is not to call into question the proportion of 15% conservation area in Germany's agricultural landscapes proposed by Heydemann (1981, 1983), converting 7.5% arable land into grassland, bushes and woodland is a high aim for the areas concerned in Querfurter Platte. This figure also corresponds to the demands of KUL (Kritische Umweltbelastungen Landwirtschaft/Critical Pollution in Agriculture) for a minimum amount of ecological priority areas in the agricultural landscape (Roth 1994, Eckert & Breitschuh 1995)

The direct change of land use in practice on the basis of landscape optimization and drafting an implementation plan with the aid of a computer algorithm is difficult for the following reasons:

- No information is provided about the creation of linear structures (hedges, rows of trees), as optimization only deals with areas.
- Some of the tiny and fragmented land use polygons proposed by optimization (e.g. small areas of arable land in woodlands) resulting from the smallest common geometry in the GIS are too small to farm economically or may be irrelevant for planning.
- Important interrelations within the cultural landscape and infrastructure aspects must be taken into account.
- Implementation may be impeded by existing ownership.
- Higher level planning may be useful if the optimization results are not sufficient, e.g. in order to distinguish between the options "woodland" and "grassland".

Therefore it makes sense to revise the land use options on the basis of the above information. If the 'rough structure' provided by optimization is to be implemented, cultural landscape information must be included in planning. Therefore the main cultural landscape elements of the study were compiled and possible planning measures incorporated (Table 8.11).

The locations for woodland, bushes and extensive grassland (meadows and pasture) calculated by the optimization and shown in Map 8.4 are concentrated on hilltops, hollows (wettish areas), steep hillsides and depth contour areas. At some areas the information was not detailed enough to differentiate between extensive grassland and woodland/bushes. Flat arable land with a homogeneous soil structure was not chosen for the creation of new landscape elements.

On this basis, that the cultural landscape information was incorporated into the results of optimization. In particular, the results shown in Map 8.5 (draft landscape plan) are produced by field division by hedges, rows of trees and grassy margins and taking into account the direction of cultivation and the pathway network.

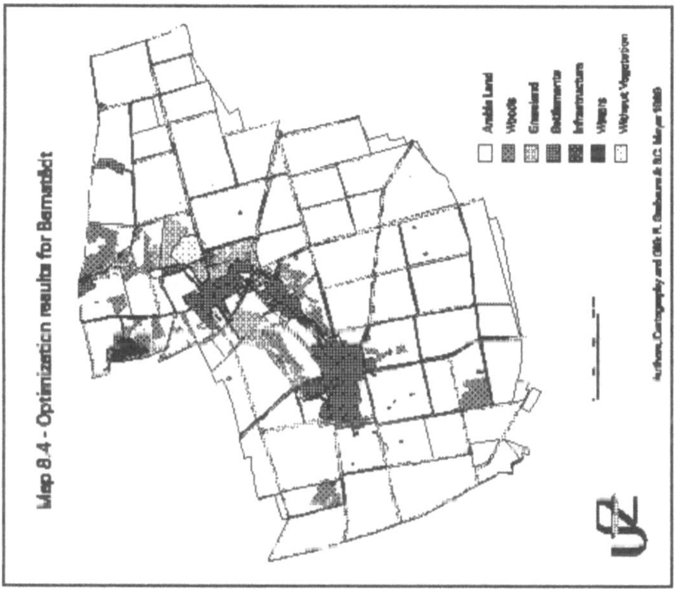


**Map 8.4:** Optimization result Barnstaedt test site

**Map 8.5:** Landscape plan Barnstaedt test site

**Table 8.11:** Cultural landscape elements, incorporated information levels and possible measures (Querfurter Platte)

| Cultural landscape elements/information | Possible measure |
|---|---|
| Hilltops, hollows, hillsides | Woodland, bushes |
| hillsides, dry valleys, wet areas, depth contours | Extensive grassland (meadows, pasture) |
| Ownership structure, tenure lots | Rebuilding old pathway networks into hedges and rows of fruit trees, taking account of lot structure |
| Historical pathway network, traditional pathways | See above; reactivating traditional pathways |
| Accessibility of fields | Connection to path network |
| Direction of cultivation of fields | Taking account of technological possibilities |
| Rivers, streams, ponds (alder edges) | Banks of water bodies (10m), renaturalisation of water bodies |
| dry valleys, railway embankments | Dry slopes, dry grassland |
| Rows of cherry trees | Planting new rows and preserving existing ones in line with historical maps and current structures |
| Diversified hedges | Planting adjoining woodland areas |
| Wind protection hedges | Planting diversified hedges perpendicular to the main wind direction asif possible adjoining exist structures and on outkirts |
| Rows of trees | Planted next to bodies of water, on paths and village outskirts |
| Individual trees, | Planting at distinct areas, next to bodies of water and ditches |
| Field margins | Layout next to hedges, paths and water bodies to separate fields (5-20 m wide) |
| Agricultural buildings/sheds, village outskirts | Enclosure with shrubs and trees, outskirts of village at areas exposed to wind |
| Animal husbandry, stables and other animal buildings | Connection to grassland |
| Landscape appearance and sight relations | Windmills, Eichstädter Warte, Querfurter Burg |
| Recreation areas | Possibly allow new path links |
| Succession areas | Allow continuously |
| Sand pits | Partly leave as open pits |

### 8.1.5 Results

The draft landscape plan is then reassessed, enabling comparison with the landscape's current condition. This comparison with the goals proves to be very positive for the draft landscape plan in the study area. These changes are summarized in Table 8.12. It should again be mentioned that the functional improvement for linear structural elements and for the reduction of areas at risk of wind erosion are incorporated via the draft landscape plan, whereas the areal elements are mainly based on the optimization results.

The development of a draft landscape plan based on the optimization results hence proves valuable from a functional viewpoint in order to arrive at a more extensive improvement in land use as proposed in the model. This draft can be effectively substantiated together with the comparisons with current use.

**Table 8.12:** Functional comparison of the current state with the draft landscape plan

| Indicator | Current state | Draft landscape plan | Change in % current state → draft landscape plan |
|---|---|---|---|
| Biodiversity: | | | |
| Arable land [%] | 86.2 | 79.6 | -6.6 % |
| Conservation land [%] (excl. structural elements) | 1.0 | 6.9 | +590 % |
| Linear structural elements | 0.9 | 1.7 | +89 % |
| Water erosion: | | | |
| Annual pot. erosion in t/ha | 1.45 | 0.96 | -33.8 % |
| Annual pot. total erosion in t | 5,231 | 3,714 | -29 % |
| Wind erosion: | | | |
| Area at risk in ha | 2,321 | 1,313 | -43.4 % |
| Retention: | | | |
| Average estimate 1-5 | 2.4 | 2.36 | 4.5-% (min. = 1.51) |
| Production : | | | |
| Average soil index of cultivated areas | 88.0 | 88.9 | +0.9 % |
| Areas with a SI > 80 in ha | 3,040 | 2,925 | -3.8 % |

### 8.1.6 Summary

All in all, there are many ways to improve the method and make it a powerful instrument for planners that can help tackle diverse requirements in a planning region. This was confirmed by the positive reactions among both the farmers in the study area and local planners and politicians. The results were presented on a number of occasions and prepared for implementation. The method was hence developed by science to the point where it was ready to be applied in practice (Mühle et al. 2000, Grabaum et al. 1999, 2000).

Functional assessments and optimization used in turn are a powerful instrument in the preparation of political decisions. This approach has already been tried out in four study areas (Table 8.13).

One particular strength of the method is its constant processing in a GIS up to the preparation of a draft landscape plan, which is produced in an area-specific manner, and every step of which is comprehensibly drawn up on the basis of vector information. This means that goals (e.g. from the conclusion of the model) can be expressed more precisely for a specific location while simultaneously taking the whole study area into account. Another major advantage is that an infinite number of scenarios can be produced by changing the weighting. Therefore, the further development of the method must focus on the problematic, normative selection of the scenarios to be calculated and the weightings to be used. In future it will be possible to use decision-support methods for the selection of functions.

One considerable weak point of the method is the incomplete inclusion of linear structural elements. Although the way in which they affect functions can be assessed, there is as yet no technical method for indicating exactly what and where structural elements should be created. This indicates that effects of lateral processes have not yet been sufficiently integrated into the method.

One future goal is to enable the closer integration of socioeconomic factors. However, the number of functions included must remain manageable. Scenarios should in future be based on the German Conservation Act using the natural functions listed here, which include a priori human land use.

**Table 8.13**: Areas where multicriteria assessment followed by optimization has been tried out

| Study area | Authors | Land use |
|---|---|---|
| Berchedesgaden National Park | Grabaum (1996) | Prototype for land use |
| Jesewitz/north Saxony | Meyer (1997); Grabaum & Meyer (1998; 1999) | Regional agricultural planning |
| Barnstädt/Saxony-Anhalt | Grabaum, Meyer Mühle (1999) | Landscape plan in an agricultural area |
| Dresden | Bobert (1999) | Development planning |

# 8.4 Polyfunctional assessment of landscapes using landscape units

## 8.4.1 The conflict between segregating and integrating the protection of natural resources

The assessment of multifunctional land use in the Halle-Leipzig-Bitterfeld district is based on the areal stocktaking of landscape-ecological information. Necessary measures for the protection of the environment and natural resources are concluded on a mesoscale scale. The starting-point is the natural functions and landscape structures which are directly or indirectly influenced by agricultural use.

At present no (regionalized) landscape indicators or systems of indicators exist which can be used to adequately measure the degree of environmental and resource protection necessary for agricultural land use and the cultural landscape shaped by agriculture. This is all the more regrettable since the theory of differentiated land use (put forward by Haber 1972) provided a basis for the functionally compensatory, multifunctional cultural landscape. A differentiated regional formulation of this theory has only so far occurred in non-transferable test areas. This formulation is beset by numerous methodological and normative problems.

On the spatial basis of landscape units, in this work natural functions are first of all individually assessed using different methods, and are then summarized using a simple polyfunctional points system. The work is structured as follows:

- The problems of environmental and resource protection in the conflict of interests between the segregation and integration of natural conservation into land use will be listed.
- The landscape units in the Halle-Leipzig-Bitterfeld conurbation will be classified by landscape type.
- Eight functions will then be assessed on the basis of landscape units and the conclusion of measures necessary for the development of their functional ability and for the development of new landscape structures.
- The results will comprise a prioritization for integrated landscape assessment based on polyfunctional assessment.

Following on from Haber & Salzwedel (1992), the main environmental problems caused by the large-scale intensive cultivation in the Halle-Leipzig-Bitterfeld district can be listed as follows in order of importance:

- The most serious effect is the reduction of quasi-natural biotopes, whose proportion within the total area of intensively used landscapes has declined to just 2-3% or even less. Moreover, biotopes low in nutrients are deteriorating as a result of fertilizers and pesticides, which has a negative impact on fauna and flora.
- Groundwater pollution owing to the input of nitrate and pesticides.

- Soil pollution owing to compaction, erosion, changes in soil microbiology, contamination, and acidification caused by fertilizers and pesticides etc.
- The impairment of surface water by drainage, piping, river correction, filling in ponds etc.
- The deterioration of nutritional quality.
- Air pollution caused by agriculture, which not only leads to the direct pollution of ecosystems (eutrophication) but also contributes to the depletion of stratospheric ozone and the greenhouse effect.

When drawing up environmental quality objectives for landscapes, the question of the segregation or integration of conservation and environmental protection into agriculture must be decided. This problem can be discussed at the level of consideration of interregional segregation, intra-regional segregation or regional integration.

- **Interregional segregation:** It is assumed that a region (e.g. owing to its fertile soil and flat conditions) is especially suitable for agricultural production and should therefore be primarily used for agriculture. The necessary measures for environmental and resource protection are then applied in other regions (e.g. peripheral low-mountain regions where cultivation is more difficult), where agricultural land use is then significantly reduced.
- **Intra-regional segregation:** The regionally specific proportion of quasi-natural biotopes and their distribution throughout the area are determined in a certain area depending on the quasi-natural features of the landscape (or its current use). One frequently specified political value is for example 15% of the area for "nature conservation land use" or the stipulation of a variable proportion of land as ecological priority areas; Eckert & Breitschuh (1995); Roth (1994). On the rest of the land, agriculture can be pursued in accordance with the requirements of "good practice", e.g. as described by Heissenhuber & Hoffmann (1992). This good practice corresponds to the minimum requirements of resource protection.

**Table 8.14:** Landscape types within the study area

| Landscape type | Abbreviation LT | Area $km^2$ |
|---|---|---|
| Agrarian landscapes | AL | 1773 |
| Agrarian landscapes with some woodland | AsW | 733 |
| Agrarian-woodland landscapes | AWL | 320 |
| Woodland-agrarian landscapes | WAL | 228 |
| Woodland landscapes with some agriculture | WsA | 19 |
| Wetland and valley landscapes | Wet | 369 |
| Post-mining landscapes | PML | 518 |
| Urban landscapes | UL | 523 |
| Insufficient data | no | 399 |

- **Regional integration:** Conservation and environmental protection are integrated as far as possible into the entire land use. All agricultural measures are to be examined and assessed with respect to impacts on environmentally sustainable development, i.e. on the ability to function of environmental aspects with a medium function (biotopes, climate, soil, water and landscape appearance/recreation; demands from e.g. Jedicke 1994; Bick 1988; Plachter 1991; Riedl 1991). Contrast to segregation, integrated conservation means nature conservation and environmental protection on 100% of the area. As a result, natural functions everywhere are protected, developed and integrated into land use.

### 8.4.2 Landscape units and landscape types

The model analysis "Conservation and Care of Environmental Media in the Leipzig-Halle-Bitterfeld region" involves the assessment of use-dependent functions on landscape units employing a small selection of functions relevant to landscape. The work is summarized in an expert report on the integrated landscape assessment of an area of 4,877 km² at a scale of 1:400,000 (Meyer and Krönert 1998). In this respect, particular attention is given to the extensive influences that agriculture (particularly arable farming) can have on functions.

The assessment of regional integration calls for suitable reference areas. Landscape units are well suited to the polyfunctional assessment of landscapes. The objective of this assessment is to identify conflicts of use and to outline ideas for protecting the environment, more specifically for preserving or reestablishing the landscape's regulation functions.

Landscape units are used as the spatial reference area in the studies carried out (Krönert 1997). These are classified into landscape types based on land use (see Table 8.14). While landscape types are taken as the landscape reference area, soil regions are used as the natural reference area, and the two are compared. However, this approach identifies an urgent need for more practical and standardized assessment procedures for mesoscale areas.

### 8.4.3 Functional assessments

In this article we do not have the opportunity to explain all the algorithms of functional assessments used for the polyfunctional assessment by Meyer & Krönert (1998). Several assessments results developed by Meyer (1997) to assess the landscape structure are used to base this macroscale investigation.

Each landscape unit is evaluated in terms of eight functions (Table 8.15), unless constrained by the available data and methods. The need for remedial action is drawn up for landscape types and assessed according to a regional rating. The assessment values are related to the whole investigation area. Thresholds, baselines and reference values are described by Meyer & Krönert (1998). The need for remedial action is evaluated on a monofunctional basis and expressed using a five-point scale (1= very low need and 5 = very high need). The results of an assessment of 53 landscape units in the Halle-Leipzig-Bitterfeld area were collated to produce an average need across all units for each of the functions (Table 8.16).

**Table 8.15:** Assessment of functions.

| Function | Method of assessment |
|---|---|
| Soil erosion due to water | Schwertmann et al. (1987) |
| Soil erosion due to wind | Capelle & Lüders (1985) |
| Soil compaction | Horn (1981) |
| Groundwater protection | Marks et al. (1989) |
| Restoration of surface waters | Meyer & Krönert (1998) |
| Suitability for recreation | Meyer & Krönert (1998) |
| Landscape diversity | Meyer & Krönert (1998) |
| Biotope diversity in agricultural areas | Meyer & Krönert (1998) |

**Table 8.16:** The need for remedial action with regard to eight functions, expressed as an average across all landscape units (highest need ranked first). Each function was assessed independently of the others.

| Landscape type | soil erosion due to water | soil erosion due to wind | soil compaction | groundwater protection | restoration of surface waters | suitability for recreation | biotope diversity | landscape diversity | assessment points | number of functions | RV |
|---|---|---|---|---|---|---|---|---|---|---|---|
| AWL | 3,2 | 3,8 | 2,0 | 4,0 | 3,0 | 1,4 | 2,2 | 3,0 | 22,6 | 7,4 | 3,1 |
| AsW | 4,3 | 3,3 | 3,3 | 2,8 | 4,3 | 2,3 | 2,5 | 3,0 | 25,5 | 7,8 | 3,3 |
| AL | 4,9 | 2,2 | 3,8 | 1,8 | 3,9 | 4,4 | 4,5 | 3,0 | 28,5 | 7,8 | 3,7 |
| Wet | 4,8 | 1,6 | 4,8 | 4,3 | 0,0 | 2,0 | 1,1 | 0,0 | 18,6 | 6,0 | 3,1 |
| PML | 4,2 | 2,8 | 3,6 | 2,8 | 0,0 | 2,7 | 4,1 | 2,6 | 22,7 | 6,8 | 3,3 |
| WAL;WsA; | 3,6 | 3,8 | 2,0 | 3,2 | 3,2 | 1,2 | 3,0 | 4,0 | 24,0 | 7,4 | 3,2 |
| UL | 4,5 | 2,9 | 3,4 | 3,3 | 5,0 | 5,0 | 3,5 | 3,8 | 31,3 | 8,0 | 3,9 |

### 8.4.4 Polyfunctionale Assessment

The polyfunctional assessment (i.e. the synoptic integration of different monofunctional assessments) is expressed for each landscape unit in the form of a standardized result value (RV), i.e. one which is independent of the number of variables used in its calculation (not all functions could be assessed on all landscape units; Table 8.16). This allows the results to be summarized at a regional level. The RVs for landscape types are expressed on a five-point scale (<2.6 = very low >3.8 = very high) and average RVs for overall remedial needs were calculated for each landscape type, allowing the latter to be ranked accordingly (Table 8.17).

The results show that there are clear differences between landscape units in the study area. Although the total size of the sample of landscape units is relatively small (53), there are also clear differences between landscape types. However, the multi-layered nature of the polyfunctional assessment approach means that the size of the RV cannot be used to draw conclusions about individual functional assessments.

In addition to the urban landscapes (which have been subjected to extreme anthropogenic influence), it is the agricultural landscapes (i.e. those dominated by arable farming) which are most in need of measures to protect the environment and natural resources.

In addition we compared the different reference areas, "landscape types" and "soil regions", according to the same eight assessments of landscape ecological functions. In this case landscape types are found to be more useful for the assessment of remedial action, as there are clearer differences between categories related to the land use. However for other applications of polyfunctional landscape assessments, the selection of the most useful topic-specific reference area should be reassessed each time a new polyfunctional assessment is carried out (Map 8. 6).

**Table 8.17:** Average values for remedial need with regard to function and landscape types in the Leipzig-Halle-Bitterfeld area .

| Objective | Average need for remedial action in the Leipzig-Halle-Bitterfeld area | Number of landscape units evaluated |
|---|---|---|
| Reduction in soil erosion caused by water | 4.4 | 53 |
| Restoration of surface water | 4.0 | 35 |
| Prevention of compaction in upper soils | 3.5 | 52 |
| Improvement in landscape diversity | 3.4 | 35 |
| Improvement in biodiversity | 3.2 | 53 |
| Improvement in the recreational potential | 3.0 | 52 |
| Safeguarding of groundwater protection | 3.0 | 52 |
| Reduction in soil erosion caused by wind | 2.7 | 53 |

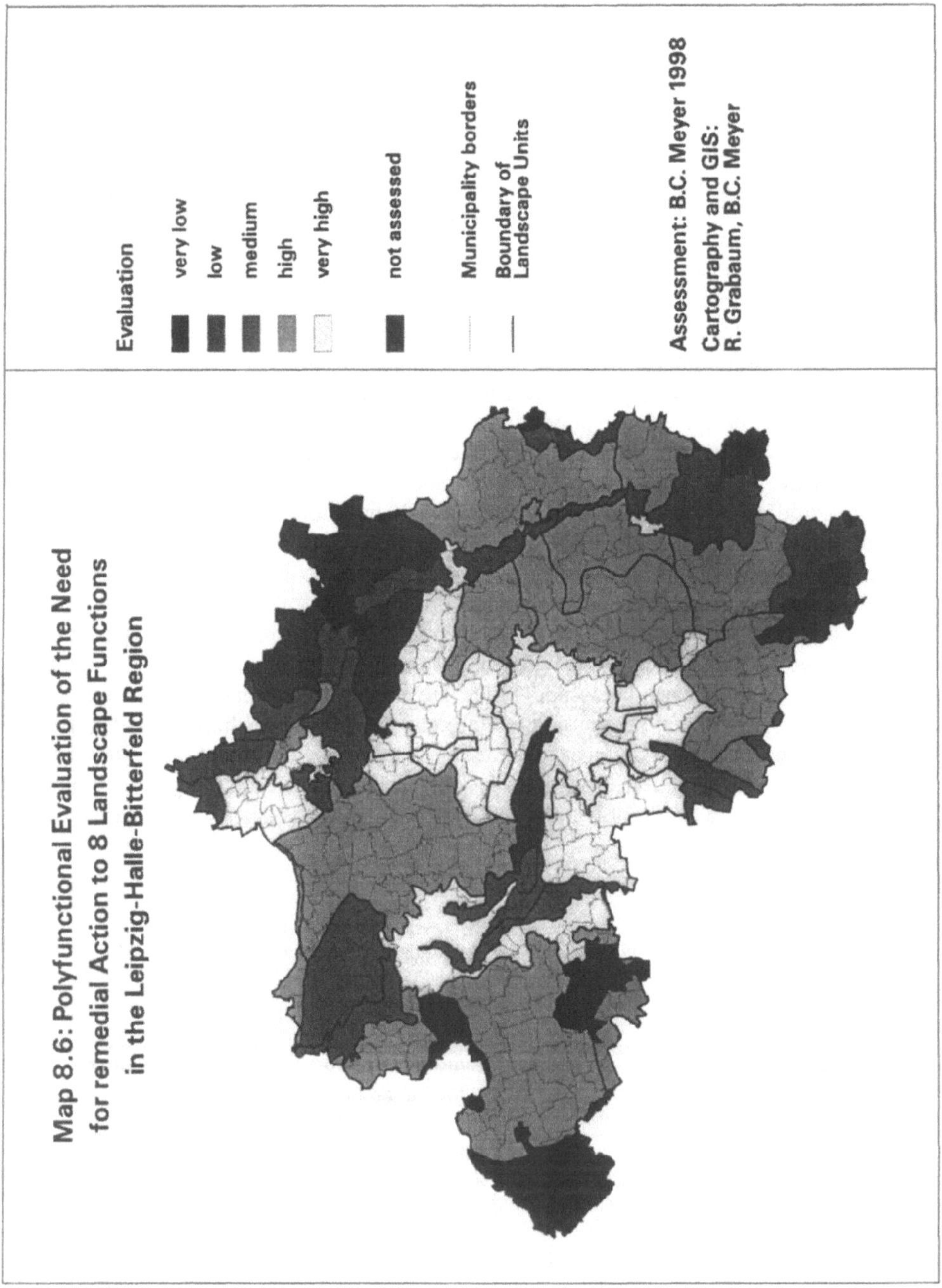


**Map 8.6:** Polyfunctional assessment of the need for remedial action to eight landscape functions in the Leipzig-Halle-Bitterfeld region

### 8.4.5 Summary

The polyfunctional categorization of remedial needs (protection of the environment and natural resources) in landscape units involves the assessment of heterogeneous areas. Individual results are used only where sensible and methodologically safe conclusions can be drawn from them (e.g. an estimate of the real danger of soil erosion is only useful for arable fields, while an estimate of the potential danger of soil erosion can be used to describe the whole area).

The procedures used in evaluating functions differ both in terms of their maturity and their validity. Each procedure is subject to different constraints on its application and is suited to different levels of scale. The procedures include validated studies of small parcels of land, relatively simple estimates based on a few indicators, and regionally valid analyses of extensive test areas (Meyer 1997). Many procedures are valid at all levels of scale (topological to mesoscale), while others have been developed only for application at mesoscales. In our opinion, the issue of the appropriateness of a procedure at different dimensions, and the choice of indicators used in an assessment, need to be reassessed for every change in the contents of an assessment; the choice of procedure or indicators is not easily transferable. Close attention must also be given to scaling-up and scaling-down issues, particularly when using GIS.

Although the validity of individual procedures certainly needs to be improved, the precision of a procedure used in a polyfunctional assessment may be less important, since the assessment of other functions may involve a smaller scale. The polyfunctional procedure used involves ordinal scales while many monofunctional assessment procedures already use cardinal scales.

## 8.5 GIS-based landscape assessment

In this chapter, three examples were used to demonstrate assessment methods which describe relevant indicators with respect to function, structure and the dynamics of change overtime.

A number of indicators of formal landscape structure and their assessment on a topological scale were presented for the conclusion of development objectives in agricultural landscapes. It is surprising that assessment techniques used for nature conservation and landscape ecology are still overwhelmingly restricted to the assessment of areal landscape functions (functional indicators). Little attention is paid to the interaction between the formal and functional structure. This is all the more astonishing since planners are increasingly calling for quantitative reasons for the development of the landscape (e.g. in ecological point models, for compensatory land registers for regulating intervention, in indicator systems for environmentally sustainable agriculture, or more generally for sustainable development etc.). Structural indicators of for instance.

- The proportion of woodlands and grassland adapted to the landscape
- Field size adapted to the landscape

- The number and length of structural elements (hedges, field margins, bodies of water etc.)
- Tolerable landscape fragmentation
- Landscape isolation
- Landscape consumption

must be regionally assessed. The concept of the "network of compensative areas as an ecological infrastructure of territories" (according to Mander et al. 1988) provides a good basis for this, followed by landscape structure assessments. According to Weiers & Meyer (2000), structural indicators (such as those discussed in Chap.5 for landscape metrics) and spatial distribution patterns of landscape elements in landscapes ought to be investigated with respect to neighbourhood relations, fragmentation, connectivity and target conclusions. This in turn should produce conclusions concerning fundamental processes and geoecological differentiation. Regionally comparative analyses of indicators in different natural areas or landscape units and multitemporal comparisons should form the basis for the aforementioned assessments. Landscape structure assessments must be applied in a strictly scale-based manner. The assessments presented in Chap. 8.2 and 8.3. use a scale of 1:10,000. In Chap. 8.4., aggregated regionalized landscape structure indicators are used together with functional assessments to validate mesoscale methods of assessment.

The example of using GIS for multifunctional landscape assessment and multicriteria optimization in scenarios of landscape development for a landscape or regional agricultural development plan describes an integrative method of landscape ecology. Different landscape-ecological functional assessments are incorporated in scenarios of landscape development by means of multicriteria optimization, which results in an optimum compromise between different area-related goals. The main advantage of this method is that the scope of the aims and weighting can be varied, hence enabling scenarios to be comparatively evaluated in terms of their functional impact. This is a fundamental requirement for application-related political advice.

One methodological problem is that separately developed monofunctional assessments must be selected for an integrated assessment (Slocombe 1998). The expert interviews used in the method could be replaced by formalized techniques for function selection, which deal in a standardized manner with the currently mostly unknown interrelations between the functional assessments. Equally, the linear landscape structures not yet integrated into optimization could be integrated into this method by means of buffer formation. Here, too, their order of magnitude can be calculated using regionalized landscape structure analysis.

In comparison to the two aforementioned examples, the polyfunctional assessment of landscape units was used on a mesoscale basis to conclude necessary measures of regional environmental and resource protection. At this scale, more highly indicated indicators should be used. The assessment of these indicators is in the range of validity for some methods (e.g. soil erosion according to Schwertmann et al. (1989)). However, there are still no methods for many functions and structures on this scale. For polyfunctional assessments, too, different monofunctional assessments were brought together using simple additive technique. This enables the areas to be qualified which have the highest need for resource protection - information which is of major importance for agricultural environmental programmes and regional planning.

Two ways of further developing the methods of polyfunctional assessment are already emerging. Firstly, coupling with multicriteria optimization is conceivable, as has already been successfully tested by Grabaum (1996). Secondly, the assessment of functions of monofunctional methods could develop into a unified consistent polyfunctional assessment technique, as outlined by Niemann (1982). However, data and GIS standardization has not yet been incorporated.

The above examples of scientific assessment techniques throw up numerous problems which need to be solved for broad practical application. These problems can be divided into the areas of digital geoinformation, scale transformation, data management and the development of new assessment methods.

Digital geoinformation is becoming increasingly important for the application of assessment methods. However, at present there are only a few attempts to simplify and standardize the provision of data. The reasons for this are the sectoral administrative structure in Germany and the resulting different ranges of competence. The result is numerous unlinked and hardly compatible databases. There is much room for improvement in the updating of important environmental data levels in all areas. Therefore much of the work required for assessment tasks involves compiling the right levels of data. For example, an informative soil map with a scale of 1:10,000 simply does not exist. Moreover, there is no uniform standard for geinformation data, another aspect which makes the integration of different data levels a job for specialists and hinders the broad land use of assessment methods. These questions must be solved independently of landscape assessment.

Scale transformations are problematic. The very simple change of data scales using GIS is due to technical reasons, not the content of data. When using assessment methods, this means that in future scale-specific techniques will need to be developed for each scalar level which use indicators which could be the same or different at different scales. As the normative component of landscape assessment depends on the subject of assessment, the question of transferring (scaling) assessment results from one scale to another is not very urgent, as different issues of assessment have to be answered at different scales. By contrast, a comparative analysis to conclude regional magnitudes for indicators and for scaling is essential. One promising approach appears to be the drafting of a scalar hierarchy of environmental indicators. These could then be used for assessment (e.g. in the form of checklists to measure sustainability; Van Mansveld & Van der Lubbe 1999). Contrary to this, the assessment of complex indices to measure sustainability is regarded as less promising (Sands & Pedmore 2000)

At present, far too much time and energy are expended on data management for assessments. Assistance could be provided by standardized expert systems, which support the process of assessment routines. Whether this should take place automatically or in dialogue with users (semi-automatically) depends on the subject of assessment. As we saw above, different information levels can be integrated by using multicriteria assessment and optimization, or polyfunctional assessment. Equally, however, further developed applications of benefit analyses and ecological risk analysis or non-formalized methods based on the planning, argumentative method (Auhagen 1998) are conceivable. For smaller areas, simple structured decision-tree analyses are conceivable (Bastian 1999). This area also covers the further development of GIS-based scenario techniques.

The development of monofunctional assessment methods will doubtless remain continuously necessary with the development of science. In this respect, special attention should be paid to establishing the interdependencies of different assessment levels. The development of assessment methods for the landscape structure should be especially emphasized.

## 8.6 References

Akademie für Natur- und Umweltschutz Baden-Württemberg (ANU 1996) Bewertung im Naturschutz. Ein Beitrag zur Begriffsbestimmung und Neuorientierung in der Umweltplanung. Beiträge der ANU Bd.23, Stuttgart

ANL (Bayrische Akademie für Naturschutz und Landschaftspflege 1994) Leitbilder - Umweltqualitätsziele - Umweltstandards. Laufener Seminarbeiträge 4/94, Laufen

Auhagen A (1998) Verbal-Argumentation oder Punkte-Ökologie - Bewertungsverfahren unter der Lupe des Planers. In: Dresdner Planergespräche Tagungsbericht, Dresden, pp 57 -109

Bastian O, Schreiber K-F (1994) Analyse und ökologische Bewertung der Landschaft, Gustav Fischer, Stuttgart

Bastian O (1999) Landschaftsfunktionen als Grundlage von Leitbildern für Naturräume. Natur und Landschaft 9: 361-373

Bick H (1988) Belastungen von Natur und Landschaft durch landwirtschaftliche Flächennutzung. In: Flächenstillegung und Extensivierung für Naturschutz. Jahrbuch für Naturschutz und Landschaftspflege 41, pp 9-16

Blaschke T (1997) Landschaftsanalyse und -bewertung mit GIS. Forschungen zur Deutschen Landeskunde Bd. 243, Trier

Bobert J (1999) Zur Herstellung der Verträglichkeit von Nutzungen nach dem Nachhaltigkeitsprinzip des Naturschutzes. GIS-Anwendung im Beispielsraum Dresden. Diplomarbeit TU-Dresden

Capelle A, Lüders R (1985) Die potentielle Erosionsgefährdung der Böden in Niedersachsen. Göttinger Bodenkundliche Berichte 83: 107-127 (quoted by Hennings 1994)

Changkong V, Haimes YY (1983) Optimization-based methods for multiobjective decision-making - an overview. Large Scale Systems 5: 1-33

Dale VH et al (1998) Assessing Land-Use Impacts on Natural Resouces. Environmental Management 22(2): 203-211

de Groot R (1992) Functions of nature. Evaluation of nature in environmental planning, management and decision making. Wolters-Noordhoff, Groningen

Dewess G (1985) Zum spieltheoretischen Kompromiss in der Vektoroptimierung. Optimization 16(1): 29-39

Dollinger F (1989) Landschaftsanalyse und Landschaftsbewertung. Mitteilungen des Arbeitskreises für Regionalforschung Sonderband 2, Wien

DUDEN - Das Fremdwörterbuch (1990) Duden-Verlag, Mannheim-Leipzig-Wien-Zürich

Duttmann R, Mosimann T (1994) Die ökologische Bewertung und dynamische Modellierung von Teilfunktionen und -prozessen des Landschaftshaushaltes - Anwendung und Perspektiven eines geoökologischen Informationssystems in der Praxis. Petermanns Geographische Mitteilungen 138: 3-17

Eckert H, Breitschuh G (1995) Kritische Umweltbelastungen Landwirtschaft (KUL). Anlage 3 zum Protokoll der gemeinsamen Fachsitzungen I/II/X des VDLUFA am 16.03.95, Leipzig-Möckern

Falkenhagen M (1989): Ein mehrkriterielles Optimierungsverfahren als Ansatz für ein Territorialmodell. Dissertation, Universität Halle

Farina A (1998) Principles and methods in Landscape Ecology. Chapman & Hall, London

Finke L (1994) Landschaftsökologie. Westermann, Braunschweig

Forman RTT (1985) Land Mosaics. The ecology of landscapes and regions. Cambridge

Grabaum R (1996) Verfahren der polyfunktionalen Bewertung von Landschaftselementen einer Landschaftseinheit mit anschließender "Multicriteria Optimization" zur Generierung vielfältiger Landnutzungsoptionen. Shaker-Verlag, Aachen

Grabaum R, Meyer BC, Mühle H (1999) Landschaftsbewertung und -optimierung. Ein integratives Konzept zur Landschaftsentwicklung. UFZ-Bericht 32/1999, UFZ, Leipzig

Grabaum R, Meyer BC (1997) Landschaftsökologische Bewertungen und multikriterielle Optimierung mit Geographischen Informationssystemen (GIS). Photogrammetrie-Fernerkundung-GIS 2: 121 - 134

Grabaum R, Meyer BC (1998) Multicriteria optimization of landscapes using GIS-based functional assessments. Landscape and Urban Planning 554: 1-14

Grabaum R, Meyer BC (1999) The application of GIS for landscape ecological assessments and multicriteria optimization for a test site near Leipzig. In: Dikau et al. (Hrsg) GIS for Earth Surface Systems. Analysis and Modelling the Natural Environment. Berlin Stuttgart, pp 91-108

Grabaum R, Meyer BC, Meyer R, Mühle H, Sauerbier H (2000) The Agricultural Landscape in the Central Region of Germany - Concepts and plans - In: Bundesministerium für Ernährung, Landwirtschaft und Forsten (Hrsg) Country Reports and Misccellaneous Contributions to rural21. International Conference on the Future and Development of Rural Areas. Bonn, pp 29-36

Gustafson E J (1998) Quantifying Landscape Spatial Pattern: What is the State of the Art? Ecosystems 1: 143-156

Haase G (Hrsg) (1991) Naturraumerkundung und Landnutzung. Geoökologische Verfahren zur Analyse, Kartierung und Bewertung von Naturräumen. Beiträge zur Geographie 34. Berlin

Haber W, Salzwedel J (1992) Umweltprobleme der Landwirtschaft - Sachbuch Ökologie. Stuttgart

Haber W (1972) Grundsätze einer ökologischen Theorie der Landnutzungsplanung. Innere Kolonisation 21: 294-298

Haber W, Riedel B, Theurer R (1991) Ökologische Bilanzierung in der Ländlichen Neuordnung. Materialien zur ländlichen Neuordnung, Heft 23, München

Heissenhuber A, Hoffmann H (1992) Überlegungen zur Realisierung einer umweltschonenden Landbewirtschaftung. In: Bay. Staatsmin. Landw. und Umweltfragen. Materialband 84, Umwelt und Entwicklung Bayern, pp 151-166

Hennings V (Koord 1994) Methodendokumentation Bodenkunde. Auswertungsmethoden zur Beurteilung der Empfindlichkeit und Belastbarkeit der Böden. Geologisches Jahrbuch, Reihe F Bodenkunde, Heft 31. Hannover

Heydemann B (1981) Zur Frage der Flächengrösse von Biotopbeständen für den Arten- und Ökosystemschutz. Jb. Natursch. Landschaftspfl. 31: 21-51

Heydemann B (1983) Die Beurteilung von Zielkonflikten zwischen Landwirtschaft, Landschaftspflege und Naturschutz aus der Sicht der Landespflege und des Naturschutzes. Schr.-R. für ländliche Sozialfragen 88, pp 51-78

Horn R (1981) Die Bedeutung der Aggregierung von Böden für die mechanische Belastbarkeit in dem für Tritt relevanten Auflastbereich und deren Auswirkungen auf physikalische Bodenkenngrößen. Landschaftsentwicklung und Umweltforschung 10, Berlin (quoted by Hennings 1994)

Jedicke E (1991, 1994) Biotopverbund: Grundlagen und Massnahmen einer neuen Naturschutz-strategie. Ulmer. Stuttgart

Koch R, Graf D, Hartung A, Niemann E, Rytz E (1989) Polyfunktionale Bewertung von Flächennutzungsgefügen. Wissenschaftliche Mitteilungen, 32, IGG, Leipzig

Krönert R (1997) Landnutzung und Landnutzungsänderungen. In: Feldmann R et al (ed) Regeneration und nachhaltige Landnutzung: Konzepte fur belastete Regionen. Springer, Berlin-Heidelberg-New York, pp 27-34

Landgesellschaft Sachsen-Anhalt mbH et al. (1995) Agrarstrukturelle Vorplanung "Querfurter Platte".

Lee JT et al (1999) The role of GIS in landscape assessment: using land-use-based criteria of the Chiltern Hills Area of Outstanding Natural Beauty. Land Use Policy 16: 23-32

Lemly AD (1997) Risk Assessmemt as an Environmental Management Tool: Considerations for Freshwater Wetlands. Environmental Management 21(3): 343-358

Leser H, Klink HJ (1988): Handbuch und Kartieranleitung Geoökologische Karte 1:25.000. Forschungen zur Deutschen Landekunde Bd. 228, Trier

Leser H (1991, 1997) Landschaftsökologie: Ansatz, Modelle, Methodik, Anwendung. Ulmer, Stuttgart

Leser H (1999) Das landschaftsökologische Konzept als interdisziplinärer Ansatz - Überlegungen zum Standort der Landschaftsökologie. In: Mannsfeld K, Neumeister H (Hrsg) Ernst Neefs Landschaftslehre heute. Pertes, Gotha, pp 65-85

Leser H et al. (1984) Diercke-Wörterbuch der Allgemeinen Geographie. Westermann, Braunschweig

Mander U et al (1988) Network of compensative areas as an ecological infrastructure of territories. In: Schreiber KF (Hrsg): Connectivity in landscape ecology. Münstersche Geogr. Arbeiten 29: 35-38

Marks R, Müller MJ, Leser H, Klink, HJ (Hrsg) (1989) Anleitung zur Bewertung des Leistungsvermögens des Landschaftshaushaltes (BA LVL). Forschungen zur dt. Landeskunde 229, Trier

Meyer BC, Grabaum R (1996) Szenarien zur Einschätzung der Bodenerosionsgefährdung durch Wasser mit GIS (ARC-INFO) - Dargestellt am Beispiel des Untersuchungsgebietes Jesewitz/Sachsen. Geoökodynamik 17: 45-67

Meyer BC (1997) Landschaftsstrukturen und Regulationsfunktionen in Intensivagrarlandschaften im Raum Leipzig-Halle. Regionalisierte Umweltqualitätsziele - Funktionsbewertungen - Multikriterielle Landschaftsoptimierung unter Verwendung von GIS. Dissertation Köln und Leipzig, UFZ-Bericht Nr 24/97, UFZ, Leipzig

Meyer BC (1998) Bewertung von Landschaftsstrukturen auf topischem Niveau zur Ableitung von Entwicklungszielen in Agrarlandschaften. In: Steinhardt U, Grabaum R (Hrsg) Landschaftsbewertung unter Verwendung analytischer Verfahren und Fuzzy Logic. UFZ-Bericht 6/1998, UFZ, Leipzig, pp 39-50

Meyer BC, Krönert R (1998) Bewertung von Maßnahmenotwendigkeiten des Umwelt- und Ressourcenschutzes im Raum Leipzig-Halle-Bitterfeld. UFZ-Bericht 15/1998, UFZ, Leipzig

Meyer BC (1996) Regionalisierte Grundlagen für Maßnahmen des Umwelt- und Ressourcenschutzes im Ballungsraum Halle-Leipzig-Bitterfeld. Projektbericht für das UFZ, Leipzig

Mosimann T (1999) Angewandte Landschaftsökologie - Inhalte, Stellung und Perspektiven. In: Schneider-Sliwa R, Schaub D, Gerold, G (Hrsg) (1999) Angewandte Landschaftsökologie: Grundlagen und Methoden. Springer, Berlin-Heidelberg-New York, pp 5-24

Mühle H (1998) Sustainable Development in Agricultural Landscapes. In: Schellnhuber H-J, Wenzel V (eds) Earth System Analysis. Integrating Science for Sustainability. Springer, Berlin-Heidelberg-New York, pp 277-287

Mühle H, Grabaum R, Meyer BC (2000) Dauerhaft umweltgerechte Landbewirtschaftung in Mitteldeutschland - Ideen, Methoden, Ergebnisse. Agrarspectrum 31: Entwicklung nachhaltiger Landnutzungssysteme in Agrarlandschaften, pp 74-80

Naveh Z, Lieberman A (1994): Landscape Ecology. Theory and Application. Springer, New York-Berlin-Heidelberg

Neef E (1965) Topologische und chorologische Arbeitsweisen in der Landschaftsforschung. Petermanns Geographische Mitteilungen 107: 249-259

Neumeister H (1989) Geoökologie - Denk- und Arbeitsweise in den Geowissenschaften (Beispiel: Geoökologische Forschung in der Agrarlandschaft). Petermanns Geographische Mitteilungen 2: 85-108

Niemann E (1977) Eine Methode zur Erarbeitung der Funktionsleistungsgrade von Landschaftselementen. In: Archiv für Naturschutz und Landschaftsforschung 17, pp 119-157

Niemann E (1982) Methodik zur Bestimmung der Eignung, Leistung und Belastbarkeit von Landschaftselementen und Landschaftseinheiten. Wiss. Mitt. Sonderheft 2, IGG; Leipzig

Plachter H (1991) Naturschutz. Stuttgart

Plachter H (1994) Methodische Rahmenbedingungen für synoptische Bewertungsverfahren im Naturschutz. Zeitschrift für Ökologie und Naturschutz 3: 87-106

PoschmannC et al. (1998) Umweltplanung und -bewertung. Gotha

Preu C, Leinweber P (1996) Konzeption und inhaltlich methodische Ansätze geowissenschaftlicher Naturraumbewertung. In: Preu C, Leinweber P (Hrsg) Landschaftsökologische Raumbewertung. Konzeptionen - Methoden - Anwendungen. Vechtaer Studien zur Angewandten Geographie und Regionalwissenschaft Bd. 16, S. 11-22

Riecken U, Ries U, Ssymank A (1994) Rote Liste der gefährdeten Biotoptypen der Bundesrepublik Deutschland. Schriftenreihe für Landschaftspflege und Naturschutz, Heft 41, Bonn-Bad Godesberg

Riedl U (1991) Integrierter Naturschutz - Notwendigkeit des Umdenkens, normativer Begründungszusammenhang, konzeptioneller Ansatz. Beiträge zur räumlichen Planung 31, Hannover

Roth D (1994) Zum Konflikt zwischen Landwirtschaft und Naturschutz sowie Lösungen für seine Überwindung. Natur und Landschaft 69: 407-411

Sachsen-Anhalt (1994) Landschaftsprogramm des Landes Sachsen-Anhalt. Ministerium für Umwelt und Naturschutz des Landes Sachsen-Anhalt

Sands GR, Podmore TH (2000) A generalized environmental sustainability index for agricultural systems. Agriculture, Ecosystems and Environment 79: 29-41

Scheffer F, Schachtschabel (1984) Lehrbuch der Bodenkunde. Enke, Stuttgart

Schellnhuber HJ et al (1997) Syndromes of Global Change. GAIA 6(1): 19-34

Schmid B, Schelske O (1997) Interdisziplinäre Forschung im "Integrierten Projekt Biodiversität" des Schweizerischen Nationalfonds: Ziele und Strukturen. Z. Ökologie u. Naturschutz 6: 247-252

Schreiber KF (1988) Connectivity in Landscape Ecology. Proceedings of the second International Seminar of the "IALE" Münster 1987. Münstersche Geographische Arbeiten 29, Paderborn

Schwertmann U, Vogl W, Kainz M (1987) Bodenerosion durch Wasser. Vorhersage des Abtrags und Bewertung von Gegenmassnahmen. Ulmer, Stuttgart

Scott MJ et al (1998) Valuation of Ecological Resources and Functions. Environmental Management 22(1): 49-68

Slocombe DS (1998) Defining Goals and Criteria for Ecossystem-Based management. Environmental management 22(4): 483-493

Smith JA, Lyon DJ, Dickey EC, Rickey P (1992) Emergency wind erosion control. University of Nebraska NebGuide Publication, G75-282-A

Sukopp H, Wittig R (Hrsg) (1993) Stadtökologie, Gustav Fischer, Stuttgart-Jena-New York

van Mansfeld JD, van der Lubbe MJ (1999) Checklist for sustainable landscape management. Elsevier, Amsterdam

Weiers S, Meyer BC (2000) Vergleichende funktionale und strukturelle Landschaftsbewertung intensiver und extensiver Agrarlandschaften zur Umsetzung in regionale Umweltqualitätsziele. Kooperationspapier UFZ-DLR, Köln 5 S. (unpubl.)

Wiegleb G (1997) Leitbildmethode und naturschutzfachliche Bewertung. Z. Ökologie u. Naturschutz 6: 43-62

Wierzbicki AP (1979) The use of the reference objectives in multiobjective optimization. IIASA wp-79-66

Zepp H, Müller MJ (1999) Landschaftsökologische Erfassungsstandards. Ein Methodenbuch. Forschungen zur Deutschen Landeskunde Bd 244, Flensburg

# 9 Landscape function assessment and regional planning: Creating knowledge bases for sustainable landscape development[1]

Daniel Petry

## 9.1 Introduction

Spatial planning has to integrate the often conflicting demands and interests stemming from the multiple functions landscapes have to fulfil. One crucial point in this context is the regional scale, where the overall principle of sustainable development needs to be diversified into spatially differentiated objectives and standards for future landscape development.

In Germany (and possibly elsewhere), where a formalized planning system on the basis of binding legal standards once hindered or at least rejected the usefulness of open and informal schemes, in recent years planning processes have become more receptive to the involvement of various interest groups and the general public. Other European countries such as the Netherlands and the United Kingdom where planning systems were less formalized and where in particular regional planning had a weaker legal position (Hendriks 2000, Cullingworth & Nadin 1997) already saw the need to develop informal ways of comprehensive planning earlier on (Dekker 2000).

However, target definition is no longer the preserve of just a few specialized experts but the outcome of discussion and consideration among planners, politicians, land users and the concerned public - who form a heterogeneous group of stakeholders with differing backgrounds of knowledge as well as often contrary interests and demands. It is within this group that the implementation of measures for sustainable landscape development needs to be established in the first place.

The success of planning and its objectives therefore largely depends on their transparency by means of scientific underpinning, methodological aspects of derivation, and the descriptive power of presentation. Therefore, the implementation of sustainable landscape development targets increasingly relies on a sound landscape-ecological knowledge base. This knowledge base needs to be a comprehensive procedure including:

---

[1] The contents of this chapter are based on the author's research at the UFZ Centre for Environmental Research, which is part of his PhD thesis at the University of Halle, Germany (Petry 2001)

- Data management, analysis and assessment tools,
- Information, presentation and visualisation techniques,
- Normative standards like environmental quality criteria as a prerequisite for evaluation, target definition and conflict solving as well as
- Relevant planning instruments and categories.

Geographical information systems (GISs) provide a powerful basis for developing such a procedure. But apart from technical aspects, using a GIS does not indicate the relevant spatial and time scales of a certain analysis and planning context.

Therefore, one important task for applied landscape ecology is to highlight the interactions between land use, landscape balance and landscape aesthetics. The relevant processes and structures must be identified in an integrative manner. The scale-dependent factors of the driving forces of groundwater recharge, soil erosion, habitat diversity, and aesthetic potential and vulnerability are represented by indicators.

This means reducing complexity to its relevant essentials and increasing transparency and handiness for the planning level. Landscape-ecological research delivers the appropriate methods and tools in this respect and creates knowledge bases for planning decisions. The planners and participants of the landscape-related decision-making processes are enabled to develop, review and weigh up their objectives and decisions in the light of interacting landscape processes and structures. These interconnect land use options, the indispensable protection of soil and water, as well as the aesthetic potentials resulting from the landscape's natural and cultural history.

## 9.2 The key position of the region for landscape analysis and spatial development

In modern planning theory, regions hold a key position for problem identification and spatial development (ARL 1995, Locher et al. 1997). At this scale, problems and conflicts are sufficiently apparent to allow the development of appropriate problem-solving strategies and measures. Moreover, regions have the advantage of dealing with a certain problem in a holistic rather than isolated manner by both spatial (local scale) and analytical (system approach) means. On a broader scale, landscape matters become increasingly complex, with the result that problem-solving is effective at the strategy level but not the result of systems analysis. Furthermore, a region is a spatial level at which a concerned public might be apparent. Ecological, economic and social interactions become obvious on a regional scale, which makes public involvement necessary and, compared to broader scales, practicable.

In landscape ecology and its neighbouring disciplines, due to their commitment to systems and hierarchy theory (Klijn 1995), the regional scale is referred to in different contexts. The analysis of water and matter fluxes in watersheds as well as of species dispersal via ecological networks requires a regional approach. This of course needs to be accomplished by scale-specific methods and techniques (see Chap. 6).

## 9.3 Multifunctional landscapes and sustainable development

The The landscape function concept is well established in planning, landscape ecology and environmental sciences. However, in line with the international literature, certain connotations in its meaning need to be mentioned and clarified. Forman & Godron (1986: p. 11) define landscape function as "the interactions among the spatial elements, that is, the flows of energy, materials, and species among the component ecosystems." In central European landscape ecology, with its strong geographical tradition, this would rather be the definition of biological, morphological and chemical *processes* in landscapes. The Dutch ecologist De Groot (1992: 7) referred to environmental functions as "the capacity of natural processes and components to provide goods and services that satisfy human needs", which is based upon the General Ecological Model for the spatial development of the Netherlands where functions are taken as an equivalent for use options of the natural environment (van der Maarel & Dauvellier 1978, p. 134). Naveh & Lieberman (1994) follow this definition as the conceptual basis for environmental assessment procedures in spatial planning. In German landscape ecology, the above cited meaning of function is covered by the term 'potential'. Both terms are important categories in planning and are sometimes difficult to distinguish. The potential of a certain nature or landscape unit is more or less equivalent with De Groot's "capacity" for human use. In this context, the functions of a landscape are defined as the actual use or meaning of the potential in an ecological, economic or social context.

The landscape function concept is the methodological framework where landscape ecological assessment meets the duties of spatial planning in steering and differentiating the societal functions of landscapes. This definition of landscape functions in landscape ecology coincides with functions in spatial planning (see Fig. 9.1). In this study, priority is therefore given to this definition of landscape function.

The landscape function concept combines landscape-ecological and planning methods within a sound theoretical framework. Environmental media such as soil and water are strongly interconnected with different types of land use by landscape processes and structures. In this context, landscape functions resulting from the landscape's physical and biotic basis as well as human interference enable a holistic yet problem-orientated view of relevant processes and structures. Here, landscape functions are taken as functions which are realized by landscapes as parts of the 'Total Human Ecosystem'. Thus it is not necessary to distinguish between functions maintaining ecosystem conditions or enabling human use of landscapes for agriculture, drinking water supply or amenity reasons. Taking the sustainability paradigm of spatial planning into account, landscapes need to be addressed with regard to their suitability to fulfil a certain function as well as their sensitivity to the deterioration of that function by human or natural impacts.

In planning terms this means that instruments are used such that the utilization of a function is enabled, and simultaneously this function is protected against internal (by usage itself) and external (by the use of other functions) impacts. Because this might seem obvious or confusing, let us clarify matters with an example. Agriculture benefits from the landscape function for biomass production. Yet by practising it with inappropriate techniques at inappropriate places, agriculture threatens its 'own' function by causing soil erosion and degradation. Additionally, agriculture threatens but also supports other functions such as groundwater recharge for the drinking water supply and ecosystem stability. On the other hand, the landscape functions for agriculture and groundwater recharge are endangered in the long run by suburbanization and the resulting surface sealing. The interactions between different functions as well as the two aspects of each function - suitability and sensitivity - need to be considered in landscape analysis and assessment aimed at sustainable landscape development.

The methodological framework presented here considers three landscape functions by way of example:

a) *Recharge of groundwater resources*
The renewal of groundwater in sufficient quantity and quality is regulated by a landscape's physical and biotic compartments, which are closely related to land use. Due to groundwater flow systems, horizontal interactions over considerable distances are characteristic. Sustainable water use is therefore not only concerned with intake rates but also needs a farsighted landscape management scheme on a broad scale. Thus, only spatial planning on a regional scale has both the appropriate scale and the integrative power to incorporate water use into sustainable landscape development.

b) *Agricultural production*
Suitability for biomass production and sensitivity to deterioration are accomplished by landscapes by their characteristic combination of climatic, morphologic and pedological factors. They determine whether a landscape delivers fertile arable land, is prone to soil erosion or has a sensitive water balance.

c) *Retention of water and matter fluxes*
Sustainability concepts for spatial development regard the retention of water, matter and energy in landscapes as one of the most important goals for the long-term stability of landscape systems (Ripl 1995). The regulation of water and matter fluxes is a landscape function which is not directly used in an economic sense, but affects ecological and economic development perspectives irreversibly.

The selected functions form a small but important part of actual landscape systems: they are closely linked to each other, they are objects of conflicting interests and demands, and they represent important - in ecological, economic and social terms - uses of the countryside.

## 9.4 Normative dimensions of landscape development

Objectives for the future development of landscapes cannot be derived from landscape-ecological research results themselves. This would mean a classic 'naturalistic fallacy': you cannot say what you want from what you have. But there are strong normative concepts and attitudes concerning the future role and functions of landscapes. Some are relatively long-established, such as the preservation of historical landscapes, and some are only just emerging, like the growing awareness of the importance of dynamics in intensively used landscapes. All of them need to be implemented into the sustainability concept as the overall guideline for future development, which has so far found its way into the legal framework of landscape-related planning.

For this study, both ecological principles and legal standards are used as a basis for the definition of strategic objectives geared towards the sustainable and environmentally sound development of landscapes. Bastian (1999) summarizes the principles for landscape development derived from the sustainability concept:

- The usage of physical resources shall not exceed their regeneration rate
- The release of substances into natural systems shall not exceed their retention capacity for these substances
- Time scales of human interference in natural systems have to follow the time scales for natural dynamics of these systems
- Minimizing irreversible water, matter and energetic losses from landscapes

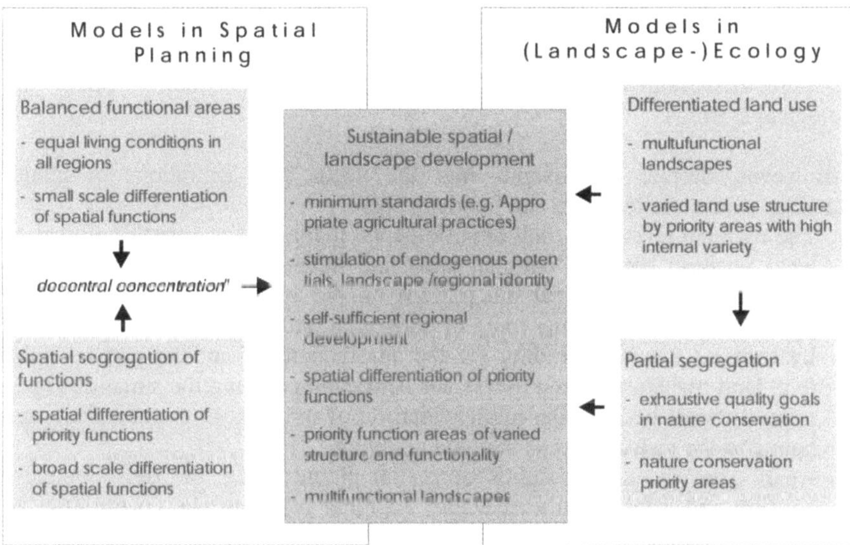


**Fig. 9.1:** Models of differentiation of spatial functions in landscape ecology and spatial planning

Since the early 1970s Haber (1972, 1990, 1998) has developed the Differentiated Land Use concept, in which he states the necessity of multifunctionality for landscapes as ecological systems. The core idea of Haber's approach is that monofunctional landscapes tend to cause more negative impacts on the environment than multifunctional ones. Therefore, he develops a concept in which a predominating function or land use needs to be combined with further functions or land uses on not less than 10-15% of a certain area. This theory can be seen as a predecessor of approaches for ecological networks (Cook & van Lier 1994), which have found their way into modern nature conservation strategies (partial segregation, Plachter 1994) and spatial planning models (Fig. 9.1).

## 9.5 Instruments and processes of landscape related regional planning

A general shift from formal planning within a strong legal framework to a more flexible, problem-orientated approach can be observed in spatial, overall and landscape planning throughout Europe. Therefore, the relevant processes and instruments for the sustainable development of landscapes are changing. The success of landscape-related planning, i.e. the implementation of goals and objectives, is increasingly relying on its communicational and persuasive skills and power rather than its legal instruments. In this context, the central feature of planning is no longer a plan but the consideration process in which the goals of all landscape-related interests and demands (e.g. agriculture, nature conservation, the control of water and material flows) are prioritized, integrated or - if the conflicts cannot be solved - spatially segregated. During this process, goals of sustainable landscape development have to be presented to a broader public - an essential condition if the necessary backing from society is to be achieved. According to Luz (2000), the latter is the crucial point in successful planning due to the fact that the implementation of planning takes place in socioeconomic rather than ecological systems.

However, despite the insights into the needs and aspirations of modern planning, several deficiencies can be identified:

- The regional level of overall environmental planning has a weak political and legal position between national and state interests as well as strong local interests (duty of vertical integration) on the one hand and also partially powerful sectoral planning (duty of horizontal integration) on the other.
- Instruments are only binding for the administration, not the general public (e.g. land owners and land users); for landscape planning the situation is even more difficult because the implementation of its goals in overall planning is bound to the instruments of regional planning.
- Apart from the classic duties of spatial planning, i.e. the development of housing and transport infrastructure, regional planning has no tradition and therefore no methods and concepts for playing an active, future-orientated role within sustainable landscape development.

In the light of the above deficiencies and changes of planning processes, some authors see the main duties and tasks of spatial planning within the following aspects:

- The identification of potential conflicts between developments proposed by sectoral planning as well as different levels of political decision-making.
- The portrayal of consequences of proposed developments for the physical and biotic basis of landscapes.
- The availability and accessibility of data, trends and scenarios for the planning and decision-making process in the relevant policy-making sectors.

From the discussion of duties, instruments and processes of landscape-related planning, some important obligations for landscape ecological research at the regional scale are concluded:

- Defining criteria from a sound landscape-ecological basis for the consideration and definition of goals of sustainable landscape development.
- Defining an appropriate spatial context for the designation of planning categories such as priority areas in order to ensure and further develop landscape functions.
- Using accessible data and workable as well as manageable, planning process orientated methods for analysis, assessment and evaluation.
- Establishing a comprehensive GIS for analysis and evaluation as well as general information and visualization as a handy tool supporting consideration, weighting and decision-making within planning processes.

The scope of application for the presented procedure covers a wide range of duties in environmental planning:

*Regional planning*

The strategic regional plan defines goals and objectives for the overall spatial development of a region. Goals are represented by spatially and functionally differentiated *priority areas* for certain functions and types of land use (agriculture, forestry, water supply, recreation, nature conservation, flood prevention, infrastructural developments) in order to prevent land use conflicts and to minimize environmental impacts. *Suitability areas* are a rather new category in planning legislation; they complement priority areas since they define zones to which a certain function is restricted. They are frequently used to direct the construction of wind turbines to less sensitive areas in terms of landscape amenity and nature conservation.

*Landscape planning*

In contrast to many other countries, Germany has a statutory landscape planning system, which has to define objectives and goals for the environmentally sound development of landscapes from the viewpoint of nature conservation. Plans are to be designed not just for nature reserves but with the whole landscape in mind. While this is of outstanding strategic importance for sustainable landscape development, the implementation of goals and measures depends on their acceptance and adoption by regional planning.

*Compensation rule*

Legal measures are designed to compensate for negative impacts on the environment from developments and projects, e.g. surface sealing by housing or infrastructure developments have to be compensated for by closing off areas functionally linked to the impact area. In recent years the compensation rule has been integrated into a regional approach, which enables the identification of impact sensitivity and suitable compensation areas on the landscape scale.

*Information and informal agreements*

Sustainable landscape development greatly depends on the success of informal and consensual agreements among potentially conflicting actors. By using GISs, the assessment procedure presented allows the spatially differentiated visualization of sensitivities, suitabilities and importances of landscapes for different functions. This ensures the transparency of goal definition in discussion, consultation, consideration and decision-making processes.

*Environmental (Impact) Assessment (EA / EIA)*

This is a well-established instrument in European and other countries. The 1985 EC Directive on EA (85/337/EEC) imposes a general obligation on the Member States to assess projects likely to have significant effects on the environment. In Germany, EA is closely linked to the application of the above-mentioned compensation rule. The assessment of landscape functions at the regional scale is a prerequisite for a problem-oriented EA setup.

*Strategic Environmental Assessment (SEA)*

Since the publication of the proposed EC Directive on environmental assessment of plans and programmes (SEA) in 1997, vehement discussions have revolved around its advances and constraints. However, one of its advances lies in the strengthening of environmental and landscape issues at the regional planning level, which is of key importance for the implementation of strategic goals in plans and programmes.

*River Basin Management (RBM)*

The forthcoming EC Water Framework Directive (WFD) on water policy sets standards for watershed management, the combined analysis of point and non-point sources of pollution, and the integrated management of groundwater and surface water. Whereas RBM is at least in part well established in some countries like the UK, in other countries such as Germany a fundamental change in water policy, management, planning, and not least perception is emerging. Since the combined analysis of land and water is a crucial point in RBM, it is a key instrument for achieving sustainable development (Newson 1992). The regional assessment of landscape functions within a GIS framework might easily be set into a watershed environment and gives substantial support to RBM.

# 9.6 A landscape function assessment procedure for the Dessau region

## 9.6.1 The Dessau region

The Dessau region is a representative part of the central European old morainic lowlands and river landscapes. As one of the three regional administrative and planning districts in the federal state of Saxony-Anhalt, the region is situated halfway between Berlin and Leipzig. It covers an area of 4,300 km² north and south of the River Elbe. Map 9.1 shows the land cover structure of the region by aggregated classes of the CORINE data.

Three dominant landscape types can be distinguished from their physical nature and land use structure.

- *Rivers and flood plains*
  The River Elbe and its major tributaries, the Mulde and Saale. The floodplains of the Elbe and the Mulde are famous for their undisturbed nature and are of great conservational value at a national and European level, which led to the designation of the first German biosphere reserve back in 1979. Characteristic features include riparian woodlands and historical cultural landscapes such as the 18th-century Wörlitzer Park near Dessau, one of the first garden landscapes in continental Europe.

- *Loess plains*
  The location of the loess plains in the lee of the Harz mountains causes a semi-continental climate with an annual precipitation of less than 500 mm. Therefore, relict chernozems dominate the western part of the region and form the fertile basis for intensive agriculture. The loess originates from the nearby moraines and was accumulated by aeolic sedimentation during the last ice-age. The landscape type is poor in natural features, and soil erosion is a widespread phenomenon due to the huge field complexes.

- *Glacial moraines*
  The centre and north of the region are comparatively hilly parts with altitudes of up to 200 m a.s.l. On these end moraines, only poor sandy soils have been allowed to develop, serving as the ground for spruce and pine forests, the predominant land cover type. Here, annual precipitation exceeds 600 mm and contributes to the renewal of groundwater resources of regional importance. The morainic landscapes fulfill considerable recreational functions for the agglomerations of Berlin, Leipzig and Magdeburg. Agriculture is prone to marginalization, which is already obvious from long-term fallows and affects the landscape's character and water balance.

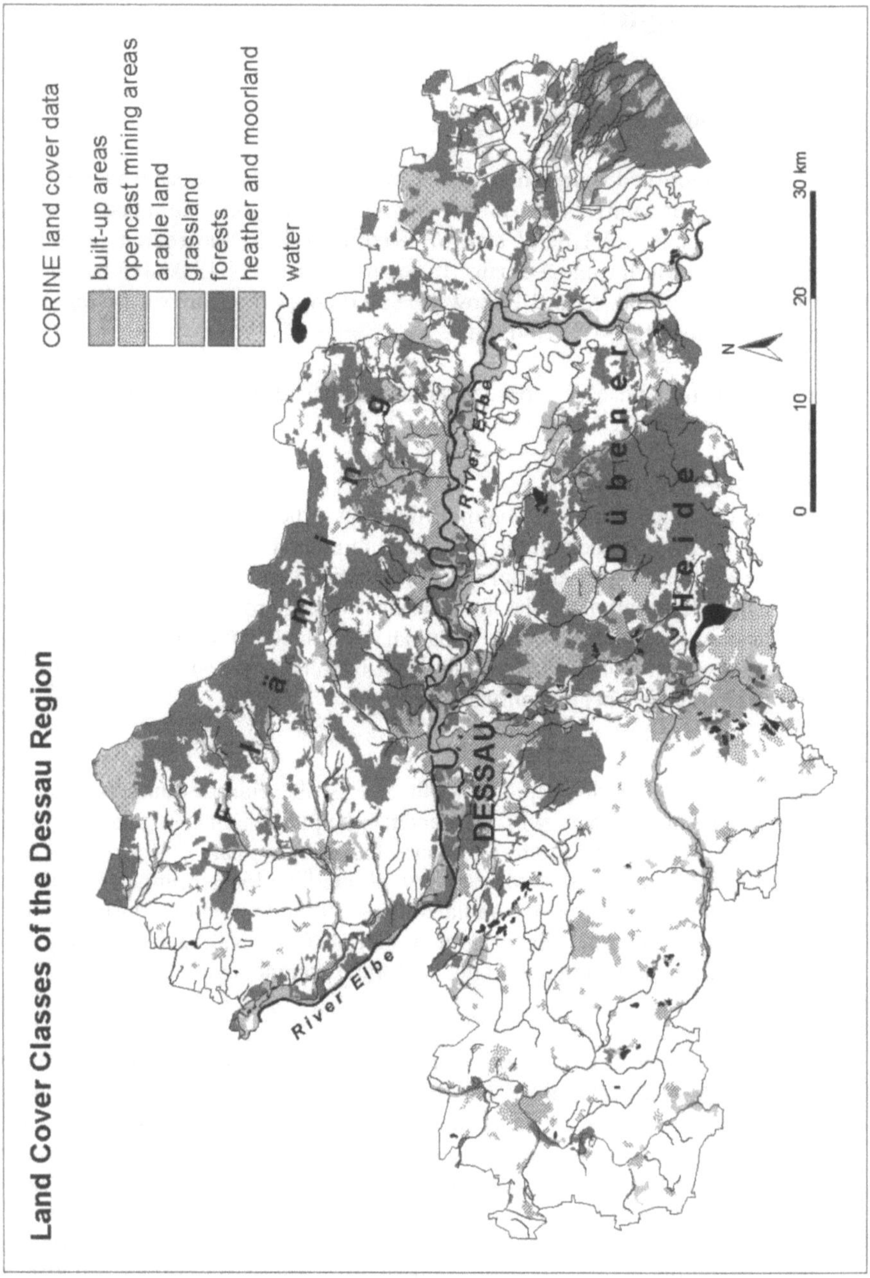


**Map 9.1:** Land cover in the Dessau region by aggregated classes of the CORINE land cover data

In addition, opencast mining areas and derelict industrial land cover a considerable portion of the area around Bitterfeld in the south of the region.

A total population of 300,000 is concentrated in the town of Dessau and smaller regional centres like Bitterfeld and Wittenberg. The countryside, especially the sandy moraines, is sparsely populated. The region has faced dramatic economic and social changes since German reunification reflected by high unemployment and a population decline of 10% within the past 10 years. Since economic development is of major concern for policy-makers, environmental planning is confronted by powerful conflicting interests and demands. Therefore, and for the variety of landscape types and dynamic land use processes, the Dessau region is chosen for the exemplary application of a methodological procedure designed to strengthen comprehensive regional planning as a vital agent of sustainable landscape development.

The assessment procedure has to ensure practicability within planning and transferability to other regions. Therefore, only available and easily accessible data were used. Table 9.1 shows the data selected for the landscape function analysis described in 9.6.2.

**Table 9.1:** Data basis for landscape function analysis

| Data | Type | Scale | Source |
|---|---|---|---|
| Land cover | CORINE land cover project | 1:100,000 | Federal Statistical Office |
| Climate | Long term means of precipitation and evaporation | 1 km² raster data | Federal Meteorological Office |
| Soil | Bodenübersichtskarte containing soil types, substrate, hydromorphic status | 1:200,000 | Geological Office of Saxony-Anhalt |
| Planning | Regional development programme for the Dessau region | 1:200,000 | Ministry for Spatial Planning and Environment of Saxony-Anhalt |

### 9.6.2 Methods of landscape analysis and evaluation

Established, suitable methods and model applications are applied and integrated into assessment methods of the planning disciplines. A key position in this context is given to the Ecological Risk Analysis (ERA, Fig. 9.2), which was developed in the late 1970s in order to promote decision-making processes in spatial planning (Bachfischer et al. 1980).

During the 1980s and 1990s, ecological risk analysis became a well-established method, especially after the introduction of the EU legislation on environmental assessment (EA) (Scholles 1997). This method enables the identification of environmentally sensitive areas (e.g. soil erosion, groundwater protection), areas confronted by conflicting demands for functional priorities (e.g. agricultural use versus drinking water supply), and areas with complementary functional characteristics (e.g. nature conservation and water retention). Objectives for the areas identified are linked to planning categories such as priority areas, and proposals for the improvement and further development of these categories aimed at strengthening sustainable development are discussed.

The assessment procedure presented in this study uses a modified version of ERA in Fig. 9.3, adopted from Kühling (1992), to evaluate landscape functions. Suitability and sensitivity are key criteria for assessing a landscape's capacity for fulfilling a certain function. For each of the selected landscape functions, indicators of suitability and sensitivity were defined and combined within an integrated assessment procedure to 'potential conservation value' as the final criteria. The term *potential* is added to indicate the lacking consideration of already existing impairments and impacts of the landscape functions considered. 'Conservation value' is a criterion frequently used in EA terminology. Besides protective and conservational values, such criteria may also indicate the need for planning and management concerned with developmental aspects of improving the fulfilment of landscape functions.

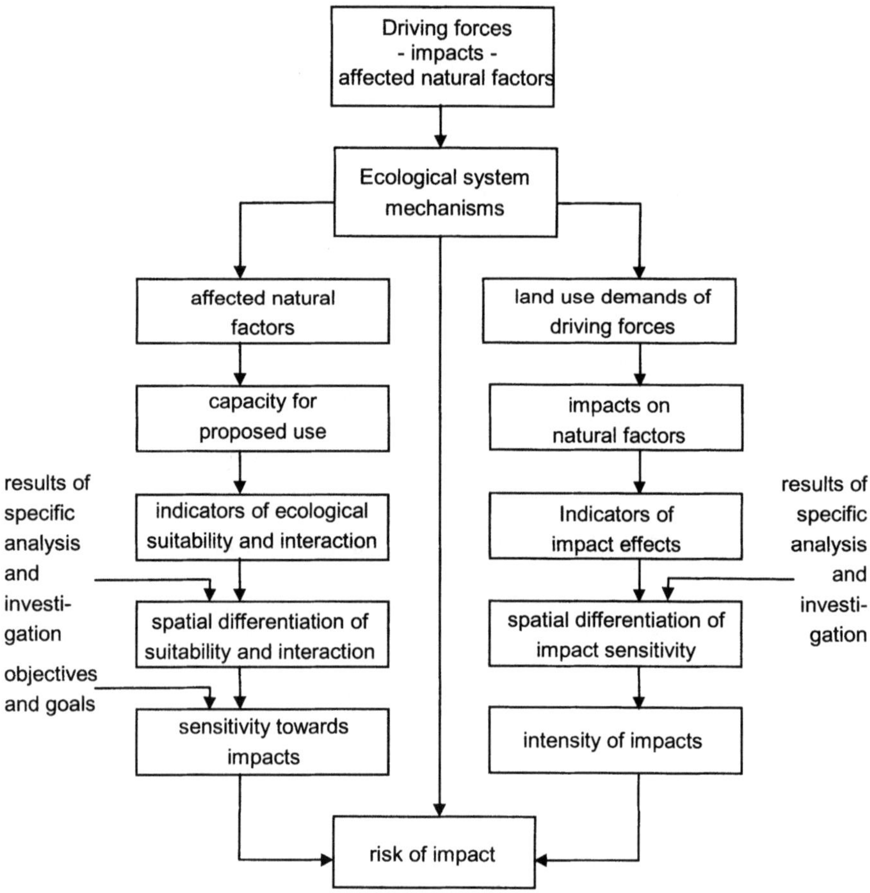


**Fig. 9.2:** Procedure of Ecological Risk Analysis (from Bachfischer 1980)

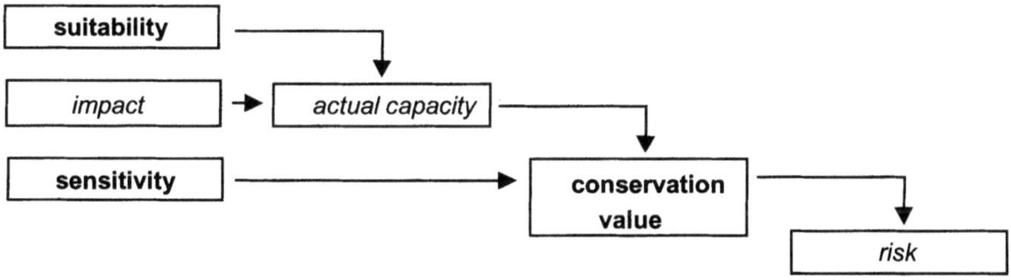


**Fig. 9.3:** Modified Ecological Risk Analysis in environmental planning (from Kühling 1992, modified)

ERA serves as a link between landscape-ecological analysis and the definition of planning targets and measures as well as their implementation with relevant instruments. The normative settings defined in 9.4 are used for the evaluation of suitabilities and sensitivities. The methodological strength of ERA lies in its integrative character, which defines a common frame for both qualitative and quantitative results from analysis.

The complete assessment procedure is described in Table 9.2, showing the stepwise operation from data management to evaluation criteria for landscape functions. The analysis and assessment of landscape function need to be addressed in the broader context of a societal, political and planning framework, which allows the constraints and chances of sustainable landscape development to be identified. Fig. 9.4 illustrates the interdependencies and feedback among the active agents of landscape development concerned.

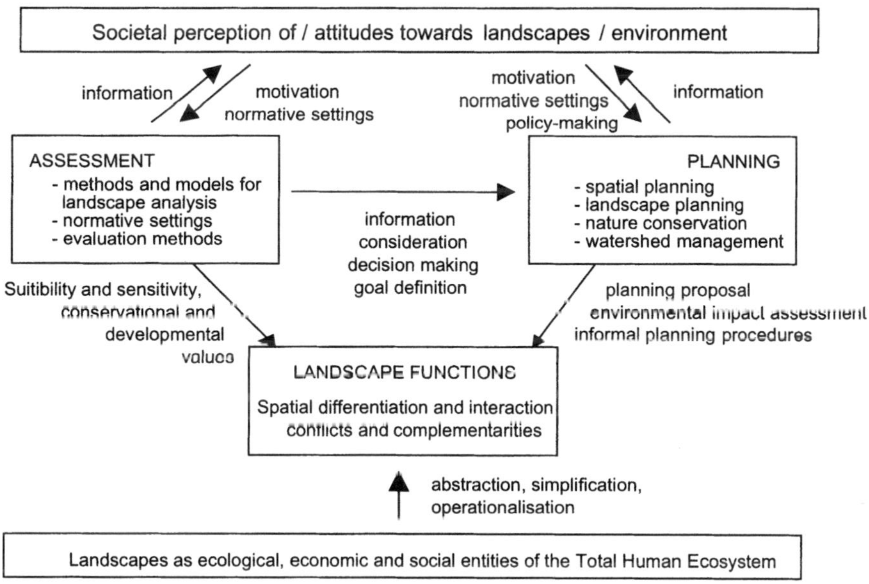


**Fig. 9.4:** Implementation of landscape functions in landscape related planning

**Table 9.2:** Integrated landscape function assessment procedure

| | Landscape functions | **Renewal of groundwater resources** | | **Capacity for agricultural production** | | **Retention of water and matter fluxes** | |
|---|---|---|---|---|---|---|---|
| Landscape evaluation | Criteria | potential conservation value for renewal of groundwater resources | | potential conservation value for agricultural production | | potential conservation value for retention of water and matter fluxes | |
| | | suitability for groundwater recharge | sensitivity against groundwater | suitability for agricultural production | sensitivity against impairments | suitability for water retention | suitability for nitrate retention |
| | Methods | Ecological Risk Analysis (ERA) | | | | | |
| | Normative standards/ objectives | ▪ minimising surface sealing<br>▪ directing urban and infrastructural developments to environmentally non sensitive areas<br>▪ promoting regional resources<br>▪ reducing nitrate leaching | | ▪ minimising surface sealing<br>▪ minimising erosion to rate of natural soil formation<br>▪ protecting rare and relict soil types<br>▪ sustaining the productive capacity for agriculture | | ▪ minimising water and matter losses<br>▪ enhancing multiple landscape functions and uses<br>▪ improving ecological water quality<br>▪ reducing nitrate leaching | |
| Landscape analysis | Indicators | mean annual ground-water recharge rate | groundwater protection ability of landscapes | biomass productive quality of landscapes | disposition to soil erosion | runoff regulation capacity of landscapes | disposition to nitrate leaching |
| | Methods | runoff model ABIMO / semi-quanti-tatve | qualitative factor combination | qualitative factor combination | semi-quantitative modelling using USLE | qualitative factor combination | runoff model ABIMO / qualitative factor |
| | Parameters | ▪ mean annual precipitation<br>▪ mean annual evapotranspiration<br>▪ land cover type<br>▪ groundwater level<br>▪ soil texture<br>▪ available field capacity | ▪ mean annual groundwater recharge rate<br>▪ soil permeability<br>▪ groundwater level | ▪ slope<br>▪ soil texture<br>▪ groundwater level<br>▪ field capacity<br>▪ mean annual temperature<br>▪ mean annual precipitation | ▪ slope<br>▪ soil texture and substrate<br>▪ mean annual precipitation<br>▪ land cover type | ▪ land cover type<br>▪ slope<br>▪ infiltration capacity<br>▪ field capacity | ▪ groundwater recharge rate<br>▪ field capacity |
| | Data | Primary data integration (climate, land cover, soil, relief, planning) and management with GIS for analysis, modelling, assessment, visualisation | | | | | |

The purpose of this chapter is not to present sophisticated, complex models and methods which are able to represent landscape systems in as much detail as possible. Instead, methods are used which provide a convenient tool for comprehensive procedures in regional planning taking into account constraints of time and money, and which provide the information required with reasonable effort.

### 9.6.3 Assessing the renewal of groundwater resources in the Dessau region

The *suitability* for the renewal of groundwater resources is indicated by the mean annual groundwater recharge rate. According to the hierarchy principles discussed by Klijn (1995) and Steinhardt & Volk (Chap. 6), in this study a top-down approach is used. In the first step, the simple numerical model ABIMO (runoff-simulation model) is used for mesoscalar water balance modelling. Developed at the East German water management authority and now established within German hydrologic monitoring at the Federal Institute of Water Research (Bundesanstalt für Gewässerkunde, see also Volk & Steinhardt in Chap.7), its input factors are climate (mean annual precipitation, potential evapotranspiration), land cover (settlement, agriculture, deciduous and coniferous forest, unvegetated land), soil (field capacity, soil texture) and the groundwater level. The output is the annual mean of actual evapotranspiration (ETa), and the total runoff can be concluded from the basic water balance equation R = P - ETa. In flat landscapes with permeable underground, the model sets total runoff equivalent to groundwater recharge. This is only valid assuming that surface runoff and interflow are not relevant, which is obviously not the case. In the second step therefore, a runoff separation divides total runoff into groundwater recharge using slope and groundwater level as criteria, and direct runoff using hydromorphic soil characteristics.

The resultant groundwater recharge rate is one input factor for assessing the groundwater protective function of a landscape, which is taken as an indicator of the *sensitivity* to impacts of the renewal of groundwater resources. The method taken for the assessment procedure was developed by Marks et al. (1992) and modified by Bastian & Schreiber (1999), is based upon a qualitative combination of the parameters soil permeability, groundwater level and groundwater recharge rate.

Finally, in accordance with ecological risk analysis, suitability and sensitivity are combined to form the landscape's *potential importance* for the renewal of groundwater resources via a simple point matrix (Table 9.3).

The resulting five classes of potential importance are reduced to three classes (low - moderate - high) within Map 9.2. This is partly due to practical reasons of clarity and highlighting major functional differences. However, it must be stressed that a qualitative evaluation of analysis which is also partly qualitative is limited in its capacity for detailed, spatially differentiated assessment. Therefore, the number of classes presented here honestly reflects the quality of data and methods used. They are selected with the intention of producing a practicable tool for strategic regional planning which must be able to identify important spatial differences in functional characteristics.

**Table 9.3:** Evaluation of the potential importance of a landscape for the renewal of groundwater resources by suitability and sensitivity criteria

| **Potential conservation value of the landscape function** ***renewal of groundwater resources*** | | **Sensitivity** *classes of groundwater protection ability* | | |
|---|---|---|---|---|
| | | low (1) *5* | moderate (2) *3 - 4* | high (3) *1 - 2* |
| **Suitability** | Low (1) *< 50* | low 1 | low 2 | moderate 3 |
| *mean groundwater recharge rate in mm/a* | Moderate (2) *50 - < 100* | low 2 | moderate 3 | high 4 |
| | high (3) *≥ 100* | moderate 3 | high 4 | high 5 |

Map 9.2 shows that a high potential conservation value is limited to the following landscape types:

- *High suitability and high sensitivity*
  Forested land on sandy soils as well as agricultural land on sandy-loamy soils in the upper parts of the end moraines with more than 600 mm rainfall per year

- *High suitability and moderate sensitivity*
  Agricultural land on sandy loess with an annual precipitation of more than 500 mm

- *Moderate suitability and high sensitivity*
  Agricultural land on sandy soils in the flood plains and lowlands along the River Elbe with varying annual precipitation and a groundwater table of more than 1m below surface

Additionally, Map 9.2 contains the priority areas for groundwater extraction in the regional development programme. It is apparent that the priority areas are not situated where the potential importance for the renewal of groundwater resources is highest. This is due to the fact that hitherto regional planning did not follow the precautionary principle inherent in the sustainability concept, since it reduced its measures for the protection of extraction wells. Using the evaluation of the selected landscape function, the assessment procedure allows the introduction of predictive, precautionary and process-oriented methods of groundwater protection, i.e. protecting the areas where groundwater is recharged, not just where it is extracted.

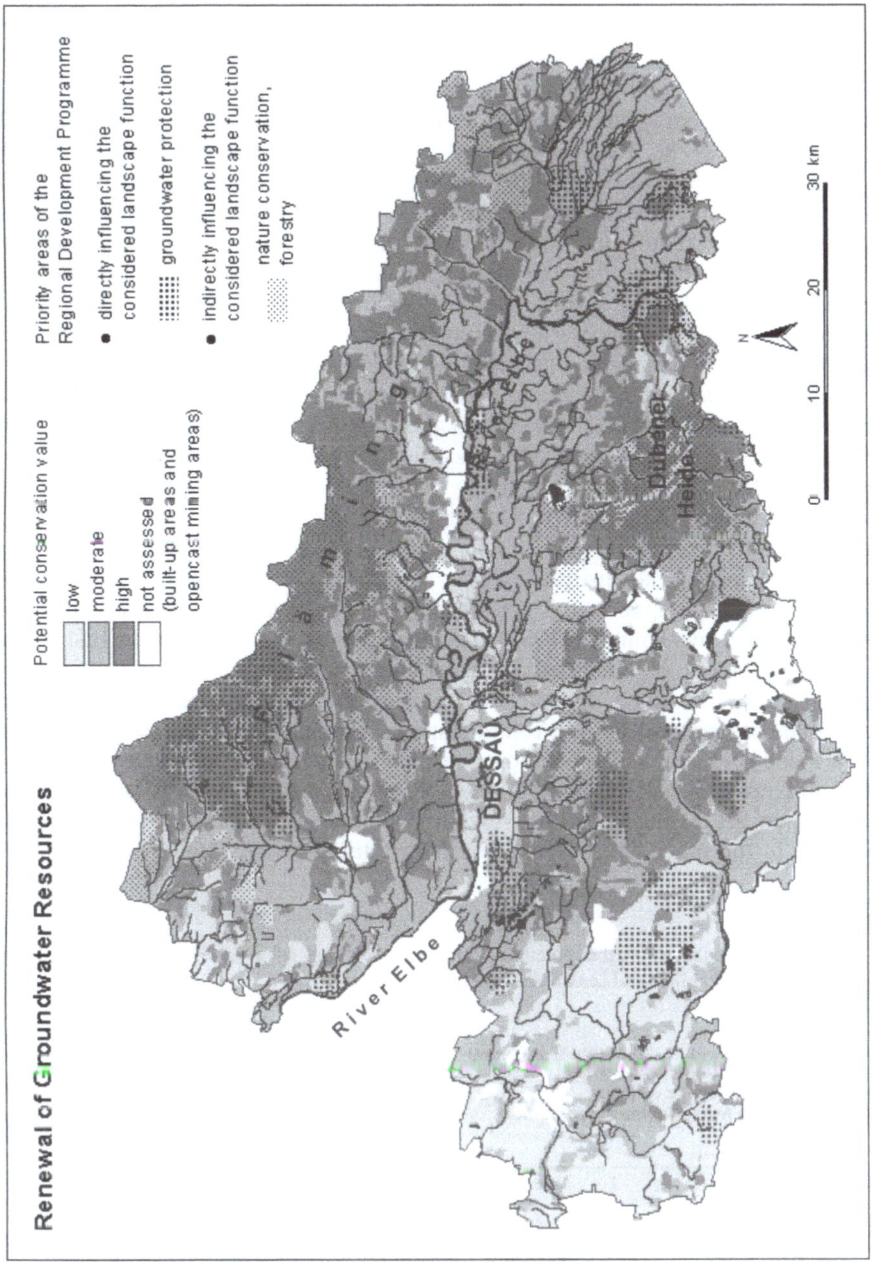


**Map 9.2:** Renewal of groundwater resources in the Dessau region - potential conservation values and actual objectives in regional planning

### 9.6.4 Assessing the capacity for agricultural production

Analysing and evaluating the landscape potential for agricultural production provides an indicator of the *suitability* of landscapes for agriculture. Marks et al. (1992) develop a point rating system in which the relevant factors are represented by one or more parameters. Bastian & Schreiber (1999) stated that the number of parameters may be reduced as long as all relevant factors are considered. Therefore, the method is simplified for this study using the parameters slope for relief, soil texture and rooting depth for soil, groundwater level, hydromorphic character and available field capacity (water storage available for plant uptake) for landscape water balance, annual mean temperature and mean annual precipitation for climate. The rating system uses five different classes from 1 to 5 for each parameter and introduces a minimum factor system for the parameter combination: the parameter classified lowest determines the result.

The *sensitivity* of landscapes to the deterioration of their productive potential is assessed via the potential risk of soil erosion. Analysis within the assessment procedure is based upon the USLE, modified for central European conditions and termed ABAG by Schwertmann et al. (1990). The stochastic model calculates the average soil loss in t/ha/a. Here, it is simplified and applied to mesoscalar analysis, even though its quantitative character limits applicability to the field scale. As transfer to mesoscalar analysis reduces accuracy, the qualitative description of erosion risk is assigned to 5 classes ranging from very low to very high as recommended by Hennings (1994). The evaluation system inherent in USLE and ABAG needs to be modified in accordance with the normative settings of sustainability presented above. The soils of the loess plains form the most important agricultural land in central Europe. Furthermore, since loess is a non-renewable resource, soil protection is of major importance. Class boundaries for the soil erosion risk evaluation are altered, so that loess soils under arable cultivation are evaluated with a high risk instead of a moderate risk for soil erosion in the procedure proposed by Schwertmann et al. (1990) and Hennings (1994).

**Table 9.4:** Evaluation of the potential conservation value of a landscape for agricultural production by suitability and sensitivity criteria

| **Potential conservation value of the landscape function** ***capacity for agricultural production*** | | **Sensitivity** *disposition to soil erosion* low (1) *5* | moderate (2) *3 - 4* | high (3) *1 - 2* |
|---|---|---|---|---|
| **Suitability** | low (1) *< 50* | low 1 | low 2 | moderate 3 |
| *potential biotic productivity* | moderate (2) *50 - < 100* | low 2 | moderate 3 | high 4 |
| | high (3) *≥ 100* | moderate 3 | high 4 | high 5 |

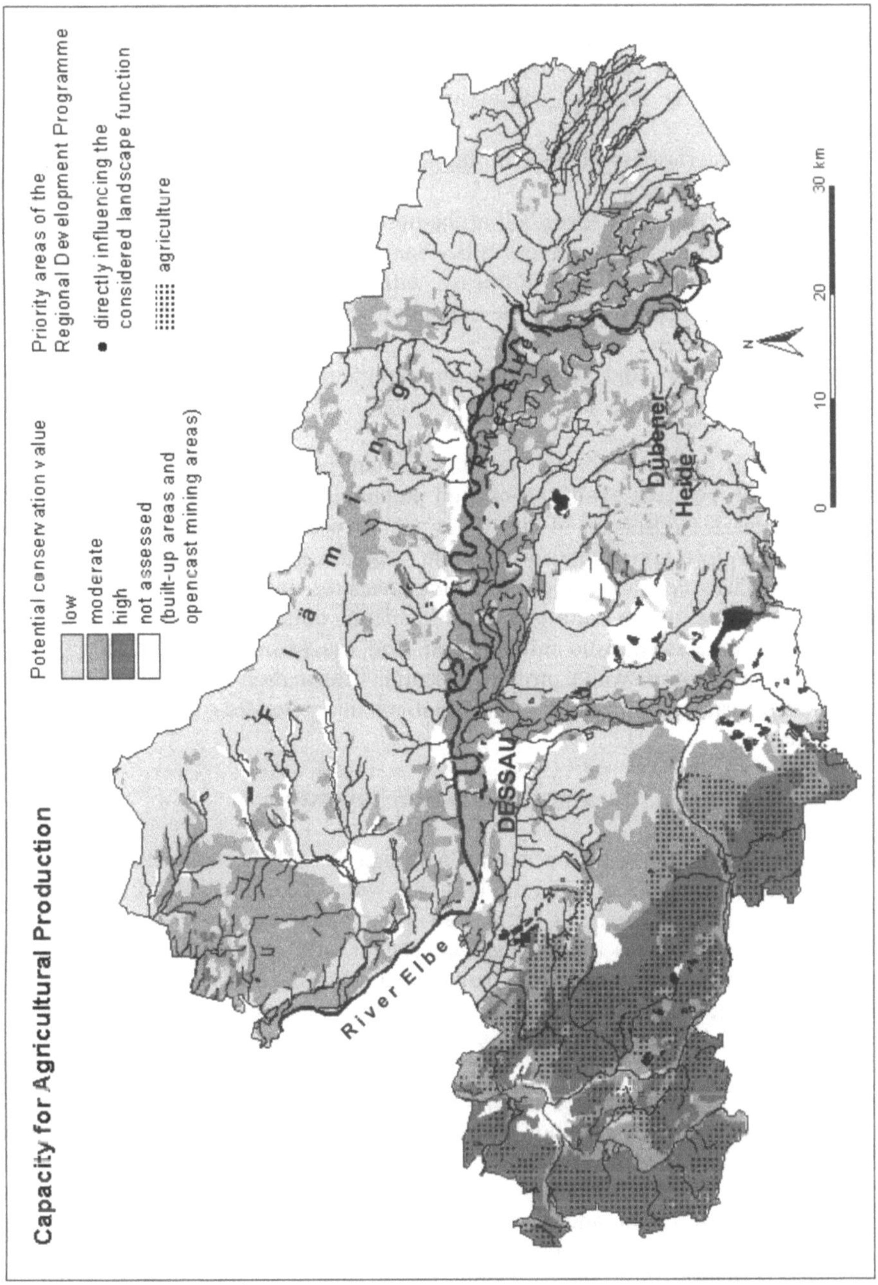


**Map 9.3:** Capacity for agricultural production - potential conservation value and actual objectives in regional planning

Table 9.4 shows the matrix combining a landscape's suitability for and sensitivity to the impairment of its function for agricultural production with its potential importance for this landscape function. The result is a clear spatial differentiation of the potential conservation values within the Dessau region (see Map 9.3):

- *High* - western loess plains with high to very high suitabilities and moderate to high sensitivities
- *Moderate* - the sandy loess belt at the north eastern fringe of the loess plains, the fluvial plains of the rivers Elbe, Saale and Mulde, as well as the loamy ground moraines of the northeast with high suitabilities and moderate sensitivities; steeper parts of the end moraines with moderate suitabilities and high sensitivities.
- *Low* - the poor sand areas north and south of the River Elbe with low to moderate suitabilities and low sensitivities

The priority areas of the regional development programme shown in Map 9.2 cover most of the areas evaluated of high potential conservation value. These are the agricultural landscapes of the loess plains with relict chernozems, where priority is given to agricultural production. Hitherto, priority has only been given to other land use types instead of protective measures minimizing soil erosion and degradation. One of the shortcomings of spatial planning in Germany is that existing instruments have no influence on land use practices. The results of the assessment procedure would provide essential information for a comprehensive and process-oriented approach of regional planning towards the future role of agriculture.

### 9.6.5 Assessing the water and matter retention

The water and matter retention within landscapes is a "regulation function" (De Groot 1992) which does not directly coincide with a certain societal use or value. The links are rather indirect: water retention contributes to the reduction of flooding since it reduces runoff peaks consistent with direct runoff components. Matter retention contributes to the maintenance of many landscape assets such as biomass productivity, and physical and biotic resources. Conversely, matter retention in landscapes is a prerequisite for reducing eutrophication in the North Sea.

Therefore, unlike other, land-use oriented functions, the evaluation of suitability and sensitivity cannot be regarded as the two sides of the landscape function. A high runoff retention, one of the indicators of water and matter retention, means a low degree of surface runoff and interflow, which stands for both high suitability for and low sensitivity against impacts of this regulation function and vice versa. Therefore, the indicators described below are directly associated with protective and developmental aspects of the *potential conservation value*.

The above-mentioned *runoff retention* is assessed using a point rating system from Marks et al. (1992). The parameters land cover (forest, shrubs, grassland, densely vegetated arable land, sparsely covered arable land), slope, infiltration capacity, and available field capacity are classified and the runoff retention class is

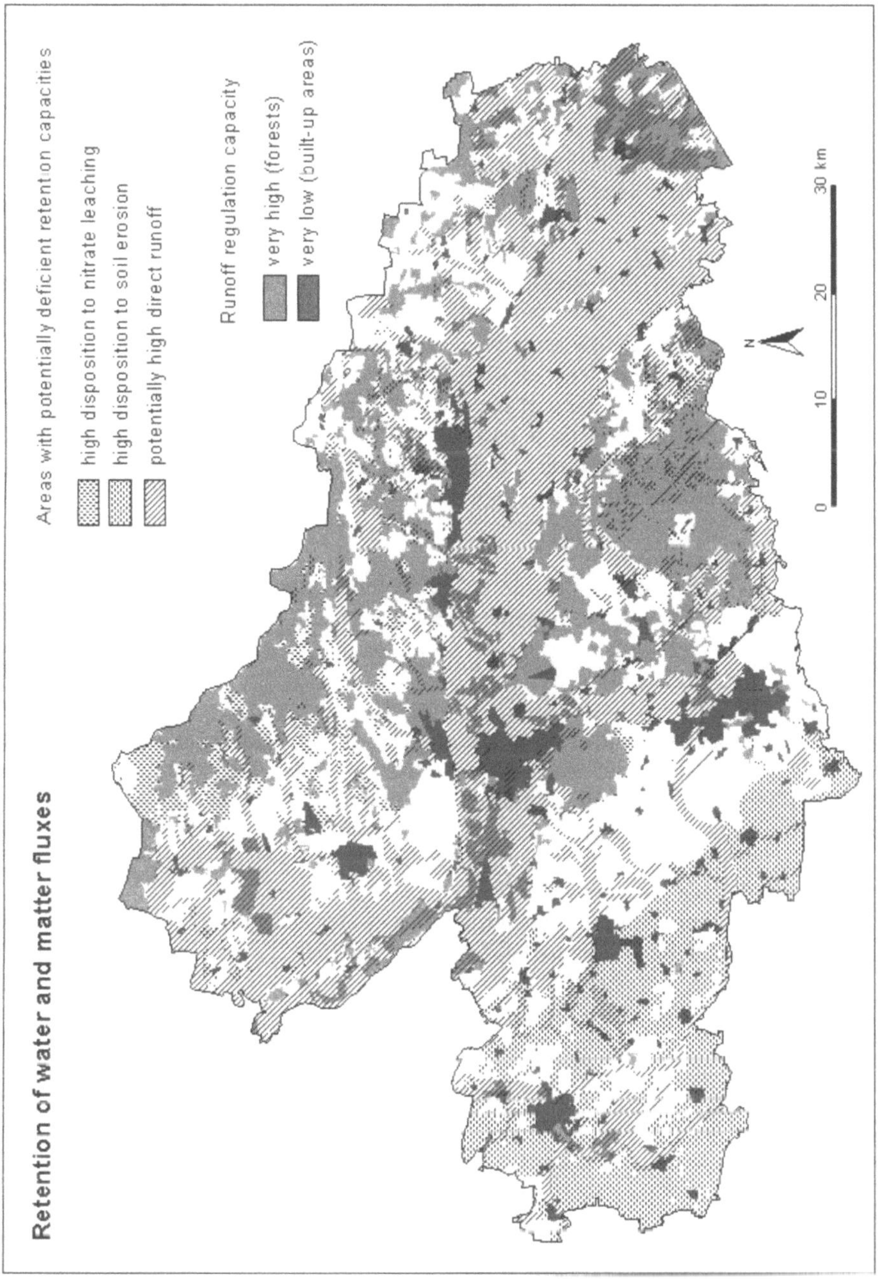


**Map 9.4:** Retention of water and mater fluxes - deficiencies and capacities of landscapes

calculated by the arithmetical mean of all parameters. In addition, the groundwater level is taken into account by a reduction of one class value. Another indicator is the *nitrate leaching risk*, which is indicated by the annual soil water turnover calculated from groundwater recharge (9.6.4) and field capacity, the total capacity for water storage of a soil (Frede & Dabbert 1998). Other indicators are the soil erosion risk (introduced above) and the runoff separation quotient.

The protective and developmental aspects of the potential conservation value are the result of the evaluation described in Table 9.5 and shown in Map 9.4:

- *Protective aspects*
  Areas with very high runoff retention capacities in combination with a low annual soil water turnover have high suitabilities for water and matter retention as well as low sensitivities against impacts. Environmental planning targets should concentrate on safeguarding and protecting this function. In the Dessau region, most parts of the forested land, which is already protected via priority areas for forestry, is part of this category. This result does not lead to satisfactory spatial differentiation. It must be questioned whether the applied method is appropriate in this context.

- *Developmental aspects*
  - Areas with low to moderate water retention capacities are evaluated with a high potential improvement value, leading to developmental rather than conservational planning targets. This is substantiated and differentiated by the further evaluation of ...
  - high annual soil water turnover, indicating the vertical leaching of polluting substances (e.g. nitrate) in those parts of the Fläming under agricultural cultivation (some parts are within priority areas for groundwater abstraction, which needs further consideration in planning);
  - high runoff separation quotient, indicating lateral water and matter fluxes in areas with high groundwater levels and hydromorphic soils (e.g. floodplains of rivers and streams);
  - high soil erosion risk indicating lateral water and matter fluxes in the loess plains and the steeper parts of the end moraines.

**Table 9.5:** Evaluation of the potential conservation value of landscapes for retention of water and matter fluxes

| **Potential conservation value** Of the landscape function ***Retention of water and matter fluxes*** | ***Importance of protective aspects*** **high** | **moderate** | **low** |
|---|---|---|---|
| Runoff retention | very high - high (1 - 2) | moderate - low (3 - 4) | very low (5) |
| | ***Importance of developmental aspects*** | | |
| | **high** | **moderate** | **low** |
| Runoff retention | very low - moderate (5 - 3) | high (2) | very high (1) |
| Annual soil water turn over (nitrate leaching risk) | > 200 % | 100 - 200 % | - |
| Runoff quotient | ≥ 2 | ≥ 1,5 - < 2 | < 1,5 |
| Disposition to soil erosion | high | - | - |

## 9.6.6 Integrating landscape functions: conflicts and complementarities of planning objectives

The isolated assessment of individual landscape functions presented above is inadequate for both the definition of a comprehensive target system and the thorough understanding of the landscape system. Decision-making in landscape-related planning and policy requires knowledge of interactions among the functions and uses of the landscape.

The GIS platform of the assessment procedure allows the spatial overlay of the information desired on the selected functions: their potential importance as well as the landscape's sensitivities and suitabilities. While technical realization is simple, extensive knowledge is required to decide what a certain functional overlay (in GIS terms) or interaction (in landscape-ecological terms) means to the functional integrity of landscapes and the need for planning action concluded therefrom. This is the crucial point of planning, where landscape-ecological knowledge has to be met by normative settings such as environmental quality standards and objectives. This is discussed and illustrated below.

Map 9.5 shows the integrated evaluation of the considered landscape functions. To identify functional conflicts and complementarities, only those functional importances which were evaluated as high in the previous step of the assessment procedure are shown. Therefore, different categories of functional interaction become apparent:

- *Groundwater recharge versus agricultural production*
  In the southern part of the sandy loess plain belt east and southeast of Köthen, conflicts between these two functions are likely to occur. This area is characterized by intensive agricultural production due to high biomass

productivity comparable to the loess chernozem region in the west. The decisive distinction comprises higher soil permeability and higher annual precipitation.

- *Water and matter retention versus agricultural production*
  The fertile loess plains covering the western part of the Dessau region exhibit high sensitivity to soil erosion and are therefore of high importance for water and matter retention within the region. Potential conflicts with agricultural production arise and are exacerbated by large-scale farming and a lack of natural features such as hedgerows or buffer strips along watercourses.

- *Water and matter retention meets groundwater recharge*
  The complementarity of targets for future development can be concluded for the end moraine landscapes covered by forests. Dübener Heide and the Fläming fulfil important functions for the renewal of regional groundwater resources as well as the retention of water and matter. Since nitrogen input from aerial deposition has become a considerable source of pollution, the importance of forests and their adequate management has grown.

The integration of assessment results into the target system of the regional plan depends on the further development planning instruments. Schmidt (1996) suggests complementing planning categories like priority areas by extending their target defining and legally binding character:

- Areas with specific demands or standards for groundwater protection
- Areas with specific demands or standards for soil protection
- Areas requiring maintenance of open landscape character
- Areas requiring enrichment with natural structures

These specifications would mean the integration of the environmentally sensitive area concept discussed extensively by landscape ecologists (recently by Steiner et al. 2000) into the regional planning framework.

The results of the assessment procedure form the knowledge base for regional planning in the Dessau region in order to develop and maintain multifunctional landscapes:

- *Priority areas for agriculture with specific demands for groundwater protection and/or soil protection* would give priority to agricultural land use yet simultaneously define environmental standards for agricultural practice. This would form a powerful instrument within not only spatial planning but also river basin management, since water policy and management lack instruments for combatting non-point source pollution (e.g. phosphorus and nitrogen) from agricultural land.

- *Areas requiring the maintenance of their open landscape character* would be a useful instrument for maintaining the agricultural enclosures of the forest landscapes. This would be important for preserving the landscape character and their important role for groundwater recharge and therefore complements priority areas for groundwater abstraction as well as nature conservation.

- *Areas requiring enrichment with natural structures* should be designated where lateral water and matter flows are high. This would be of significant importance for areas prone to soil erosion and the river and stream floodplains.

Realizing these specifications would drastically widen the scope of regional planning instruments. Regional planning would have the ability to play an active role in predicting landscape development. Strategic targets and long-term objectives are spatially differentiated and form a strategic framework for small-scale local action and planning. In particular, the quality of compensation rule measures could be improved substantially. Currently, measures are carried out in a local framework, where practicability rather than the functionality of landscape systems is the guiding principle. A regional context is the prerequisite for integrating local measures into a strategic perspective, e.g. the enrichment of the loess plains with natural structures in order to minimize soil erosion and to improve water and matter retention. The multifunctionality of agricultural landscapes could thus be strengthened, since the financial aspects of compensating farmers for reduced profits often limit planning action in such areas.

Yet regardless of widening the scope of legally binding planning instruments, the GIS-based assessment procedure gives planners and decision-makers easily accessible information on landscape functions. Compared to the information traditional regional plans rely on, this is a new quality and a powerful instrument in itself. It was stated above that the role of planning is changing from a system of legally binding instruments to a more flexible system of consideration processes and agreements. As a result, information on spatial interactions and the transparency of considered targets are becoming key features of planning processes and decision-making.

The assessments and spatial specifications presented are the result of normative settings of priorities. They are not definitive and do not represent an optimal situation. However, every step of the assessment procedure is transparent - and so are the normative settings. They can be altered by increasing knowledge about landscape systems or the need to compromise within a decision-making process. But the available information about suitabilities and sensitivities, the concluded importance and the identified functional conflicts and complementarities form a sound knowledge base for planning - and enabling the effect of any planning decision on the landscape to be judged.

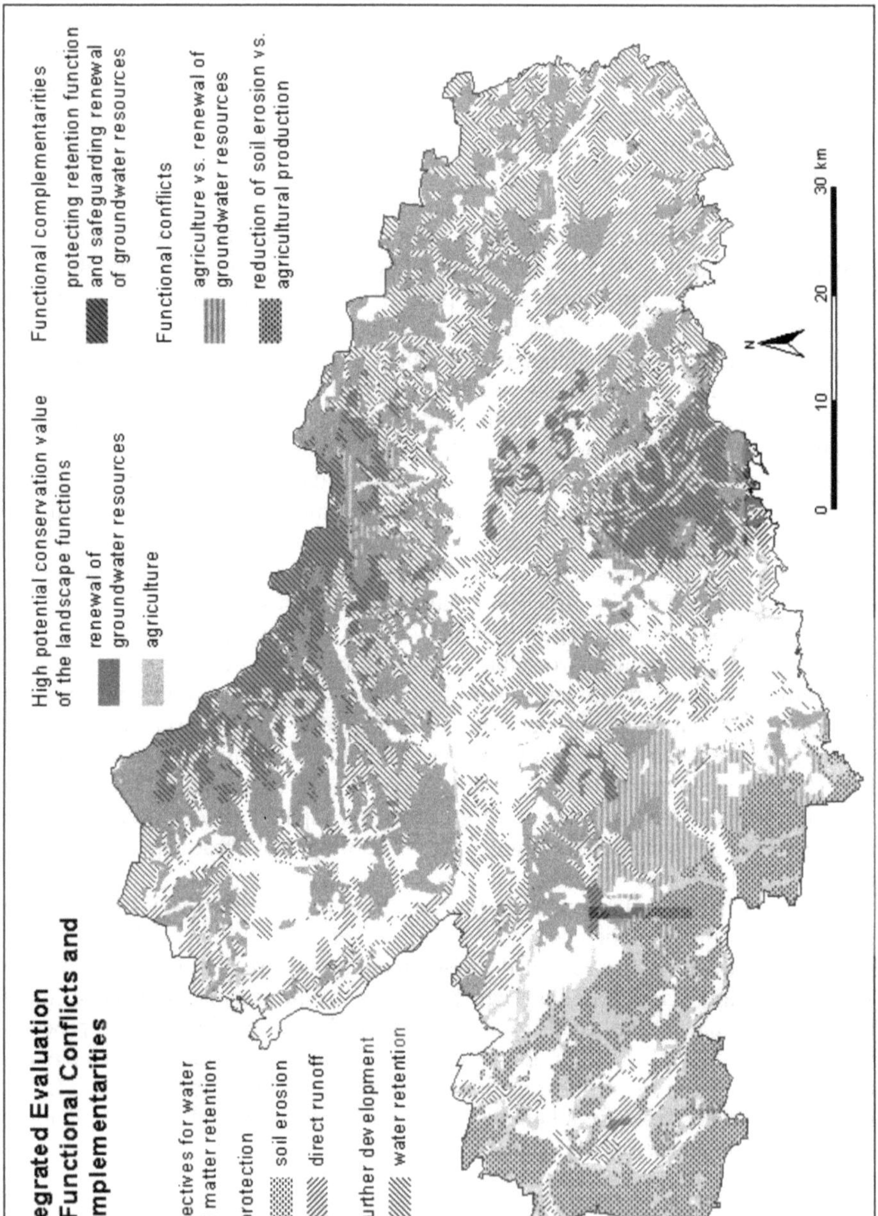


**Map 9.5:** Integrated evaluation of landscape functions - potential conservation values, functional conflicts and complementarities

## 9.7 Conclusions

Growing scepticism surrounds the feasibility of environmental planning in view of the complexity of landscapes and the relevant steering factors. The German landscape planner Jessel (1995) even provocatively asks whether landscape planning really is feasible.

The answer would be no, if sustainable landscape development needed to be planned down to the last detail to obtain optimum results and exactly what was initially intended - which is doubtless the case for planning and building a house. But even there the planner is confronted by uncertainties necessitating planning alteration during construction. Landscapes are not constructed, since landscapes are dynamic and permanently changing systems, whose functional interactions are not understood sufficiently and probably never will be, and the active agents, be they physical, biotic or human, are manifold in their system behaviour.

But the answer is definitely yes if uncertainty, dynamics, and permanent, often unpredictable change are accepted as inherent characteristics of a landscape's identity (see also Hobbs 1997) - and if there is not necessarily a need for control. Environmental planning has an active role in initiating and supporting processes identified as contributing to sustainable landscape development. Of course, this is not sufficient for achieving environmentally sound development. Critical loads for polluting substances need to be defined in a legally binding way, as is already the case (or is at least proposed) for improving water quality. The EC nitrate directive sets standards for drinking water quality, while the EC directive on water policy puts forward standards for many important substances like pesticides and heavy metals. But the sound environmental development of landscapes needs support at a regional and strategic level, where legally binding measures are much harder to define and objectives are the result of considering different interests and demands. Therein lies the strength of the presented assessment procedure, which supports modern planning in the following ways:

- *Multifactoral analysis and assessment using GIS*

The complexity of landscapes is represented by the landscape function concept. Landscape functions are the result of interacting factors such as climate, soil, land use and land cover. The assessment procedure defines parameters which enable the analysis of landscape functions with a set of established data, methods and models. GIS serves as a platform for primary data integration, the linkage of methods and models to form a coherent system. Therefore, it is possible to select the best practicable data, methods and models from case to case. This has proven to be much more flexible and practicable than the development of some kind of 'supermodel' which can answer any question. The assessment procedure provides information on a broad range of landscape matters and serves as a tool for conflict analysis and decision-making

- *Flexibility in spatio-temporal contexts*

Primary data integration resulted in a huge database of nearly 18,000 polygons containing all the relevant information. They form quasi-homogenous base units of landscape information. Therefore, spatial analysis is easily adopted to very different planning aspects in spatial planning, landscape planning or river basin management. Predefined spatial units such as hydrologic response units in watershed analysis or landscape units in landscape planning have their strength in certain contexts. They are derived from a distinct factor combination to deal with a certain phenomenon. The way chosen here allows the definition of specific spatial units but does not restrict analysis to them. Another aspect of flexibility is that of monitoring change and using scenario techniques to evaluate future developments. For mesoscalar analysis, data with an adequate spatiotemporal resolution are often lacking, but the methodical structure of the assessment procedure allows such applications.

- *Integrating land use and environmental planning*

Planning is divided into sectors. Spatial planning is very often land use planning influencing the spatial distribution and pattern of settlements, infrastructure, open space etc. On the other hand landscape planning concentrates on physical and biotic landscape features for their importance for species and populations. The landscape function concept of the assessment procedure integrates both landscape views, since landscape functions combine land use and land resources aspects. Therefore it becomes possible to define priority areas for agriculture by means of productivity and soil fertility criteria, as is currently solely done, and by means of sensitivity to soil degradation, which is neglected in current practice.

- *Backing participative and open planning processes*

The assessment procedure serves as an information system. It describes certain landscape factors and processes, evaluates suitabilities and sensitivities, and shows where certain functions are of regional importance. The methods used for analysis and the normative settings behind evaluations are apparent and may be altered if they prove insufficient. All steps of the procedure are transparent to everyone involved in planning and decision-making processes, be it the planner who has to identify suitable areas for an ecological network and wants to evaluate potential conflicts he may be confronted by, be it the politician, who is under considerable pressure to create jobs and is seeking suitable sites for industrial development, or the concerned public, or a land user who wants to know why certain restrictions inhibit his interests.

## 9.8 References


ARL (Hrsg) ( 1995): Zukunftsaufgabe Regionalplanung. In: Forschungs- und Sitzungsberichte der ARL 200.

Bachfischer, R., David, J. & H. Kiemstedt (1980): Die ökologische Risikoanalyse als Entscheidungsgrundlage für die räumliche Gesamtplanung - dargestellt am Beispiel der Industrieregion Mittelfranken. In: Buchwald, K. & W. Engelhardt (eds): Handbuch für Planung Gestaltung und Schutz der Umwelt, Band 3: Die Bewertung und Planung der Umwelt, München: 524-545.

Bastian, (1999b): Das Nachhaltigkeitsprinzip als Leitbild der Landschaftsentwicklung. In: Böhm, H.-P, Dietz, J. & H. Gebauer (eds): Nachhaltigkeit - Leitbild für die Wirtschaft? TU Dresden, Zentrum für Interdisziplinäre Technikforschung: 159-170.

Bastian, O. & K.-F. Schreiber (Hrsg) (1999): Analyse und ökologische Bewertung der Landschaft. Heidelberg, Second edition. Berlin.

Cook, E.A. & H.N. van Lier (Hrsg) (1993): Landscape planning and ecological networks. Developments in Landscape Management and Urban Planning 6F. Amsterdam.

Cullingworth J.B. & V. Nadin (1997): Town and country planning in Britain. 12th edition. London, New York.

De Groot, R.S. (1992): Functions of Nature - Evaluation of nature in environmental planning, management and decision making. Groningen.

Dekker, A. (2000): Aufgaben und Verfahren der Regionalplanung zwischen formellen und informellen Arrangements. In: Planung in den Niederlanden - anders als bei uns? ILS-Schriften 163: 19-27.

Frede, H.-G. & S. Dabbert (Hrsg) (1998): Handbuch zum Gewässerschutz in der Landwirtschaft. Landsberg.

Forman, R.T.T. & M. Godron (1986): Landscape Ecology. New York.

Haber, W. (1972): Grundzüge einer ökologischen Theorie der Landnutzungsplanung. - In: Innere Kolonisation 21 (11): 294-299.

Haber, W. (1990): Using Landscape Ecology in Planning and Management. In: Zonneveld, I.S. & R.T.T. Forman (eds): Changing Landscapes: An Ecological Perspective. New York 217-232.

Haber, W. (1998): Reflections on the Ecological Role of Agriculture. In: Barron, E.M. & I. Nielsen (eds): Agriculture and Sustainable Land use in Europe. Papers from Conferences of European Environmental Advisory Councils. London, 147-160.

Hendriks, J.A.M. (2000): Niederländischer Verwaltungsaufbau und niederländisches Planungssystem. In: Planung in den Niederlanden - anders als bei uns? ILS-Schriften 163: 10-15.

Hennings, V. (1998): Einfluß verschiedener Verfahren der Aggregierung auf die Güte bodenkundlicher Auswertungskarten. In: Mitt. Dtsch. Bodenkund. Ges. 88: 417-420

Hobbs, R. (1997). Future landscapes and the future of landscape ecology. - In: Landscape and urban Planning 37: 1-9.

Jessel, B. (1995): Ist Landschaft planbar? Möglichkeiten und Grenzen ökologisch orientierter Planung. In: Laufener Seminarbeiträge 4/95: 7-10.

Klijn, J.A. (1995): Hierarchical concepts in landscape ecology and its underlying disciplines. In: Report 100, DLO Winand Staring Centre, Wageningen.

Kühling, W. (1992): Notwendige Anmerkungen zum Entwurf der allgemeinen Verwaltungsvorschrift zur Ausführung des Gesetzes über die Umweltverträglichkeitsprüfung. In: UVP-report 1/92: 2-6.

Locher, B., Hülsmann, W., Schablitzki, G., Dickow-Hahn, R. & C. Wagener-Lohse (1997): Zum Stellenwert der Regionalplanung in der nachhaltigen Entwicklung. In: UBA-Texte 31/97.

Luz, F. (2000): Participatory landscape excology - A basis for acceptance and implementation. In: Landscape and Urban Planning 50 (1-3):157-166.

Marks, R. et al. (eds) (1992): Anleitung zur Bewertung des Leistungsvermögens des Landschaftshaushaltes (BA LVL). Second edition. econd edition. In: Forschungen zur deutschen Landeskunde 229.

Naveh, Z. & A.S. Lieberman, A.S. (1994): Landscape ecology - Theory and application. New York.

Newson, M.D. (1992): Land, water and development: river basin systems and their sustainable management. London.

Petry, D. (2001): Landschaftsfunktionen und planerische Umweltvorsorge auf regionaler Ebene: Entwicklung eines landschaftökologischen Verfahrens am Beispiel des Regierungsbezirks Dessau. PhD-thesis, Universität Halle.

Plachter, H. (1994): Methodische Rahmenbedingungen für synoptische Bewertungsverfahren im Naturschutz. - In: Z. f. Ökologie und Naturschutz 3: 87-106.

Ripl, W. (1995): Management of Water Cycle and Energy Flow for Ecosystem Control - The Energy-Transport-Reaction (ETR) Model. In: Ecological Modelling 78: 61-76.

Schmidt, C. (1996): Beitrag zur regionalplanerischen Umweltvorsorge - unter besonderer Berücksichtigung ökologischer Wechselwirkungen zwischen Fließgewässern und Einzugsgebieten. PhD-thesis. Weimar.

Schwertmann, U., Vogl. W. & M. Kainz (1990): Bodenerosion durch Wasser - Vorhersage des Abtrags und Bewertung von Gegenmaßnahmen. Stuttgart.

Steiner, F., McSherry, L. & J. Cohen (2000): Land suitability analysis for the upper Gila River watershed. In: Landscape and Urban Planning 50 (4): 199-214.

Steinhardt, U. & M. Volk (2001): Scales and spatio-temporal dimensions in landscape research. - (in this book)

Scholles, F. (1997): Abschätzen, Einschätzen und Bewerten in der UVP - Weiterentwicklung der Ökologischen Risikoanalyse vor dem Hintergrund der neueren Rechtslage und des Einsatzes rechnergestützter Werkzeuge. In: UVP-Spezial 13. Dortmund.

Van der Maarel, E. & P.L. Dauvellier (1978): Naar en Globaal Ecologisch Model (GEM) voor de Ruimtelijke Ontwikkeling van Nederland (deel 1 en 2). 's-Gravenhage.

Volk, M. & U. Steinhardt (2001): Landscape balance and landscape assessment. - (in this book).

# 10 Decision support for land use changes - A combination of methods for policy advising and planning

Bernd Klauer, Burghard Meyer, Helga Horsch, Frank Messner, Ralf Grabaum

## 10.1 Introduction

The cultural landscape is the result of the anthropogenic usage of the natural environment. According to Leser et al. (1984), "The cultural landscape ensues from human groups and societies permanently influencing the natural landscape and in particular using it for economic purposes and settlement as they exercise their basic functions of existence." Whereas geofactors such as the climate, soil and topography are affected by mankind only indirectly and apparently very slowly (e.g. by means of climatic change or soil erosion), mankind shapes the landscape within much shorter spaces of time by means of land use. Consequently, land use neEds. to be planned responsibly. In particular, the various groups responsible must anticipate the various effects land use will have, and then only choose those forms which will have the 'best impact' from among the possible usage options. This article deals with methods which can be used for decision support for land uses changes.

Changes in land use usually exert different, sometimes even contradictory effects on the landscape functions. For example, planting a field bush will affect a landscape's water erosion, wind erosion and runoff regulation, or its suitability for recreational purposes. Moreover, changed habitats encourage different communities of animals to live. Even the landscape's economic functions may be affected; for instance, the newly planted bush will mean the agriculture once pursued on that particular parcel of land must of course be discontinued. Hence many different criteria have to be taken into account and weighed up against each other when a change of land use is assessed.

At the UFZ Centre for Environmental Research Leipzig-Halle, two different methods for decision support for land uses changes have been developed in recent years. They can be classified on different levels of decision-making and decision implementation. One of them is the "method for multifunctional assessment and multicriteria optimization for the generation of diverse land use options (LNOPT)", which was used in the Joint Project ("Impact of Land Use on Landscape Balance and Biodiversity in Landscapes Dominated by Agriculture" in an area-specific manner[1] (Grabaum 1996, Meyer 1997, Meyer, Grabaum 1998, Grabaum, Meyer 1997, 1998, Grabaum, Meyer, Mühle 1999). It gives grounds for 'optimal land use patterns' as a basis for land use changes by evaluating landscape functions and landscape elements. The result is an optimal compromise between different objectives.

The other method is the integrated ecological-socioeconomic technique used in the Joint Project "Conserving Natural Resources and Economic Development" (Klauer, Messner, Drechsler, Horsch 2001, Horsch, Ring, Herzog 2001, Klauer, Messner, Herzog 1999, Horsch, Ring 1999, Drechsler 1999, 2001). Its aim is to assess and prioritise possible options in conflict situations (in the case study for example between the protection of natural resources and economic development), and hence provide decision support for land use changes at a regional level. It does not take an area-specific approach.

Although these two decision support methods were developed for specific problems, their methodology can be transferred to many other situations. This article describes the two methods, and examines the determinants for their transfer to other areas. Of key interest is the question of how the two methods can enhance and augment each other.

We shall proceed as follows. We shall start in Section 2 by briefly characterizing the two methods and their application to date. As the multifunctional optimization of land use options has already been explained in Chapter 8, we can largely concentrate here on the integrated method for the evaluation of conflicting land uses. In the third section, we shall show how the two techniques could augment each other, and then discuss how this combined technique could be used.

## 10.2 Characterization of the two methods with an integrated concept

In the following we shall outline the structure and the main features of the two methods developed at the UFZ which are to be compared in this article. For simplicity's sake, we shall name the two techniques the Querfurt method and Torgau method after the study areas for which they were originally developed.

### 10.2.1 The Querfurt method - A concept for landscape assessment and optimization

The method for multifunctional landscape assessment and optimization was developed with the goal of proposing land use changes in the form of "optimal land use patterns". The different ways in which a landscape can be used are integrated within a framework of certain political goals to create optimal compromises for landscape development. The different land uses are assessed with the help of land use functions (Chap. 8).

[1] An area-specific approach means a reference to homogeneous areas of land use or to the "smallest common geometries". In the study area of Joint Project, Querfurter Platte these areas had a size of at least 1 ar.

In the case for which this method was developed, a landscape planning proposal for restructuring a cleared landscape which greatly increased the proportion of conservation areas in the agricultural landscape was substantiated on the basis of optimal land use patterns. The study area comprises the cadastral districts of four municipalities in Querfurter Platte, an agricultural area with a size of about 4,200 hectares in the German state of Saxony-Anhalt. Querfurter Platte is a slightly hilly, cleared landscape resulting from land redistribution and the industrialization of agriculture. It is categorized by fertile loessy black earth (the cultivated areas have an average soil index of 88), large field plots, and a very few structure-forming elements (bushes, hedges, field margins, ponds and other surface waters). Market farming prevails, with ever-shrinking crop rotation involving just a few different crops owing to economic pressures and technological possibilities.

As shown in Fig. 10.1 (Chap. 8), the Querfurt method encompasses the following steps:

1st step: *Goal definition and general assessment.* The goals are concluded from landscape analysis, discussions with stakeholders, and regional planning.
2nd step: *Data input and the selection of assessment algorithms and methods.* The choice of assessment methods to be used depends on the landscape analysis. The data required depend in turn on the assessment techniques used. The data are recorded in a geographic information system (GIS).
3rd step: *Initial assessment, assignment into assessment classes of function fulfillment.* Assessment is performed with a GIS, with validated assessment techniques being used.
4th step: *Landscape optimization, resulting in optimal landscape patterns.* Landscape optimization is used to calculate optimal land use compromises between the individual aims on the basis of envisaged types of land use.
5th step (optional): *Development of a landscape plan (or draft) into which cultural landscape information and structures can be incorporated.* As some cultural landscape information and linear landscape structures cannot yet be taken into account in optimization, they must be incorporated into a landscape plan if they are relevant to the overall aim.
6th step: *Second assessment to measure the changes between potential improvement and goal.* This assessment is carried out using the aforementioned methods. This makes the potential improvement of functional classification measurable and serves as argumentation for the decision-makers.
7th step: *Conclusion of action recommendations.* The degree of function fulfillment is determined. This provides a measure of the compromises proposed as optimal land use patterns for the land use scenarios considered. The scenarios and their functional fulfillment can be presented in maps, text form and statistics.

To sum up, the Querfurt method is a broadly structured method for integrative landscape development which has already been used in five different example areas, and which gives objective reasons for land use patterns which are optimal for the specific landscape regardless of scale.

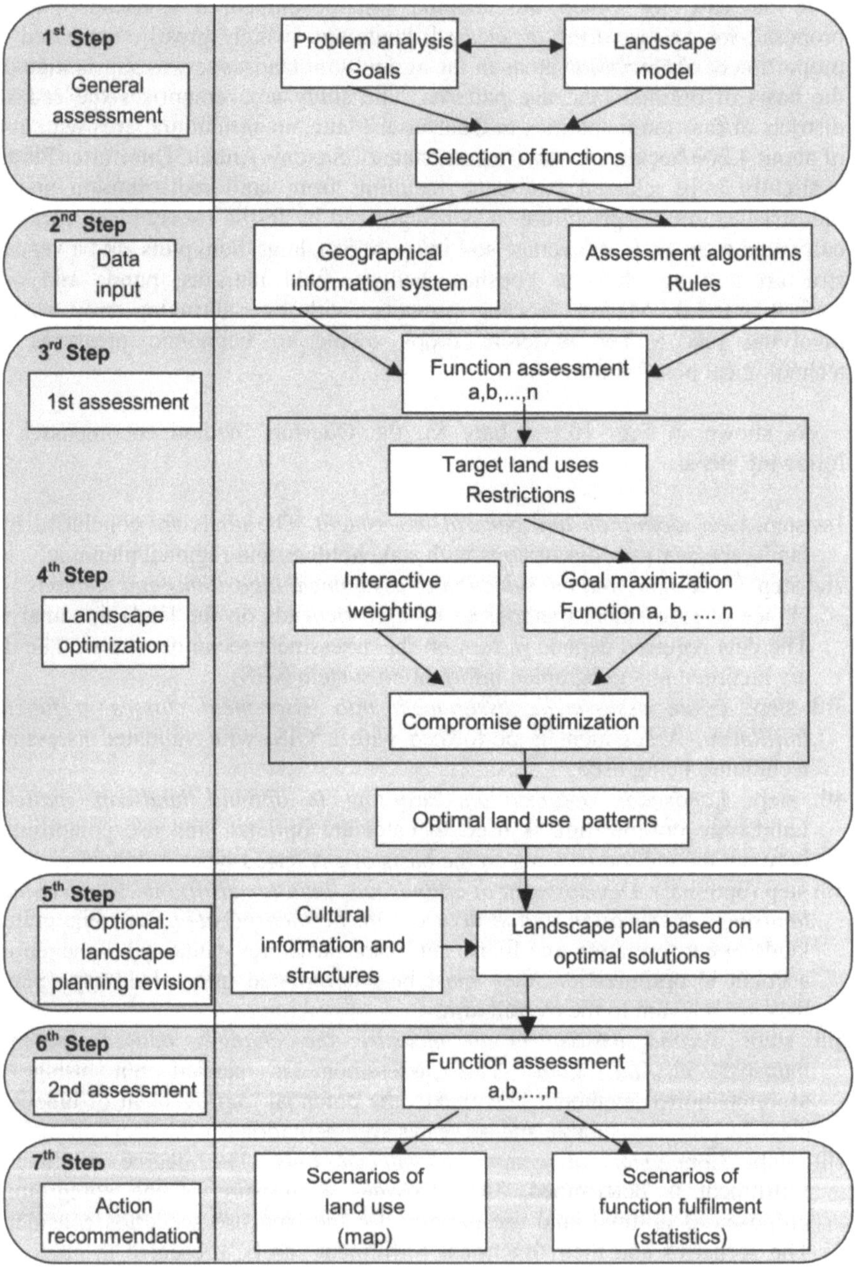


**Fig. 10.1:** Structure of the method for multifunctional landscape assessment and optimization (Querfurt method)

### 10.2.2 The Torgau method - A concept for assessing alternative options in conflict situations

The aim of the integrated assessment procedure developed during the Joint Project "Sustained Water Management and Land Use in the Elbe Catchment Area" is to provide support in selecting one of a number of alternatives in the case of land use conflicts. In the project study area, the Torgau district in north-east Saxony measuring around 700 km², a sharp conflict has emerged between the conservation of natural resources and economic development. The long-distance water supply company Fernwasserversorgung Elbe-Ostharz GmbH extracts drinking water in the Torgau area, through which the River Elbe flows. The company supplies some 3.5 million people with approximately 600,000 m³ of drinking water every day, and the waterworks in the Torgau district have the capacity to provide about 40% of this amount. To protect the groundwater, at the start of the project 33% of the study area had already been designated wellhead protection areas. In addition 52% of the area was designated landscape protection areas (albeit partly overlapping with the wellhead protection areas). In the protection zones, economic activities (especially agriculture, industrial activities and gravel quarrying) are bound by certain conditions. In view of the dramatic decline in drinking water demand since German reunification, certain parties called for the wellhead protection areas to be cut back since the elimination of the restrictive conditions in force would encourage economic development.

Another controversy arose concerning the question whether additional gravel quarries should be developed. Numerous applications to quarry gravel were received, prompted by the high demand for gravel in the first few years following reunification and the large stocks of high-quality gravel found in the Torgau Elbe floodplain. Beneficial though this may be for the local economy, an increase in gravel extraction would permanently reduce the amount of land which could be used for agriculture, diminish natural groundwater recharge, and increase pollution and noise emissions owing to the higher transport volumes.[2] Moreover, wet extraction would cause more pollutants to penetrate the soil.

The integrated assessment procedure was designed to provide decision support concerning the various options of reducing wellhead protection areas and developing new gravel quarries. The method comprises four steps (Fig. 10.2):

Step 1 :Working out possible scenarios
Step 2: Selecting problem-specific assessment criteria
Step 3: Assessment and criterion related assessment of the scenario effects
Step 4: Multicriteria analysis (MCA)

**Step 1: Working out possible scenarios**
At first, the prevailing land use conflicts are analysed in the study area. In a participatory process, the situation is discussed and possible strategies examined at an initial meeting attended by public authorities and lobbyists from the region. The fields of regional action are outlined and discussed with a view to solving the conflict, and relevant options within these areas are worked out. In the Torgau

[2] In the Torgau district, the gravel extracted is transported solely by lorry, not by ship or rail.

project, for example, one field is "drinking water protection", the basic choice being whether or not to curtail certain wellhead protection areas (specifically Zone 3b on both sides of the Elbe and Zone 3a east of the Elbe in the Mockritz wellhead protection area). The other action field is "gravel extraction", the choice being whether or not to develop new gravel quarries. The two basic options in each field result in a total of four options or action alternatives for solving the problem. These action alternatives initially concluded during a meeting attended by both public authorities and local lobbies were later stated more precisely in individual discussions.[3]

Since the benefit of each individual alternative varies depending on the future general environment, a number of different development frameworks have to be defined so that the uncertainty concerning the future general environment can be taken into account. A development framework comprises concrete future conditions and development trends in socioeconomic and ecological systems, which cannot (at least directly) be influenced by the decision-makers and which must therefore be treated as external factors. The alternative options and decision frameworks are then combined to form scenarios.[4] As three development frameworks were put forward for the Torgau project, this resulted in twelve scenarios (Messner et al. 2001).

**Step 2: Selection of problem-specific evaluation criteria**

Working towards the overall concept of sustainability, an interdisciplinary committee whose members include local lobbyists works out the problem-specific economic, ecological and social assessment criteria which can then be used to determine the expected effects of the various scenarios as quantitatively as possible. The criteria are chosen:

1. Such that they appropriately capture the scenarios‘ economic, social and ecological effects;
2. Depending on the models and methods available for estimating the scenarios‘ effects.

In the Torgau project, six economic, social and ecological criteria were chosen (Klauer, Messner, Herzog 2001):

- *Economic criteria*: 1) Net benefits; 2) Gross value added in the Torgau district
- *Social criterion*: 3) Employment in the Torgau district
- *Ecological criteria*: 4) The level of nitrate level in leachate, 5) The impact of gravel extraction and the resulting gravel lakes on nature conservation; 6) The ratio between natural groundwater recharge and extraction.

---

3 In Fig. 10.2, all the methodological steps in which such participation processes take place are indicated by the communication symbol.

4 Note that although this definition of the term scenario corresponds to that used in literature on decision theory (cf. e.g. Veeneklaas and Van den Berg 1995: 11), it differs from that used in the Querfurt method. The scenarios used in the Querfurt method are known in the nomenclature of the Torgau method as parameters of development works.

If the criteria are inserted into the columns of a matrix and the scenarios into the rows, the result is a multicriteria matrix (Fig. 10.2). For instance, entry in the second line and the column A indicates how the second scenario is assessed with respect to the criterion A. The multicriteria matrix forms the basis for multicriteria analysis in Step 4. Placement in the multicriteria matrix results from economic and ecological modelling and estimates of the scenario effects carried out in Step 3.

**Step 3: Evaluation and criterion related assessment of the scenario effects**
The third step of the Torgau method can be subdivided into two parts:

1. *Evaluation of the scenario effects*: The effects of the scenarios are evaluated using models or other methods.
2. *Criterion related assessment*: The scenario effects are evaluated in terms of the six criteria defined in Step 2.

In the Torgau project, economic analysis of the scenario's effects focused on sectoral effect analyses and a dynamic input-output model designed to reflect the interconnections between the various economic sectors. The input-output model enables the direct and indirect effects of land use changes on the economy and society (gross value added, employment figures) to be assessed (Klauer 2001). The input-output model was fed with the results of economic sectoral studies in which the development of individual economic sectors under the conditions of the various scenarios had been worked out.

From an ecological viewpoint, the main indicators are effects on the water balance and the pollution of the groundwater with nitrate. The hydrological effects were simulated with a hydrological-ecological runoff model (ABIMO), which was used to model data for natural groundwater recharge (Volk, Herzog, Schmidt 2001). This was then taken as a basis in conjunction with area-related nutrient balances to estimate nitrogen discharge (Franko, Schmidt, Volk 2001).

Criterion related assessment is used to evaluate the various scenario effects with respect to a single criterion. The most complicated stage is assessment regarding the criterion "net benefit" with cost-benefit analysis. The net benefit is the difference between the benefits and costs of an action alternative (compared to a basic alternative). In the cost-benefit analysis, the scenario effects are 'monetarized', i.e. they are measured in monetary units. Cost-benefit analysis is the standard procedure used for decision support in economics. Owing to the various methodological weaknesses (see Hanley, Spash 1993; Klauer 1999; Messner 2000; Geyler, Messner 2001; Messner, Geyler, Drechsler 2001) especially concerning monetarisation and the consideration of long-term effects, multicriteria analysis was applied for the Torgau project (see Step 4).

Assessment with respect to the other criteria seem to be less complicated at a first glance. For example, input-output analysis directly provides the values for the economic criterion "gross value added" and the social criterion "employment". In fact, assessment with respect to these two criteria actually takes place implicitly when these criteria are selected during modelling.

Extensive sensitivity analyses were carried out for all the modelling and forecasting methods, so that not only the most probable values of the criteria but also the probability distributions of the criteria values were available as input parameters for multicriteria assessment for each scenario in Step 4.

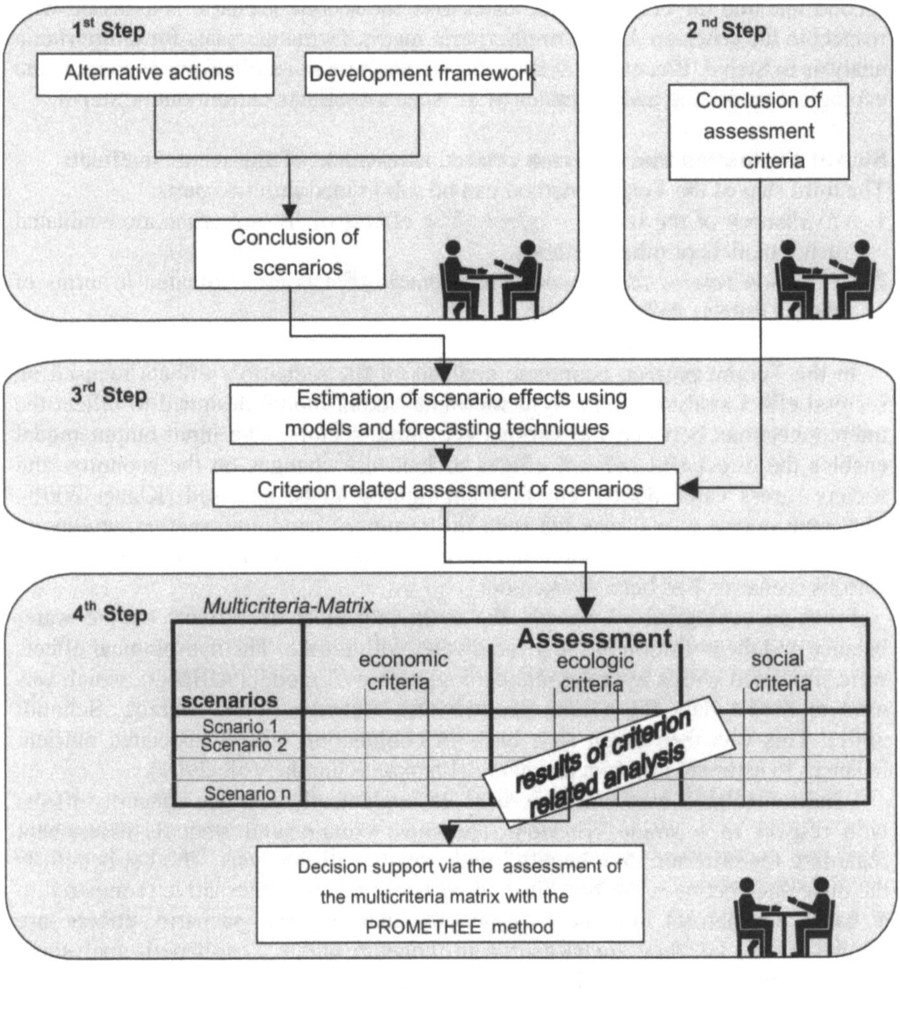


**Fig. 10.2:** Structure of the integrated method for the assessment of conflict situations (Torgau method)

**Step 4: Multicriteria analysis**
Once the multicriteria matrix (with the most likely criteria values and the probability distributions) has been filled in, the corresponding data undergo multicriteria decision analysis using an outranking procedure (PROMETHEE, Brans, Vincke 1985). The method used and further developed in the Torgau project enables the probability distribution of the model data values to be explicitly taken into account (Drechsler 2001). In the PROMETHEE method, pairs of alternatives are compared with respect to each criterion for each development framework, so that dominances and equalities can be determined. At the end of analysis, the individual results of the criteria are aggregated in two steps. First of all, in each criterion and for each alternative the number of dominances is compared to other alternatives. These figures are then added up, and if necessary the criteria are weighting differently. The alternatives are then ranked by means of the weighted totals.

Weighting reflects the value-judgements about the significance of the criteria, and they obviously affect the resulting ranking. The selection of weightings is hence an important preliminary decision which has to be taken together with the decision-makers. However, as deciding weighting is an abstract choice which may well be difficult for decision-makers, various approaches are discussed in the Torgau method.

One possibility is to analyse the influence of randomly chosen weightings. Using a specially developed computer program, results can be calculated for thousands of randomly generated weightings. These calculations provide information about how often what positions are taken by a scenario within the ranking of results, indicating initial trends within the results. If for example an action alternative is never top of the list, it would seem reasonable to concentrate on the other alternatives during further decision-making.

Another approach is to confront lobbyists and decision-makers (at a meeting or in individual discussions) with the results of the multicriteria matrix, and to have them determine criteria weighting. Whichever procedure for weighting and the inclusion of the stakeholders is chosen, the results can ultimately either be submitted to the decision-making body for decision, or a round table comprising representatives of all the interest groups can be set up in order to negotiate a compromise. This type of multicriteria analysis is called "multicriteria decision support", since instead of producing a single, unambiguous result, it simply provides scientific assistance for the decision-making process.

An exact description of the Torgau method and more detailed discussion of the results of multicriteria analysis are contained in Horsch, Ring and Herzog (2001). However, an outline of the approach in the Querfurt and Torgauer methods suffices in order to be able to compare the two methods.

## 10.3 Combining the Querfurt and Torgau methods

The Querfurt and Torgau methods differ in many respects. This is not surprising, for the two methods were developed for different problems and the developers come from different disciplines. Nevertheless, comparison of the differences as well as the features they have in common may be very fruitful, as the authors see various ways in which the two methods could augment each other.

Although both methods concentrate on decisions about land use changes, the land use changes to be decided upon by the Querfurt and Torgau methods have a different quality. The intended result for Querfurter Platte is a landscape plan in which each area is to be occupied by a certain landscape element. The plan describes land use which on the one hand complies with certain framework goals, and on the other must ultimately fulfil the landscape functions of these goals. A very large number of different land use options which fulfil all the goals has to be compared with each other.

By contrast, in Torgau not only microscale but also mesoscale and macroscale land use changes are compared. To be more precise, political action alternatives which will entail changed land use need to be assessed. In the Torgau study, one action option comprised whether or not to reduce wellhead protection areas, and the other whether or not to develop new gravel quarries. Therefore the action alternatives consisted in framework goals for the exact formulation of land use; the area-specific formulation itself was not the subject of the decision process, but was instead left up to the owners. Hence instead of a large number of detailed land use changes, only four alternative framework goals for the formulation of land use were compared. The two methods are compared in Table 10.1.

In the following we develop a structure in which the two methods complement and augment each other. Using the example of the case study in Querfurter Platte, we then explain one way in which this combined method can be used. Comparison of the Querfurt method with the Torgau method reveals the former to be particularly suitable for optimally organizing and assessing the area-specific implementation of landscape development models. The framework conditions of landscape development stem from the landscape model (e.g. from landscape framework planning or the federal state's landscape programme) and are decided by experts in discussion with local lobbyists. The process in which these framework conditions are created is the precondition for (rather than the subject of) the Querfurt method. In contrast, the Torgau model is especially suitable for weighing up different political goals for landscape development - for social conflicts concerning land use typically take place at the level of decisions over political framework goals, where specific-area usage only plays a minor role.

The obvious step is to additively combine the two methods. Using the Torgau method, the framework goals for land use changes are first decided. Then the optimal area-specific formulations are decided for the selected alternatives using the Querfurt method. However, this approach is beset by the disadvantage that oversimplifying assumptions must be made concerning the area-specific formulation of land use changes in the selection of framework goals. Moreover, the land use changes are not assessed in terms of their functional, landscape effects or analysed in terms of planning.

Yet exactly this point there is an interface via which the two techniques can be linked up in a process which is integrative rather than simply additive. Whereas the techniques in the Torgau method can be used for decision support for selecting landscape development goals moulded by political, economic and legal aspects,

**Table 10.1:** Comparison of the Querfurt and Torgau methods for decision support for land use changes

| Characteristic | Querfurt method | Torgau method |
|---|---|---|
| Goal | Optimal compromises to calculate "optimal land use pattern" | Assessment of different framework goals for land use changes |
| Spatial orientation | Area-specific for all scales | Regional (mesoscale and macroscale) |
| Subject of comparison | A large number of alternative, area-specific land use patterns are compared and optimised | Only a small number of actions can be compared |
| Methodological bases and subjects of assessment | Landscape model for planning, experts, lobbies; only decision support | Model of sustainability, methodological individualism,[5] preliminary decisions by experts involving decision-makers and lobbies; only decision support |
| Criteria of assessment | Multidimensional with ecological focus | Multidimensional incorporating socioeconomic assessment criteria |
| Evaluation of scenario effects | Scientifically substantiated evaluation procedure for each function | Various models and methods of estimation (e.g. input-output analysis, soil model) |
| Time relation | Long-term considerations | Long-term considerations taking into account time preferences |
| Consideration of uncertainty | Uncertainty ignored | Sensitivity analyses are carried out for the models and evaluation methods; the results are subjected to MCA |
| Decision support method | Multicriteria optimization for decision support (developed based on benefit analysis) | MCA using the PROMETHEE outranking method for decision support (based on pair comparison) |
| Evaluation | Integrated into multicriteria optimization | Integrated into MCA, greater possibilities for weighted analysis |
| Usage of results | Policy advising and planning | Policy advising and planning, conflict advising |

[5] Methodological individualism is a fundamental postulate of economics. It says that assessments and decisions are taken not by a state whole, a joint entity or collective, but exclusively by individuals, and consequently social value estimates are to be substantiated with individual value assessment (Petersen 1996: 85).

the Querfurt method of landscape optimization can be used to determine the area-specific optimal land use change for each alternative framework condition. Thus the Querfurt method of landscape optimization can be used as a model as it were to predict area-specific, 'tope'[6] land use changes and simultaneously to assess the alternatives concerning the evaluated landscape functions. In other words, the Querfurt method can be used within the framework of the Torgau method to predict area-specific formulation for each alternative political goal (including its functional evaluation). Such a forecast can if necessary be supplemented by the estimates of ecological, economic and social effects previously used.[7]

The structure of the combined method is shown in Fig. 10.3. The overall structure can be seen to comprise the four steps of the Torgau method. The Querfurt method enables the qualified evaluation of the effects of action alternatives, and is reflected in the step comprising the modelling and evaluation of the scenarios' effects.

For the concrete case of restructuring the cleared landscape in Querfurter Platte, the first stage of the combined method could be used for the important issue of selecting scenarios: what percentage of the landscape area should be set aside for nature conservation, and what crop rotations should be cultivated on the agricultural areas? Previous investigations left the exact crop rotation unanswered, while the proportion of conservation areas was decided by experts together with local farmers; this approach is perfectly adequate for the problems of Querfurter Platte, since the local farmers are indeed the main stakeholders, and there are no virulent social conflicts concerning land use. The farmers are aware of the necessity of intensified landscape conservation, above all to safeguard soil fertility in the long term - an objective which does not clash with economic goals. In the end, the discussion simply boiled down to the optimal localization of the conservation areas.

However, the problem of, say, establishing commercial or industrial areas or large conservation zones on areas previously used for agriculture in a regional plan is more complicated, since different interest groups directly conflict with each other and the political aims for landscape development have to be chosen amidst opposing interests. However, the combined method can provide valuable information to aid the process of social discussion, and reaching compromises and a decision. The area-specific prediction of land use changes as a result of the choice of objectives improves the method's clarity and accuracy. It is suitable for planning purposes if the optimization scenarios are taken into account for land use planning and landscape planning.

---

[6] In geoecology, a 'tope' is a homogeneous basic unit with geographical homogeneity in terms of characteristics and hence with homogeneous geoecodynamics (Leser et al. 1984: headword "Tope").

[7] If the Querfurt method is used to forecast area-specific land use changes and the effects on landscape functions, this assumes that the land use implemented corresponds to the optimal land use. This assumption appears justified if the area-specific land use can be administratively laid down or if those cultivating the land implement optimal land use of their own accord. This assumption becomes critical if the optimal land use conflicts with the interests of those cultivating the land.

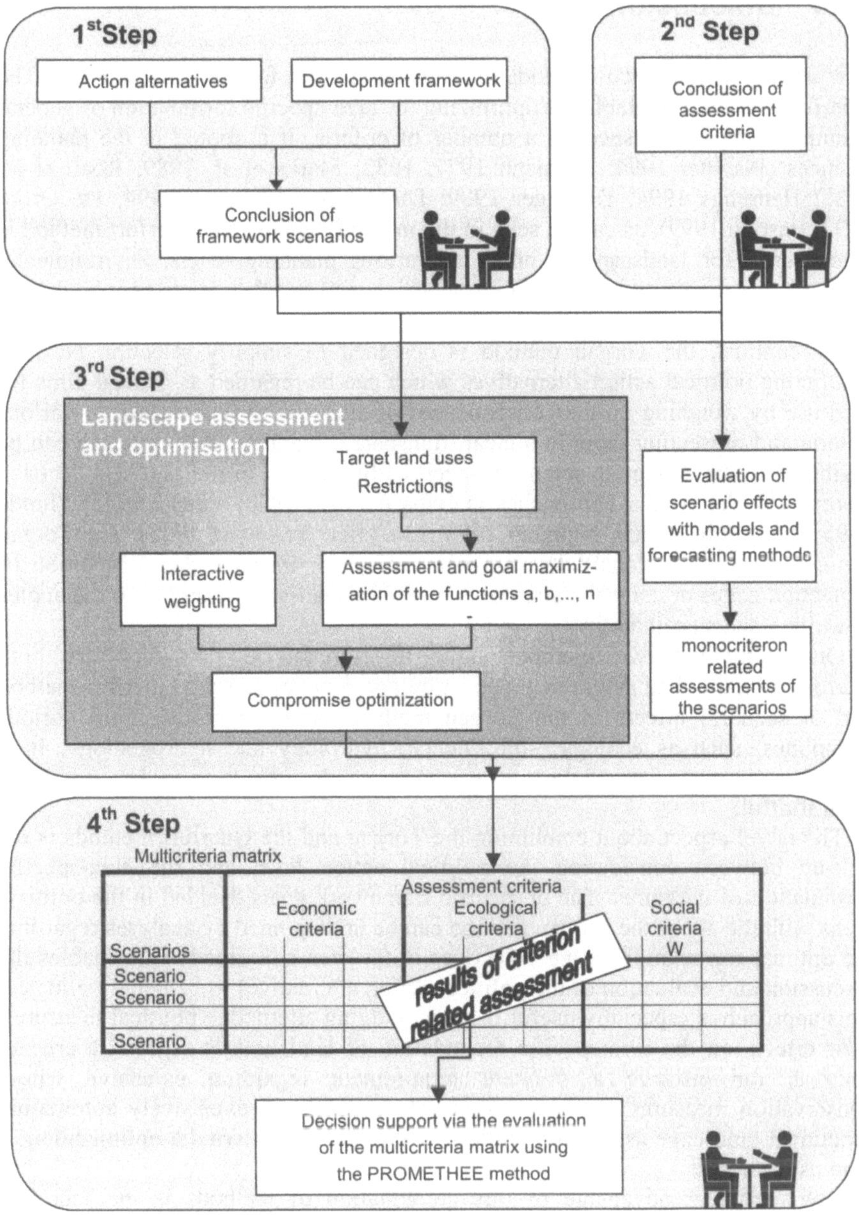


**Fig. 10.3:** Structure of the combined method

## 10.4 Discussion

This article presents two methods of decision support for land use changes. The Querfurt method is suitable for optimizing the area-specific formulation of general planning aims with respect to a number of criteria. It is rooted in the planning sciences (Plachter 1994; Niemann 1977, 1982; Marks et al. 1989; Koch et al. 1989; Hennings 1994; Dollinger 1989; Duttmann, Mosimann 1994; De Groot 1992; Bastian 1999), as can be seen in the method's goals: the Querfurt method is a tool used for landscape planning at various planning levels, environmental impact assessments in regional policy schemes, and to substantiate new landscape structures.

By contrast, the Torgau method is designed to simplify selection between conflicting political action alternatives which can be regarded as general aims for land use by weighing up the consequences of alternatives with regard to various criteria and presenting them in a clear, transparent manner. This approach can be attributed to the economic sciences, specifically the area of multicriteria decision theory (e.g. the similar approaches in Bana e Costa 1990; Vincke 1992; Munda 1995; Roy 1996; Beinat, Nijkamp 1998; El-Syaify, Yakowitz 1998). The Torgau method supports (rational) decision concerning for instance the designation of protection zones or control instruments by public decision-makers from the angles of welfare and sustainability.[8]

One thing the two methods have in common is that they are both interdisciplinary. The assessments of landscape functions in the Querfurt method and of scenario effects in the Torgau method require expertise from various disciplines, such as ecology, soil science, hydrology and hydrogeology. It is precisely this openness of the two methods which appears to make combining them fruitful.

The novel aspect about combining the Torgau and the Querfurt methods is the link-up between considering the political action level and the area-specific formulation of measures. The unspecific framework goals decided in the political arena with the aid of the Torgau method can be underpinned by analyses regarding the optimal formulation of the objectives in the area concerned. This enables the discussion and evaluation of the political action alternatives to be better validated. This approach is especially useful if, when judging alternative political measures, their effects on the area-specific formulation of land uses is especially crucial. This is for instance a frequent requirement regarding extensive nature conservation measures. Such a link is only possible by extensively automating functional landscape assessment with the subsequent multicriteria optimization of land use patterns.

One particular advantage of this combination of methods is the fact that economic and spatial-planning components can be coupled in land use decisions and alternatives.

---

[8] The traditional instrument of economic decision support, cost-benefit analysis, is integrated in the Torgau method by the net benefit (hence the difference between the economic benefit and costs) entering into the assessment of action alternatives as a *single* assessment criterion (in addition to others).

The combined method appears suitable for complex land use decision situations surrounding sustainable regional and landscape development. Land use strategies based on environmental quality target concepts and political action alternatives should be evaluated using the area-specific analysis, evaluation and solution of land use conflicts. There is no denying that applying the combined method is costly. Large quantities of data have to be collected and processed in procedures and models which entail plenty of work. Moreover, close liaison with the decision-makers and lobbyists is essential if the conflicts and interests at stake are to be properly appreciated, as well as to communicate the results of the analyses. This level of work only appears justified for important problems covering complex issues. Areas for which the combined method is suitable include:

- Riverine catchment area management, especially in connection with the new EU Water Framework Directive, which calls for action plans to improve water quality. The method would be suitable as a source of aid in assessing and deciding for various instruments and management measures. Alternative actions plans could be compared with respect to technical, economic and structure aims, as well as pollutant balances;
- The assessment of different development alternatives for sustainable urban development, especially in suburban areas. For example, investigations could be carried out into whether compactly or diffuse development patterns (which are strongly controlled by systems of political incentives) are better for sustainable development;
- The comparative assessment of alternative action plans concerning sustainable land use systems under the changing conditions of global change, for which the area-specific functional capacity of the natural balance is decisive;
- The assessment of conservation instruments such as a normative stipulation governing the proportion of priority nature conservation areas in a landscape. Although the selection of such instruments is a framework decision, its effectiveness chiefly depends on the area-specific formulation. This in turn provides better validation for landscape planning.

To sum up, therefore, combining the two methods presented here would create a possibility for making decisions about land use changes more objective and comprehensible than is possible today. The combined method could become a valuable instrument for decision support in politics, planning and conflict management.

## 10.5 References

Bana e Costa CA (ed) (1990) Readings in Multiple Criteria Decision Aid. Springer-Verlag, Berlin- Heidelberg-New York

Bastian O (1999) Landschaftsfunktionen als Grundlage von Leitbildern für Naturräume. Natur und Landschaft 9: 361-373

Beinat E, Nijkamp P (eds) (1998) Multicriteria Analysis for Land-use Management. Kluwer Academic Publishers, Dordrecht

Brans JP, Vincke P (1985) A preference ranking organization method. Management Science 31(6): 647-656

De Groot R (1992) Functions of nature. Evaluation of nature in environmental planning, management and decision making. Wolters-Noordhoff

Dollinger F (1989) Landschaftsanalyse und Landschaftsbewertung. Mitteilungen des Arbeitskreises für Regionalforschung Sonderband 2, Wien

Drechsler M. (1999) Verfahren zur multikriteriellen Analyse bei Unsicherheit. In: Horsch H, Ring I (eds) Naturressourcenschutz und wirtschaftliche Entwicklung: Nachhaltige Wasserbewirtschaftung und Landnutzung im Elbeeinzugsgebiet. UFZ-Bericht 16/1999, UFZ, Leipzig, pp. 187-214

Drechsler M (2001) Multikriterielle Bewertungsverfahren unter Berücksichtigung von Unsicherheit. In: Horsch et al. (eds) Nachhaltige Wasserbewirtschaftung und Landnutzung. (in prep.)

Duttmann R, Mosimann T (1994) Die ökologische Bewertung und dynamische Modellierung von Teilfunktionen und -prozessen des Landschaftshaushaltes - Anwendung und Perspektiven eines geoökologischen Informationssystems in der Praxis. Petermanns Geographische Mitteilungen 138: 3-17

El-Syaify SA, Yakowitz DS (eds) (1998) Multiple Objective Decision Making for Land, Water, and Environmental Management. Lewis Publishers, Boca Raton et al.

Franko U, Schmidt T, Volk M (2001): Modellierung des Einflusses von Landnutzungs-änderungen auf die Nitrat-Konzentration im Sickerwasser. In: Horsch et al. (eds) Nachhaltige Wasserbewirtschaftung und Landnutzung. (in prep.)

Geyler S, Messner F (2001) Die Kosten-Nutzen-Analyse von Landnutzungsänderungen als integrierter Bestandteil der multikriteriellen Entscheidungsanalyse. In: Horsch et al. (eds) Nachhaltige Wasserbewirtschaftung und Landnutzung. (in prep.)

Grabaum R (1996) Verfahren der polyfunktionalen Bewertung von Landschaftselementen einer Landschaftseinheit mit anschließender „Multicriteria Optimization" zur Generierung vielfältiger Landnutzungsoptionen. Diss. Universität Leipzig

Grabaum R, Meyer BC (1997) Landschaftsökologische Bewertungen und multikriterielle Optimierung mit Geographischen Informationssystemen (GIS). Photogrammetrie-Fernerkundung-GIS 2/97: 121 - 134

Grabaum R, Meyer BC (1998) Multicriteria optimization of landscapes using GIS-based functional assessments. Landscape and Urban Planning 554 (1998): 1-14

Grabaum R, Meyer BC, Mühle H (1999) Landschaftsbewertung und -optimierung. Ein integratives Konzept zur Landschaftsentwicklung. UFZ-Bericht 32/1999, UFZ, Leipzig

Hanley N, Spash CL (1993) Cost-Benefit Analysis and the Environment. Edward Elgar, Cheltenham

Hennings V (Koord 1994) Methodendokumentation Bodenkunde. Auswertungsmethoden zur Beurteilung der Empfindlichkeit und Belastbarkeit der Böden. Geologisches Jahrbuch, Reihe F Bodenkunde, Heft 31, Hannover

Horsch H, Ring I, Herzog F (eds) (2001): Nachhaltige Wasserbewirtschaftung und Landnutzung: Methoden und Instrumente der Entscheidungsfindung und -umsetzung. Metropolis-Verlag, Marburg (in prep.)

Horsch H, Ring I (eds) (1999) Naturressourcenschutz und wirtschaftliche Entwicklung - Nachhaltige Wasserbewirtschaftung und Landnutzung im Elbeeinzugsgebiet. UFZ-Bericht 16/1999, UFZ, Leipzig

Klauer B (1999) Nachhaltigkeit und Naturbewertung. Physika-Verlag, Heidelberg

Klauer B (2001) Modellierung von Wertschöpfungs- und Beschäftigungseffekten mittels Input-Output-Analyse. In: Horsch et al. (eds) Nachhaltige Wasserbewirtschaftung und Landnutzung. (in prep.)

Klauer B, Messner F, Drechsler M, Horsch H (2001) Das Konzept des integrierten Bewertungsverfahrens. In: Horsch et al. (eds) Nachhaltige Wasserbewirtschaftung und Landnutzung. (in prep.)

Klauer B, Messner F, Herzog F (1999) Szenarien für Landnutzungsänderungen im Torgauer Raum. In: Horsch H, Ring I (eds) Naturressourcenschutz und wirtschaftliche Entwicklung: Nachhaltige Wasserbewirtschaftung und Landnutzung im Elbeeinzugsgebiet. UFZ-Bericht 16/1999, UFZ, Leipzig, pp. 77-87

Klauer B, Messner F, Herzog F (2001) Die Ableitung von Bewetungskriterien. In: Horsch et al. (eds) Nachhaltige Wasserbewirtschaftung und Landnutzung. (in prep.)

Koch R, Graf D, Hartung A, Niemann E, Rytz E (1989) Polyfunktionale Bewertung von Flächennutzungsgefügen. Wissenschaftliche Mitteilungen, 32, IGG Leipzig

Leser H, Haas H-D, Mosimann T, Paesler R (1984): Diercke Wörterbuch der allgemeinen Geographie. Diercke, Braunschweig

Marks R, Müller MJ, Leser H, Klink HJ (eds) (1989) Anleitung zur Bewertung des Leistungsvermögens des Landschaftshaushaltes (BA LVL). Forschungen zur dt. Landeskunde 229, Trier

Messner F (2000) Ansätze zur Bewertung von Naturqualitäten im regionalen Entwicklungsprozess. UFZ-Diskussionspapier 5/2000, UFZ, Leipzig

Messner F, Klauer B, Geyler S, Volk M, Herzog F (2001) Die Ableitung von Szenarien für Szenarioanalysen: Methodik und beispielhafte Anwendung. In: Horsch et al. (eds) Nachhaltige Wasserbewirtschaftung und Landnutzung. (in prep.)

Messner F, Geyler S, Drechsler M (2001) Monetäre versus multikriterielle Bewetung. In: Horsch et al. (eds) Nachhaltige Wasserbewirtschaftung und Landnutzung. (in prep.)

Meyer BC (1997) Landschaftsstrukturen und Regulationsfunktionen in Intensivagrarlandschaften im Raum Leipzig-Halle. Regionalisierte Umweltqualitätsziele - Funktionsbewertungen - Multikriterielle Landschaftsoptimierung unter Verwendung von GIS. Dissertation Köln und UFZ-Bericht Nr 24/97, UFZ, Leipzig

Meyer BC, Grabaum R (1997) Multifunktionale Bewertung und multikriterielle Optimierung von Landschaftsausschnitten. In: Feldmann et al. (eds) Regeneration und nachhaltige Landnutzung - Konzepte für belastete Regionen. Springer, Heidelberg, pp. 236-243

Munda G (1995) Multicriteria Evaluation in a Fuzzy Environment. Physika-Verlag, Heidelberg

Niemann E (1977) Eine Methode zur Erarbeitung der Funktionsleistungsgrade von Landschaftselementen. Archiv für Naturschutz und Landschaftsforschung 17: 119-157

Niemann E (1982) Methodik zur Bestimmung der Eignung, Leistung und Belastbarkeit von Landschaftselementen und Landschaftseinheiten. Wiss. Mitt. IGG Leipzig, Sonderheft 2, Leipzig

Plachter H (1994) Methodische Rahmenbedingungen für synoptische Bewertungsverfahren im Naturschutz. Zeitschrift für Ökologie und Naturschutz 3: 87-106

Roy B (1996) Multicriteria Methodology for Decision Aiding. Kluwer Academic Publishers, Dordrecht

Veeneklaas FR, Van den Berg LM (1995) Scenario building: Art, craft or just a fashionable whim? In: Schoute JFT, Finke PA, Veeneklaas FR, Wolfert HP (eds) Scenario Studies for the Rural Environment. Kluwer, Dordrecht, pp. 11-13

Vincke P (1992) Multicriteria Decision-aid. John Wiley and Sons, New York

Volk M, Herzog F, Schmidt T (2001) Modellierung des Einflusses von Landnutzungsänderungen auf die Grundwasserneubildung. In: Horsch et al. (eds) Nachhaltige Wasserbewirtschaftung und Landnutzung. (in prep.)

# Index

MIX
Papier aus verantwortungsvollen Quellen
Paper from responsible sources
FSC® C105338

If you have any concerns about our products,
you can contact us on
**ProductSafety@springernature.com**

In case Publisher is established outside the EU,
the EU authorized representative is:
**Springer Nature Customer Service Center GmbH**
**Europaplatz 3, 69115 Heidelberg, Germany**

Printed by Libri Plureos GmbH
in Hamburg, Germany